Algebraic Topology

A Structural Introduction

Algebraic Topology

A Structural Introduction

Marco Grandis

Università di Genova, Italy

World Scientific

NEW JERSEY · LONDON · SINGAPORE · BEIJING · SHANGHAI · HONG KONG · TAIPEI · CHENNAI · TOKYO

Published by

World Scientific Publishing Co. Pte. Ltd.

5 Toh Tuck Link, Singapore 596224

USA office: 27 Warren Street, Suite 401-402, Hackensack, NJ 07601

UK office: 57 Shelton Street, Covent Garden, London WC2H 9HE

British Library Cataloguing-in-Publication Data
A catalogue record for this book is available from the British Library.

First published 2022 (hardcover)
Reprinted 2025 (in paperback edition)
ISBN 978-981-98-1405-3 (pbk)

ALGEBRAIC TOPOLOGY
A Structural Introduction

ISBN 978-981-124-835-1 (hardcover)
ISBN 978-981-124-836-8 (ebook for institutions)
ISBN 978-981-124-837-5 (ebook for individuals)

For any available supplementary material, please visit
https://www.worldscientific.com/worldscibooks/10.1142/12586#t=suppl

To my colleagues and friends

Preface

Algebraic Topology is a system and strategy of partial translations, aiming to reduce difficult topological problems to algebraic facts that can be more easily solved. This interface role, between topology and algebra, is the source of its power and attraction.

Each of these translations views a topological space in a particular light, and codifies this picture into an algebraic structure — typically a group, or a graded abelian group, or a graded ring. In this view, a huge amount of information is dropped, and a manageable part is kept. Choosing an appropriate translation, we can hope to solve the topological problem we are analysing.

The main subject of this book is singular homology, the simplest of these translations. Studying this theory and its applications we also investigate parts of other disciplines, which form its underlying structural layout (and in part grew out of it):

- Homological Algebra, for tensor products, the Hom functor and their derived functors,

- Homotopy Theory, for the homotopy groups and fundamental groupoid,

- Category Theory, the grammar of translations between mathematical fields.

This book is an introduction to a complex domain, with references to its advanced parts and ramifications. It is written with minimal prerequisites — basic general topology and little else — and a moderate progression, starting with a very elementary beginning. A consistent part of the exposition is organised in the form of exercises, with suitable hints and solutions.

It can be used as a textbook for a first course on Algebraic Topology or self-study, and a guidebook for further study.

Contents

Introduction

0.1 Investigating spaces with algebraic structures

Algebraic Topology studies topological spaces and continuous mappings, transforming spaces into algebraic structures (like groups and rings) and maps into homomorphisms. The general aim is to reduce difficult topological problems to simpler algebraic facts.

For instance, the following figure shows two surfaces, the sphere $\mathbb{S}^2$ and the 'torus' $\mathbb{T}$ (the surface of a lifebuoy)

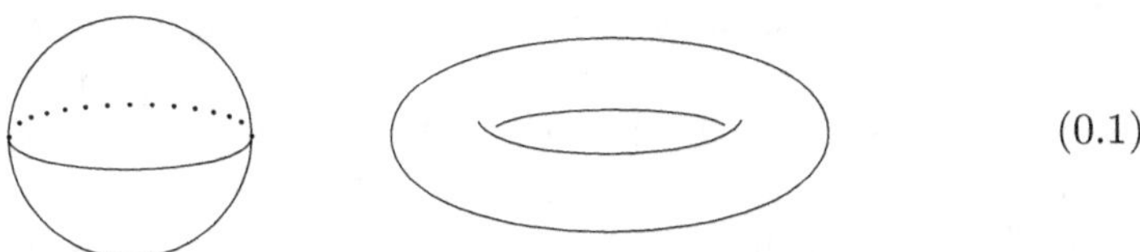

$$(0.1)$$

We have also drawn a great circle on the sphere, to suggest its curvature.

Plainly, the sphere and the torus have a different shape: we cannot continuously deform one space into the other. However, proving that these subspaces of $\mathbb{R}^3$ are not homeomorphic is not an elementary thing, like many seemingly obvious facts, in Topology.

To distinguish these spaces, we can think of a 'loop' (a closed path) on these surfaces. On the sphere, every loop can be continuously shrunk to a point, but a tight string around the lifebuoy cannot be further contracted — staying in the surface we are considering.

All this can be made precise with elementary results of Algebraic Topology: the homology group $H_1(\mathbb{S}^2)$ is trivial, while $H_1(\mathbb{T})$ is not. This is sufficient to prove that the sphere and the torus are not homeomorphic: two homeomorphic spaces X, Y have isomorphic homology groups $H_n(X) \cong H_n(Y)$, in every degree $n \geqslant 0$.

The homology groups of a space have a geometrical meaning, which is often fairly intuitive: for instance, $H_1(\mathbb{T})$ is a free abelian group with two generators, which can be realised as the homology classes $[a]$, $[b]$ of two loops a and b, around the 'equator' and a 'meridian' of the torus (with respect to the axis of rotation)

$$(0.2)$$

The homology class $3[a] + 2[b]$ can be realised as a loop which — globally — turns thrice around the equator and twice around a meridian. Negative coefficients turn up by reversing a loop.

Loosely speaking, the homology group $H_1(X)$ explores 1-dimensional properties of a space, and can detect 1-dimensional holes, as we have seen. Similarly, the homology group $H_2(X)$ is related with 2-dimensional features. For the sphere and the torus, the groups $H_2(\mathbb{S}^2)$ and $H_2(\mathbb{T})$ are infinite cyclic, detecting a 2-dimensional 'cavity' in each space; moreover, the choice of a generator in $H_2(\mathbb{S}^2)$ or $H_2(\mathbb{T})$ amounts to choosing an orientation, in each of these manifolds.

The n-dimensional sphere $\mathbb{S}^n$ (for $n > 0$) has only two non trivial groups, namely $H_0(\mathbb{S}^n)$ and $H_n(\mathbb{S}^n)$, both isomorphic to the additive group of the integers. This shows that all the spheres are really distinct spaces: $\mathbb{S}^m$ and $\mathbb{S}^n$ are only homeomorphic if $m = n$, and justifies our idea of topological dimension. It is again an important fact, which looks obvious in low dimension, but is not easily proved with ordinary topological tools (after the easy cases, for $m \leqslant 1$ or $n \leqslant 1$).

There are diverse homology theories of topological spaces, but all of them coincide on 'reasonably good' spaces. More precisely, all homology theories satisfying Eilenberg–Steenrod axioms, with the same coefficient group, give the same result on every triangulable space, up to isomorphism (see Section 3.4).

0.2 Homology and cohomology theories

Basically, a homology theory is a sequence of transformations $H_n(-)$; each of them turns a topological space X into an abelian group $H_n(X)$, and a continuous mapping $f\colon X \to Y$ into a homomorphism $H_n(f)\colon H_n(X) \to H_n(Y)$. In this transformation, composition of maps and identity maps are preserved:

- for consecutive maps $f\colon X \to Y$ and $g\colon Y \to Z$, we have: $H_n(gf) = H_n(g)H_n(f)$,

- for a space X we have: $H_n(\mathrm{id}\,X) = \mathrm{id}\,H_n(X)$.

In the basic terminology of category theory, each H_n is a covariant functor, from the category **Top** of topological spaces (and continuous mappings) to the category **Ab** of abelian groups (and homomorphisms). As an obvious consequence, each H_n preserves invertible arrows: it transforms a homeomorphism of spaces into an isomorphism of groups.

In fact, a homology theory turns up in a richer form: for every 'relative pair' (X, A) of topological spaces (where A is a subspace of X), and a fixed abelian group G, we have a sequence of groups $H_n(X, A; G)$ of relative homology, with coefficients in G. The basic case is absolute homology with integral coefficients: $H_n(X) = H_n(X, \emptyset; \mathbb{Z})$.

On the other hand, a cohomology theory $H^n(-)$ takes a continuous mapping $f\colon X \to Y$ to a homomorphism $H^n(f)\colon H^n(Y) \to H^n(X)$, reversing the direction of arrows, reversing composition and preserving identities. Here also we have an enriched version, for relative pairs of topological spaces and a coefficient group G.

An important fact appears: the cohomology groups with coefficients in a ring R have a graded multiplication, and their family forms a graded R-algebra. We shall study singular cohomology (a by-product of singular homology), and briefly review other theories, namely Alexander–Spanier cohomology and de Rham cohomology, which naturally arise in contravariant form.

0.3 An outline

Chapter 1 is an introduction to the goals and methods of Algebraic Topology, with a brief analysis of the elementary issues of category theory involved in the interface between topology and algebra.

Chapter 2 introduces singular homology, the simplest homology theory of general topological spaces, showing how to compute the homology groups, and get information on spaces and maps. Section 2.4 is devoted to the homology groups of the spheres, and their consequences: for instance, the Theorem of Topological Dimension, the topological degree of a map $\mathbb{S}^n \to \mathbb{S}^n$, and the theorem about vector fields on even-dimensional spheres. Several computations of homology groups can be found in Sections 2.5, 2.6.

Singular homology is constructed in the cubical form, that — in the author's opinion — gives a simpler approach. The more usual simplicial form is briefly presented in Section 2.8; their equivalence is proved in Section 5.6.

In Chapter 3 we extend singular homology to relative pairs (X, A). Section 3.3 shows how relative homology is able to investigate local features, for instance the local and global orientation of topological manifolds. The Eilenberg–Steenrod axioms for relative homology theories are listed in Section 3.4, and verified for relative singular homology. Alexander–Spanier cohomology, described in Section 3.5, satisfies the dual axioms, for a relative cohomology theory; its graded product is dealt with in Section 3.6. Similarly, de Rham cohomology of differentiable manifolds gives a graded algebra on the real field (see Section 3.7).

Singular homology and cohomology with a coefficient group is studied in Chapters 4 and 5. Varying the coefficient group we may be able to get results that the ordinary theory (with integral coefficients) cannot obtain, as shown in Subsection 4.3.7. All this requires some tools of Homological Algebra, namely the tensor and torsion products for homology, or the Hom and Ext functors for cohomology (in Sections 4.1 and 5.1–5.3).

Chapter 5 also explores the multiplicative structure of singular cohomology with coefficients in a ring (in Section 5.5), and the homology groups of a product of spaces (in Section 5.6). The latter are detemined by Eilenberg–Mac Lane's Acyclic Model Theorem, which is also used to prove the equivalence of the cubical and simplicial constructions of singular homology.

Chapter 6 is an introduction to the fundamental groupoid and homotopy groups, with their relationship to singular homology.

Chapter 7 briefly reviews issues that have already appeared in the previous chapters, on category theory (categorical limits and adjoint functors) and general topology (the compact-open topology). Finally Chapter 8 gathers most solutions to the exercises of the previous chapters.

0.4 Translations and underlying affinities

As already said, Algebraic Topology is a 'discipline of translations', from topology to algebra. Each translation gives a partial view of spaces and maps, in a particular perspective. This simplified picture may be able to solve the problem we are studying; otherwise, another translation might do.

The grammar of this discipline is provided by elementary category theory, reviewed here in some sections of Chapter 1, as far as this makes clear how the translation works.

As a simple instance, the fact that the functor H_n transforms a disjoint union of spaces $\bigcup X_i$ into a direct sum $\bigoplus H_n(X_i)$ of abelian groups is the outer appearance of an underlying property, the preservation of categorical sums: a disjoint union of spaces and a direct sum of abelian groups satisfy the same categorical universal property (see 1.6.2).

The effectiveness of a translation likely depends on the affinity between the two languages involved, in our case the source category of topological spaces versus the algebraic category of destination. In this regard, the category Ab is sufficiently flexible to offer 'covariant translations', by homology theories, and 'contravariant translations', by cohomology theories.

Less flexible categories, like those of rings and graded algebras, can only give a 'mirror-image', by a cohomology functor H^*. But this translation is far richer than viewing $H^*(X)$ as a mere graded abelian group.

The roots of this phenomenon can be found in an affinity between Top and Rng^{op}, the dual of the category of rings; an affinity expressed in theorems, like Gelfand duality, and in formal categorical aspects sketched in 5.5.9.

0.5 An inductive approach on structural bases

Notions will be presented in a concrete, 'inductive' way, starting from elementary examples. Whenever possible, the reader will be guided to build the theory through a series of exercises.

On the other hand, the roots of the interplay between Topology and Algebra, formalised in Homological Algebra and Category Theory, will be investigated more deeply than usual in an introductory book.

Algebraic Topology is a complex domain, with ramifications in diverse fields. After covering the basic parts, and some more advanced ones, we shall sketch several developments that do not have a place here, giving extensive references for further study.

The exposition will be particularly elementary and detailed in Chapter 1 and the first part of Chapter 2: this is not a book on general topology or abelian groups, but the reader will be invited (and guided) to check the topological or algebraic ground on which we are building. This point made, we shall go on more quickly.

Chapters 1 and 2 form a basic introduction to singular homology; Chapters 1 and 6 play the same role for homotopy groups; the three of them can give a preliminary view of Algebraic Topology.

0.6 Prerequisites and literature

The only prerequisites are general topology and the very basic theory of groups, abelian groups and rings. Categories and functors are introduced when their need arises. The same holds for the part of Homological Algebra used here: exact sequences, chain complexes of abelian groups, tensor products and Hom-functors, their derived functors.

For the fascinating field of Algebraic Topology there are elementary textbooks, like Vick [Vi] and Massey [Mas3], and more advanced ones, like Hilton–Wylie [HiW], Spanier [Sp2] and Hatcher [Ha]; the last is freely downloadable. The history of this discipline is dealt with in Dieudonné [Di].

For Homological Algebra we shall refer to Cartan–Eilenberg [CE] and Mac Lane [M1]; for General Topology to Kelley [Ke], Munkres [Mu] and Bourbaki [Bou2].

Category Theory is exposed in well-known books, like Mac Lane [M2], Borceux [Bo1, Bo2, Bo3], Adámek, Herrlich and Strecker [AHS]. The author's [G5] is a textbook for beginners, also devoted to applications in Algebra, Topology, Algebraic Topology and Homological Algebra.

0.7 Notation and conventions

Weak inclusion of sets is denoted by the symbol $\subset$, instead of the more usual $\subseteq$.

As usual, the symbols $\mathbb{N}$, $\mathbb{Z}$, $\mathbb{Q}$, $\mathbb{R}$, $\mathbb{C}$ denote the sets of natural, integral, rational, real or complex numbers; $\mathbb{N}^*$ is the subset of the positive integers. Open and semi-open real intervals are denoted as $]a, b[$, $[a, b[$, etc. — a Bourbaki's notation. The standard euclidean interval $[0, 1]$ is written as $\mathbb{I}$, the standard n-sphere as $\mathbb{S}^n$ (see 1.4.1).

The identity mapping of a (possibly structured) set X is written as $\mathrm{id}\, X$. A singleton set is generally written as $\{*\}$. The equivalence class of an element x with respect to an assigned equivalence relation is denoted as $[x]$, by default. A bullet in a diagram stands for an object. The two-valued index α (or β) varies in the set $\{0, 1\}$, also written as $\{-, +\}$.

A topological space is generally called a 'space', provided no ambiguity can arise. In this domain, a 'map' is a continuous mapping, and 'nbd' stays for neighbourhood.

In a space X, the interior of a subset A is written as $\mathrm{int}\, A$ or $\mathrm{int}_X A$, the closure as $\mathrm{cl}\, A$, or $\mathrm{cl}_X A$, or $\overline{A}$; the subset $\underline{\partial} A = \mathrm{cl}\, A \setminus \mathrm{int}\, A$ will be called the *border* of A in X. (The more usual term 'boundary' is here reserved for chain complexes and manifolds with boundary.)

A euclidean space is any subspace of some $\mathbb{R}^n$ with euclidean topology. A ring is assumed to be unital, by default.

A part marked with * is out of the main line of exposition. It may refer to issues dealt with further on, or be addressed to readers with some knowledge of the subject which is being discussed, or give references for higher topics. Marginal remarks are written in small characters.

Each chapter and section has its own introduction. Many subsections end with a list of exercises, whose solution is generally deferred to Chapter 8; but the solution can follow the exercise, when it is important for the sequel. Easy exercises are generally left to the reader.

0.8 Acknowledgements

This book is dedicated to the colleagues and dear friends with whom I have worked and am working.

In particular to the memory of Gabriele Darbo, my professor and friend. To Tim Porter and Ettore Carletti, with whom I discussed many topics which are developed here. To Bob Paré, George Janelidze and many others, recalling the happy hours we spent together when travelling was easier than it is today.

Diagrams and figures are composed with 'xy-pic', by K.H. Rose and R. Moore — a free package.

The cover comes out of a collaboration. A spider prepared the installation; the night dew arranged crystal-clear balls, in rhythmic progressions; the early morning light drew bright spheres out of them; I took the picture; Mr Loo Chuan Ming gave the final design.

I would also like to thank Dr Lim Swee Cheng and Ms Liu Yumeng, at World Scientific Publishing Co., for their kind help in the publication of this book.

1

Introducing Algebraic Topology

Algebraic Topology studies spaces and continuous mappings, transforming them into algebraic structures and homomorphisms, in order to reduce difficult topological problems to algebraic facts that can be more easily solved.

For instance, proving that the homology group H_1 of the circle is the additive group $\mathbb{Z}$ of the integers has various consequences, from straightforward ones (the circle is not a retract of the plane) to others, requiring additional work, as the Fundamental Theorem of Algebra. Computing the homology groups of the spheres has far-reaching results; several of them can be found in 1.1.5 and Section 2.4.

This introductory chapter is written in a very elementary way, and also reviews well-known issues about topological spaces and abelian groups. The reader will know which parts to skip, or browse, or read.

1.1 Classifying spaces and maps

We begin by considering the shape and the 'homotopy shape' of a space, hinting at their investigation with homology groups. The homotopy relation will be further examined in Section 1.4.

1.1.1 The shape of a space

Two topological spaces X, Y are homeomorphic (written as $X \cong Y$) if there exists a homeomorphism $X \to Y$, i.e. a continuous mapping which has a continuous inverse. We also say that X, Y *have the same (topological) shape*, or *the same homeomorphism-type.*

To prove that two given spaces have the same shape can require long computations, in order to build a homeomorphism between them, but is generally a 'confined' problem.

To prove that they are not homeomorphic can be quite difficult, and enticing, even when we clearly 'see' that they have a different shape: we must prove that *there cannot exist* a homeomorphism between them.

The proof is generally based on a 'topological property' (invariant up to homeomorphism) which holds in one of them but not in the other: connectedness, compactness, separation or countability axioms, higher forms of connectedness, etc. When these topological properties are explored throughout algebraic structures we enter in the domain of Algebraic Topology.

The following exercises show two families of spaces which is important to classify; the first classification can be achieved with the usual means of General Topology, the second will be completed later, using singular homology.

Exercises and complements. The solutions to these exercises can be found in Chapter 8.

(a) (*The shape of the intervals, I*) Any non-degenerate interval of the euclidean line is homeomorphic to $]0, 1[$, or $[0, 1[$, or $[0, 1]$.

Hints. It is an easy exercise of Calculus, which can be solved using affine linear functions and some elementary transcendental functions: see 8.1.1. (One can also use rational functions and their pastings, but the argument would be longer.)

(b) (*The shape of the intervals, II*) These three homeomorphism-types are distinct. Adding the degenerate cases, the intervals of $\mathbb{R}$ are thus classified in five types of homeomorphism:

(i) open non-degenerate intervals,

(ii) semiopen non-degenerate intervals,

(iii) compact non-degenerate intervals,

(iv) singletons,

(v) the empty space.

Hints: consider the connected components of an interval deprived of a point.

(c) (*Topological Dimension in low degree*) Classifying the euclidean spaces $\mathbb{R}^n$, by proving that they represent distinct homeomorphism-types, is a crucial fact in Topology, without which there would be no notion of topological dimension.

This point will be reviewed in 1.1.5(b), but the reader can solve as of now the lowest degrees: the singleton $\mathbb{R}^0 = \{0\}$ and the euclidean line $\mathbb{R}$ are not homeomorphic to any higher euclidean space $\mathbb{R}^n$, for $n > 1$.

(d) An interval of the line is not homeomorphic to any sphere $\mathbb{S}^n$ $(n \geqslant 0)$.

1.1.2 Classifying maps

Studying the maps $f\colon X \to Y$ between two topological spaces, we are also interested in classifying them, 'up to continuous deformation'.

Let us recall that two maps $f, g\colon X \to Y$ are said to be *homotopic* (written as $f \simeq g$) if there is a map $\varphi\colon X \times \mathbb{I} \to Y$ defined on the *cylinder* $I(X) = X \times \mathbb{I}$ (with the product topology), that coincides with f on the lower basis of the cylinder and with g on the upper one

$$\varphi\colon X \times \mathbb{I} \to Y,$$
$$\varphi(x,0) = f(x), \quad \varphi(x,1) = g(x) \qquad (\text{for } x \in X), \tag{1.1}$$

forming a continuous deformation of f into g. This is easily proved to be an equivalence relation (see 1.4.4). The homotopy will also be written as $\varphi\colon f \simeq g\colon X \to Y$.

A point $x \in X$ can be identified with the corresponding map $x\colon \{*\} \to X$, defined on the singleton space. A homotopy $a\colon x \simeq x'\colon \{*\} \to X$ is then the same as a *path* in X from x to x', i.e. a map such that

$$a\colon \mathbb{I} \to X, \qquad a(0) = x, \quad a(1) = x'. \tag{1.2}$$

On the other hand, a homotopy $\varphi\colon f \simeq g\colon X \to Y$ gives a family of paths $\varphi(x,-)\colon \mathbb{I} \to Y$ from $f(x)$ to $g(x)$, indexed by $x \in X$ and varying continuously on X.

Exercises and complements. (a) The classification of the maps $\{*\} \to X$ up to homotopy amounts to the partition of X in path components.

(b) We have two maps $f, g\colon X \to Y$ with values in a euclidean space Y (i.e. a subspace of some $\mathbb{R}^n$).

If Y is a convex subset of $\mathbb{R}^n$ (i.e. for all $y, y' \in Y$, the line segment from y to y' is contained in Y), the maps f and g are always homotopic. More generally, the same holds if, for every $x \in X$, the line segment from $f(x)$ to $g(x)$ is contained in Y.

Hints. We can use the linear structure of $\mathbb{R}^n$ to define the *affine homotopy*

$$\varphi\colon f \simeq g\colon X \to Y, \qquad \varphi(x,t) = (1-t)f(x) + tg(x), \tag{1.3}$$

which describes, at each $x \in X$, the line segment from $f(x)$ to $g(x)$, and is indeed an affine linear map in the variable $t \in \mathbb{I}$. Now we only have to verify the continuity of the mapping φ on $X \times \mathbb{I}$.

1.1.3 Classifying the homotopy type

The homotopy of maps leads to a possibility of deforming spaces, called homotopy equivalence, which is far wider than homeomorphism.

Namely, a map $f\colon X \to Y$ is a *homotopy equivalence* if it admits a *weak inverse* $g\colon Y \to X$, up to homotopy equivalence: this means that $gf \simeq \operatorname{id} X$ and $fg \simeq \operatorname{id} Y$. Then we say that the spaces X, Y are *homotopy equivalent*, or that they have *the same homotopy type*; it is again an equivalence relation between spaces (see 1.4.4), written as $X \simeq Y$.

A space is said to be *contractible* if it has the homotopy type of the singleton.

A subspace $A \subset X$ is a *deformation retract* of X if there is a map p such that

$$p\colon X \to A, \qquad\qquad pm = \operatorname{id} A, \qquad mp \simeq \operatorname{id} X, \qquad\qquad (1.4)$$

where $m\colon A \to X$ is the inclusion. Forgetting the homotopy $mp \simeq \operatorname{id} X$, we have a simple retract; the map p is called a retraction of m, and m is called a section of p.

We speak of a *strong deformation retract* if in (1.4) we can find a homotopy which does not move the points of A

$$\varphi\colon mp \simeq \operatorname{id} X, \qquad \varphi(x,t) = x \qquad\qquad (x \in A,\ t \in \mathbb{I}). \qquad (1.5)$$

Exercises and complements. (a) Every euclidean space $\mathbb{R}^n$ is contractible, and admits $\{0\}$ as a deformation retract. (Any other point would do, but this is a general fact: see 1.4.3(b).)

(b) Every non-empty convex subspace $X \subset \mathbb{R}^n$ is contractible. More generally, the same is true of every subset $X \subset \mathbb{R}^n$ which is *starred* with respect to some $x_0 \in X$; this means that: if $x \in X$, the line segment from x_0 to x is contained in X.

(c) Let X be a non-empty space. The following conditions are equivalent:

(i) X is contractible,
(ii) $\operatorname{id} X$ is homotopic to a constant map,
(iii) for any space Y, each map $Y \to X$ is homotopic to a constant map,
(iv) for any space Y, each map $X \to Y$ is homotopic to a constant map.

(d) The sphere $\mathbb{S}^n$ is a deformation retract of the pierced space $X = \mathbb{R}^{n+1} \setminus \{0\}$. *Hints:* use the normalisation mapping N

$$N\colon X \to \mathbb{S}^n, \qquad N(x) = x/\|x\| \qquad\qquad (x \neq 0). \qquad (1.6)$$

(e) The subspace $A \subset X$ is a deformation retract of X if and only if there is a map φ such that

$$\begin{aligned}
&\varphi\colon X \times \mathbb{I} \to X, \quad \varphi(x,0) \in A, \\
&\varphi(x,1) = x, \qquad \varphi(a,0) = a \qquad\qquad (\text{for } x \in X,\ a \in A).
\end{aligned} \qquad (1.7)$$

For a strong deformation retract we replace the last condition with:

$$\varphi(a, t) = a \qquad (\text{for } a \in A,\ t \in \mathbb{I}).$$

1.1.4 Hints at singular homology

Singular homology will be defined in Chapter 2. The construction requires some work, and we try to explain why it is worthwhile.

Singular homology associates:

- to a topological space X an abelian group $H_k(X)$, for each $k \geqslant 0$,
- to a map $f\colon X \to Y$ a homomorphism $H_k(f)\colon H_k(X) \to H_k(Y)$ of groups, for each $k \geqslant 0$,

so that composition and identity mappings are preserved:

  (i) $H_k(gf) = H_k(g)H_k(f)$, for all maps $f\colon X \to Y$ and $g\colon Y \to Z$,

  (ii) $H_k(\mathrm{id}\, X) = \mathrm{id}\, H_k(X)$, for any space X.

Because of these properties the 'transformation' H_k is called a 'functor', from spaces and maps to abelian groups and homomorphisms. $H_k(X)$ is called the *k-th group of singular homology* of the space X; $H_k(f)$ is called the *associated homomorphism*, and is usually written as $f_*\colon H_k(X) \to H_k(Y)$, or f_{*k}.

As a consequence of (i), (ii), if the spaces X, Y are homeomorphic, the homology groups $H_k(X)$ and $H_k(Y)$ are isomorphic, for all $k \geqslant 0$.

In fact we have two maps, inverse to each other

$$f\colon X \rightleftarrows Y\colon g, \qquad gf = \mathrm{id}\, X, \quad fg = \mathrm{id}\, Y,$$

transformed by H_k into homomorphisms $f_*\colon H_k(X) \rightleftarrows H_k(Y)\colon g_*$ which are again inverse to each other.

As a typical property of the transformations of Algebraic Topology, we shall see that the homology functors are *homotopy invariant*: this means that whenever two maps $f, f'\colon X \to Y$ are homotopic, then $H_k(f) = H_k(f')$, for every $k \geqslant 0$.

Therefore, a homotopy equivalence $f\colon X \rightleftarrows Y\colon g$, with $gf \simeq \mathrm{id}\, X$ and $fg \simeq \mathrm{id}\, Y$, is also transformed by H_k into an isomorphism $H_k(X) \cong H_k(Y)$ of abelian groups.

The homology groups are thus *invariants of the homotopy type of a space*; directly, they cannot distinguish homotopy equivalent spaces. In particular, all the spaces $\mathbb{R}^n$ have the same homology groups as the singleton $\mathbb{R}^0 = \{*\}$, which will be (easily) computed as:

$$H_0(\mathbb{R}^n) \cong \mathbb{Z}, \qquad H_k(\mathbb{R}^n) = 0 \qquad (k > 0), \qquad\qquad (1.8)$$

the trivial group being (always) written as 0.

Yet, the distinction of the diverse $\mathbb{R}^n$ can be achieved, by suitable modifications of the spaces (which only preserve the homeomorphism type).

1.1.5 The homology of the spheres

Developing the theory, we shall learn to compute the homology groups and the associated homomorphisms. For instance, on the standard spheres $\mathbb{S}^n$ we get (in 2.4.1 and Theorem 2.4.2):

$$H_0(\mathbb{S}^0) \cong \mathbb{Z}^2, \qquad\qquad H_k(\mathbb{S}^0) = 0 \quad (k > 0),$$
$$H_0(\mathbb{S}^n) \cong H_n(\mathbb{S}^n) \cong \mathbb{Z}, \quad H_k(\mathbb{S}^n) = 0 \quad (n > 0,\, 0 \neq k \neq n). \tag{1.9}$$

Roughly speaking, a non-trivial group $H_k(X)$ detects 'holes' of dimension k in the space X.

This has many important consequences, and we begin by presenting some straightforward ones.

(a) First, these results readily prove that no sphere is contractible. Moreover the sphere $\mathbb{S}^n$ is not homeomorphic, nor homotopy equivalent, to any other sphere $\mathbb{S}^m$ (for $m \neq n$).

In fact, if the space X is homotopy equivalent to $\mathbb{S}^n$ then $H_k(X) \cong H_k(\mathbb{S}^n)$ (for all $k \geqslant 0$), and the integer n is determined by the homology groups of X as: $\max\{k \in \mathbb{N} \mid H_k(X) \neq 0\}$.

(b) (Theorem of Topological Dimension) *If the euclidean spaces* $\mathbb{R}^m$ *and* $\mathbb{R}^n$ *are homeomorphic, then* $m = n$.

In fact, the one-point compactifications $\mathbb{S}^m$ and $\mathbb{S}^n$ are homeomorphic.

Let us note the analogy with the theorem of *linear* dimension of vector spaces.

(c) We already remarked that we cannot get this result applying the homology functors, directly, to $\mathbb{R}^m$ and $\mathbb{R}^n$.

Loosely speaking, *the n-dimensional character of the topology of* $\mathbb{R}^n$ *appears more clearly in its Alexandrov compactification* — a transformation which preserves homeomorphism but does not preserve homotopy equivalence.

(d) The argument of (b) can also be achieved replacing Alexandrov compactification with a more elementary procedure, still invariant up to homeomorphism and not to homotopy equivalence: taking out a point.

Suppose for simplicity that $m, n > 0$. If the euclidean spaces $\mathbb{R}^m$ and $\mathbb{R}^n$ are homeomorphic, the pierced space $\mathbb{R}^m \setminus \{0\}$ is homeomorphic to a pierced space $\mathbb{R}^n \setminus \{x_0\}$, whence $\mathbb{S}^{m-1}$ and $\mathbb{S}^{n-1}$ are homotopy equivalent (by 1.1.3(d)), and again $m = n$.

For $m = 1$ we are reviewing the solution to Exercise 1.1.1(c): connectedness does not distinguish $\mathbb{R}$ from any other $\mathbb{R}^n$, but does distinguish the pierced line $\mathbb{R} \setminus \{0\}$ from any higher pierced space $\mathbb{R}^n \setminus \{x_0\}$ with $n > 1$.

(e) We already remarked, since the beginning of the general Introduction, how the homology groups of the sphere and the torus show that these spaces are not homeomorphic; we can now add that they are not even homotopy equivalent.

1.1.6 Studying retracts

Showing that a subspace $A \subset X$ is a retract of X is a 'confined' problem: we are to find a (continuous) retraction $p\colon X \to A$, such that $p(x) = x$ for all $x \in A$. Showing that it is *not*, i.e. that there is no (continuous) retraction, can be an important result.

For instance, saying that the subspace $\mathbb{S}^0 = \{-1, 1\}$ of the euclidean line $\mathbb{R}$ is not a retract is an obvious consequence of the Intermediate Value Theorem, and can be viewed as a way of stating the theorem. (The general statement is an elementary consequence of the previous property.)

More generally, saying that the sphere $\mathbb{S}^n$ is not a retract of $\mathbb{R}^{n+1}$ is intuitively clear (at least in low dimension), but not easy to prove with the usual tools of general topology. This fact has important consequences, like the Brouwer fixed-point theorem (see 2.4.4(a)).

Exercises and complements. (a) For $n > 0$, $\mathbb{S}^n$ is not a retract of $\mathbb{R}^{n+1}$, nor of any contractible space. *Hints:* use the functor H_n and its values on our spaces, mentioned in (1.8) and (1.9).

(b) These facts can be easily extended: for a subspace $A \subset X$, the inclusion $m\colon A \to X$ gives a homomorphism $m_*\colon H_n(A) \to H_n(X)$ (for every $n \geqslant 0$), and:

- this homomorphism need not be injective,

- if A is a retract of X, then $m_*\colon H_n(A) \to H_n(X)$ has a left inverse, and is injective.

1.1.7 Topological embeddings and projections

Even though general topology is a prerequisite of this book, we end this section by reviewing a few elementary issues of frequent use, to fix our terminology.

A map $f\colon X \to Y$ between topological spaces has a canonical factorisation

$$
\begin{array}{ccc}
X & \xrightarrow{\ f\ } & Y \\
{\scriptstyle p}\downarrow & & \uparrow{\scriptstyle m} \\
X/R_f & \xrightarrow[\ g\ ]{} & \operatorname{Im} f
\end{array}
\qquad f = mgp, \qquad (1.10)
$$

where:

- p is the canonical projection onto the topological quotient of X modulo the equivalence relation associated to f ($xR_f x'$ means that $f(x) = f(x')$),

- m is the inclusion of the subspace $\operatorname{Im} f = f(X)$ into the codomain Y,

- $g\colon X/R_f \to \operatorname{Im} f$ is the bijective map defined by $g[x] = f(x)$ ($x \in X$).

If f is open (or closed), the same holds for g, which is a homeomorphism. Here we are interested in the following cases where g is a homeomorphism.

(a) The map $f\colon X \to Y$ is said to be a *topological embedding* if it is an injective mapping and X has the initial topology for f: the open sets of X are (precisely) the preimages $f^{-1}(V)$ of the open sets of Y. Then f restricts to a homeomorphism $g\colon X \to f(X)$ between X and a subspace of Y

$$
\begin{array}{ccc}
X & \xrightarrow{\ f\ } & Y \\
 & {\scriptstyle g}\searrow & \uparrow{\scriptstyle m} \\
 & & \operatorname{Im} f
\end{array}
\qquad\qquad (1.11)
$$

In this case we may want to 'identify' X with its image in Y, so that it becomes a subspace. Plainly, topological embeddings are stable under composition. An injective map which is either open or closed is always a topological embedding.

(b) The map $f\colon X \to Y$ is said to be a *topological projection* if it is a surjective mapping and Y has the final topology for f: the open sets of Y are (precisely) the subsets V whose preimage $f^{-1}(V)$ is open in X. Then f induces a homeomorphism $g\colon X/R_f \to Y$ between a quotient of X and the space Y

$$
\begin{array}{ccc}
X & \xrightarrow{\ f\ } & Y \\
{\scriptstyle p}\downarrow & \nearrow{\scriptstyle g} & \\
X/R_f & &
\end{array}
\qquad\qquad (1.12)
$$

Here we may want to identify Y with this quotient of X. Topological projections are stable under composition, an advantage with respect to 'strict' topological quotients. (Note that, in Set Theory, a quotient of a quotient of the set X is not a quotient of X.)

A surjective map which is either open or closed is always a topological projection. In particular, a surjective map from a compact space to a Hausdorff space is closed, and a topological projection.

1.1.8 Cover Lemmas

Let $(X_i)_{i \in I}$ be a cover of the space X, that is $X = \bigcup X_i$. The continuity of a mapping $f \colon X \to Y$ with values in a topological space Y can be reduced to the continuity of its restrictions $f_i \colon X_i \to Y$ in two main cases.

Exercises and complements. (a) (The Open Cover Lemma) *Let $(X_i)_{i \in I}$ be an open cover of the space X (i.e. all X_i are open subsets of X). Then $f \colon X \to Y$ is continuous if and only if all its restrictions $f_i \colon X_i \to Y$ are.*

(b) (The Finite Closed Cover Lemma) *Let $(X_i)_{i \in I}$ be a finite closed cover of the space X (i.e. all X_i are closed subsets of X, and I is finite). Then $f \colon X \to Y$ is continuous if and only if all its restrictions $f_i \colon X_i \to Y$ are.*

(c) The general situation is a space X and a cover $(X_i)_{i \in I}$ such that X has the finest topology that makes all inclusions $u_i \colon X_i \to X$ continuous.

Categorically, this means that X is the colimit of the inclusions $X_i \to X$.

1.2 Transforming problems

We want to transform topological spaces into algebraic structures (typically groups) and continuous mappings into homomorphisms. Technically, this means to construct *functors*, from a category of topological spaces to a category of algebraic structures.

Here we give a brief presentation of these topics, making informal use of issues that will be made precise in the sequel.

The set of maps $X \to Y$ between two topological spaces is written as $\mathsf{Top}(X, Y)$.

1.2.1 Hints at categories and functors

We transform topological problems into algebraic ones setting up a procedure F that transforms

- a topological space X into an abelian group $F(X)$ (or some other algebraic structure),

- a map $f \colon X \to Y$ into a homomorphism $F(f) \colon F(X) \to F(Y)$,

so that composition and identities are preserved — as in 1.1.4, in the case of singular homology.

Such a procedure is denoted as $F\colon \mathsf{Top} \to \mathsf{Ab}$ and called a (covariant) *functor*, from the *category* Top *of topological spaces and continuous mappings* to the *category* Ab *of abelian groups and homomorphisms*. These issues will be defined in Section 1.3.

Here it will be sufficient to say that a category C consists of *objects* $X, Y, \ldots$ and *morphisms* $f\colon X \to Y, \ldots$ between objects; the latter have a (partial) *composition law*: for each pair of 'consecutive' morphisms $f\colon X \to Y$ and $g\colon Y \to Z$ there is a composed morphism $gf\colon X \to Z$. This partial operation is associative (whenever composition is legitimate) and every object X has an *identity*, written as $\operatorname{id} X\colon X \to X$, or 1_X, which acts as a unit for legitimate compositions. The notation $\mathsf{C}(X, Y)$ denotes the set of morphisms $X \to Y$ in C.

The prime example is the category Set *of sets* (and mappings), where:

- an object is a set,

- the morphisms $: X \to Y$ between two sets are the (set-theoretical) mappings from X to Y, and form the set $\mathsf{Set}(X, Y)$,

- the composition law is the usual composition of mappings: $(gf)(x) = g(f(x))$.

We also need the following categories of structured sets and structure-preserving mappings (with the usual composition):

- the category Top *of topological spaces* (and continuous mappings),

- the category Ab *of abelian groups* (and their homomorphisms),

- the category Gp *of groups* (and their homomorphisms),

- the category $R\,\mathsf{Mod}$ *of R-modules* on a commutative ring R (and their homomorphisms),

- the category Rng *of rings* (and their homomorphisms),

- the category $\mathsf{Set}_\bullet$ *of pointed sets* (and pointed mappings),

- the category $\mathsf{Top}_\bullet$ *of pointed topological spaces* (and pointed maps).

Other categories of structured sets will be introduced later on.

A functor $F\colon \mathsf{Top} \to \mathsf{Ab}$ is defined as above. More generally, we define in a similar way a functor $F\colon \mathsf{C} \to \mathsf{D}$, from a category C to a category D. (Contravariant functors are introduced in 1.2.2(iii).)

Remarks and complements. (a) By default, a ring R is assumed to be associative and unital; an R-module is assumed to be unitary.

(b) We recall that a *pointed set* is a pair (X, x_0) consisting of a set X and a *basepoint* $x_0 \in X$; a pointed mapping $f\colon (X, x_0) \to (Y, y_0)$ is a mapping $f\colon X \to Y$ such that $f(x_0) = y_0$; the composition is the usual one.

Similarly, a *pointed topological space* (X, x_0) is a space with a basepoint, and a pointed map $f\colon (X, x_0) \to (Y, y_0)$ is a continuous mapping from X to Y such that $f(x_0) = y_0$. (We shall see that the fundamental group $\pi_1(X, x_0)$ is defined for a pointed topological space.)

(c) When a category is named after its objects alone (e.g. the 'category of topological spaces'), this means that the morphisms are understood to be the obvious ones (in this case the continuous mappings), with the obvious composition law. Of course, different categories with the same objects are given different names.

1.2.2 Exploring connectedness

We begin by constructing some elementary functors $\mathsf{Top} \to \mathsf{D}$, with values in a category of algebraic structures. The important ones, the homology and homotopy functors, will require more work, but will appear to be 'higher dimensional versions' of some of the present ones, related to properties of 'higher connectedness'.

As in 1.1.4, a functor $F\colon \mathsf{Top} \to \mathsf{D}$ is *homotopy invariant* if, for every pair of homotopic maps $f \simeq g$, we have $F(f) = F(g)$.

(i) A set has a trivial algebraic structure, with no operations under no axioms, and all mappings preserve this structure: Set is itself a category of algebraic structures, the simplest.

There is an obvious *forgetful functor*

$$U\colon \mathsf{Top} \to \mathsf{Set}, \qquad U(X) = |X|, \qquad U(f) = |f|, \qquad (1.13)$$

that sends each space X to its underlying set $|X|$, and each map $f\colon X \to Y$ to the underlying mapping $|f|\colon |X| \to |Y|$.

Each category of structured sets has a similar *forgetful functor* with values in Set (see 1.3.7). Such functors are so obvious that we tend to forget *them!* The notation $|-|$ will be used when useful.

(ii) We also consider two less obvious functors with values in Set

$$\mathrm{Cc}\colon \mathsf{Top} \to \mathsf{Set}, \qquad \Pi_0\colon \mathsf{Top} \to \mathsf{Set}. \qquad (1.14)$$

The former sends a space X to the set $\mathrm{Cc}(X)$ of its connected components, and a map $f\colon X \to Y$ to the mapping $\mathrm{Cc}(f)\colon \mathrm{Cc}(X) \to \mathrm{Cc}(Y)$ which sends a component A of X to the component of Y containing $f(A)$.

Π_0 is constructed in the same way, using the path components; it will be reviewed in (1.69).

(iii) (*A contravariant functor*) Here $\mathbb{Z}$ is the ring of integers equipped with the discrete topology. We form a *contravariant functor* (denoted by a dashed arrow)

$$\mathcal{A}\colon \mathsf{Top} \dashrightarrow \mathsf{Rng}, \qquad \mathcal{A}(X) = \mathsf{Top}(X, \mathbb{Z}). \tag{1.15}$$

It sends a space X to the (commutative) ring of all maps $a\colon X \to \mathbb{Z}$, endowed with the pointwise operations:

$$(a+b)(x) = a(x) + b(x), \quad (ab)(x) = a(x).b(x), \quad 1(x) = 1_{\mathbb{Z}}. \tag{1.16}$$

A map $f\colon X \to Y$ is sent to a homomorphism of rings

$$\mathcal{A}(f)\colon \mathcal{A}(Y) \to \mathcal{A}(X), \quad \mathcal{A}(f)(a\colon Y \to \mathbb{Z}) = (af\colon X \to \mathbb{Z}). \tag{1.17}$$

The reader will note that $\mathcal{A}$ *reverses the direction* of the map f; consistently, this transformation preserves the identities but reverses the composition: a map $g\colon Y \to T$ will give a homomorphism $\mathcal{A}(g)\colon \mathcal{A}(T) \to \mathcal{A}(Y)$, and

$$\begin{aligned}
\mathcal{A}(gf) = \mathcal{A}(f)\mathcal{A}(g)\colon \mathsf{Top}(T, \mathbb{Z}) &\longrightarrow \mathsf{Top}(X, \mathbb{Z}), \\
(a\colon T \to \mathbb{Z}) &\mapsto (agf\colon X \to \mathbb{Z}).
\end{aligned} \tag{1.18}$$

Exercises and complements. (a) The forgetful functor $U\colon \mathsf{Top} \to \mathsf{Set}$ is not homotopy invariant.

(b) The functors $\mathrm{Cc}\colon \mathsf{Top} \to \mathsf{Set}$, $\Pi_0\colon \mathsf{Top} \to \mathsf{Set}$ and $\mathcal{A}\colon \mathsf{Top} \dashrightarrow \mathsf{Ab}$ defined above are homotopy invariant.

*(c) More generally, in the construction of $\mathcal{A}$ one can replace $\mathbb{Z}$ with any ring, equipped with the discrete topology.

1.2.3 Other exercises

The reader may find it interesting to explore the contravariant functor $\mathcal{A}$ (in (1.15)), either independently or with the following exercises.

(a) A function $a\colon X \to \mathbb{Z}$ is continuous if and only if each preimage $X_k = a^{-1}\{k\}$ (for $k \in \mathbb{Z}$) is open in X, if and only the function a is locally constant (i.e. constant on some neighbourhood of each point).

This implies that a is constant on each connected component of X; the converse certainly holds if the connected components of X are open, but plainly fails for the rational line $\mathbb{Q}$ whose connected components are singletons.

(b) A map $a\colon X \to \mathbb{Z}$ amounts to a partition $X = \bigcup_{k \in \mathbb{Z}} X_k$ in disjoint open subsets (automatically closed in X). But this description is ineffective with respect to the operations of the ring $\mathcal{A}(X)$, defined in (1.16).

(c) The space X is empty if and only if the ring $\mathcal{A}(X)$ is trivial.

(d) The following conditions on the space X are equivalent:

(i) X is non-empty and connected,

(ii) $\mathcal{A}(X) \cong \mathbb{Z}$ (as rings),

(iii) $\mathcal{A}(X)$ is an integral domain (i.e. a non-trivial ring without proper zero divisors),

(iv) the underlying additive group of $\mathcal{A}(X)$ is infinite cyclic.

(e) Let $X = \bigcup_{i \in I} X_i$ be the partition of X in connected components, and suppose that each of them is open in X (which is certainly true if I is finite). Then the ring $\mathcal{A}(X)$ is canonically isomorphic to the cartesian-power ring $\mathbb{Z}^I = \prod_{i \in I} \mathbb{Z}$. (Note that we can identify the index-set I with the set $\mathrm{Cc}(X)$ of connected components of X.)

*(f) For the euclidean space

$$X = \{x \in \mathbb{R} \mid x = 1/n, \text{ for some } n \in \mathbb{N}^*\} \cup \{0\}, \tag{1.19}$$

the ring $\mathcal{A}(X)$ is isomorphic to a countable subring of $\mathbb{Z}^{\mathbb{N}}$.

*(g) The ring $\mathcal{A}(\mathbb{Q})$ of the rational line (with euclidean topology) is uncountable. (In fact, it has the cardinal of the continuum, i.e. the cardinal of $\mathbb{R}$.)

1.2.4 Distinguishing shape and homotopy type

In a category C, an *isomorphism* is a morphism $f\colon X \to Y$ which has an inverse, i.e. a morphism $g\colon Y \to X$ such that $gf = 1_X$ and $fg = 1_Y$. The inverse is determined by f (by the usual verification) and written as f^{-1}.

We say that the objects X, Y are isomorphic, written as $X \cong Y$, if there is an isomorphism $X \to Y$ (in the category that we are considering).

In the categories listed in 1.2.1 this definition gives the usual isomorphisms of each structure: bijective mappings in Set, homeomorphisms in Top, isomorphisms of groups in Gp and Ab, etc.

Plainly:

(i) a functor $F\colon \mathsf{C} \to \mathsf{D}$ takes any isomorphism of C to an isomorphism of D,

(ii) a homotopy invariant functor $F\colon \mathsf{Top} \to \mathsf{D}$ transforms a homotopy equivalence $f\colon X \to Y$ into an isomorphism of the category D; in other words, $X \simeq Y$ implies $F(X) \cong F(Y)$.

Therefore, a functor $F\colon \mathsf{Top} \to \mathsf{D}$ can be used *to distinguish spaces*, up to homeomorphism: two spaces X, Y cannot be homeomorphic if $F(X)$ and

$F(Y)$ are not isomorphic in D. A homotopy invariant functor can be used *to distinguish homotopy types.*

Important results in this sense will come out of complex functors, like singular homology. The elementary examples of 1.2.2 give trivial or elementary results.

Exercises and complements. (a) Applying principle (i) to the functors U, Cc, Π_0 of 1.2.2 gives obvious facts, for which there would be no need of categories and functors: two homeomorphic spaces have equipotent underlying sets, the same number (or cardinal) of connected components, and the same number (or cardinal) of path components. We shall see that Π_0 is closely related to singular homology in degree zero.

(b) Applying principle (ii) to the homotopy invariant functors Cc and Π_0 gives elementary, less obvious facts: two homotopy equivalent spaces have the same number (or cardinal) of connected components, and the same number (or cardinal) of path components.

*(c) The functor $\mathcal{A}$ is able to distinguish the euclidean space

$$X = \{x \in \mathbb{R} \mid x = 1/n, \text{ for some } n \in \mathbb{N}^*\} \cup \{0\}, \tag{1.20}$$

of Exercise 1.2.3(f) from any infinite discrete space (while the functors U, Cc and Π_0 give isomorphic results on X and the discrete $\mathbb{N}$).

1.2.5 Free abelian groups

We recall here some points about abelian groups, that will be frequently used throughout this book. (Everything can be extended to modules on a commutative ring R, that replaces the ring $\mathbb{Z}$ of integers: see 4.1.1.)

(a) In an abelian group A, a finite linear combination $\sum_{i \in I} \lambda_i x_i$ of elements $x_i \in A$ with integral coefficients λ_i makes sense, without any ordering of the set I of indices, because of the associative and commutative property of addition.

More generally, a linear combination $\sum_{i \in I} \lambda_i x_i$ makes sense for an arbitrary set of indices I, if the family of coefficients $(\lambda_i)_{i \in I} \in \mathbb{Z}^I$ is *essentially finite*: there are finitely many indices i such that $\lambda_i \neq 0$; then the sum is computed on these indices.

(b) A *basis* of an abelian group A, is a family $(x_i)_{i \in I}$ of elements of A such that every $x \in A$ can be uniquely written as a linear combination with essentially finite integral coefficients

$$x = \sum_{i \in I} \lambda_i x_i. \tag{1.21}$$

The abelian group A is *free* (on the ring $\mathbb{Z}$) if it has a basis. The latter is an injective family, and we can embed I in A, identifying $i = x_i$.

In fact, if $i \neq j$ in I, the difference $x_i - x_j = 1.x_i + (-1).x_j$ cannot be $0_A = 0.x_i + 0.x_j$.

The *universal property* of the basis says that every mapping $f \colon I \to B$ with values in an abelian group can be uniquely extended to a group-homomorphism $g \colon A \to B$

$$g(\textstyle\sum_{i\in I} \lambda_i x_i) = \sum_{i\in I} \lambda_i f(i), \tag{1.22}$$

computed by an (essentially finite) linear combination in B.

(c) Every set I is the basis of a free abelian group. We construct the latter as the direct sum

$$\mathbb{Z}I = \textstyle\bigoplus_{i\in I} \mathbb{Z} \subset \mathbb{Z}^I, \tag{1.23}$$

namely the subgroup of $\mathbb{Z}^I$ formed by the essentially finite families $(\lambda_i)_{i\in I}$ of integers.

In fact $\mathbb{Z}I$ has a canonical basis e_i $(i \in I)$, given by the Kronecker families

$$e_{ii} = 1, \qquad e_{ij} = 0 \qquad (j \neq i), \tag{1.24}$$

since the general element $(\lambda_i)_{i\in I} \in \mathbb{Z}I$ coincides with the essentially finite linear combination $\sum_{i\in I} \lambda_i e_i$.

Identifying any $i \in I$ with the element e_i of the canonical basis, the elements of $\mathbb{Z}I$ can be written as *formal linear combinations* $\sum_{i\in I} \lambda_i i$ of the elements of I, with essentially finite integral coefficients.

(d) Finally we have the *free abelian group* functor

$$F \colon \mathsf{Set} \to \mathsf{Ab}, \qquad F(I) = \mathbb{Z}I = \textstyle\bigoplus_{i\in I} \mathbb{Z},$$
$$F(f) \colon \mathbb{Z}I \to \mathbb{Z}J, \qquad F(f)(i) = f(i) \quad (\text{for } f \colon I \to J), \tag{1.25}$$

where the homomorphism $F(f)$ is defined on the canonical basis I; its value on a general essentially finite family of $\mathbb{Z}I$ is:

$$F(f)(\lambda_i)_{i\in I} = \textstyle\sum_i \lambda_i f(i) = (\sum_{f(i)=j} \lambda_i)_{j\in J}. \tag{1.26}$$

The functorial properties of F are readily verified on the bases: for instance, for a composed mapping $gf \colon I \to J \to K$

$$F(gf)(i) = (gf)(i) = g(f(i)) = F(g)F(f)(i).$$

(e) We have studied in 1.2.2 and 1.2.3 the contravariant functor $\mathcal{A} = \mathsf{Top}(-, \mathbb{Z}) \colon \mathsf{Top} \dashrightarrow \mathsf{Rng}$.

Forgetting topology, we have a contravariant functor $P = \mathsf{Set}(-, \mathbb{Z})$ defined on sets:

$$P = \mathsf{Set}(-, \mathbb{Z}) \colon \mathsf{Set} \dashrightarrow \mathsf{Rng}, \qquad P(I) = \mathbb{Z}^I = \Pi_{i \in I}\, \mathbb{Z},$$

$$P(f) \colon P(J) \to P(I), \qquad P(f)(\lambda_j)_{j \in J} = (\lambda_{f(i)})_{i \in I}, \tag{1.27}$$

where $f \colon I \to J$ is a mapping. The functor P is the same as the composite of the discrete-topology functor $D \colon \mathsf{Set} \to \mathsf{Top}$ with $\mathcal{A} \colon \mathsf{Top} \dashrightarrow \mathsf{Rng}$.

1.2.6 A structure theorem

We shall also use some other classical results on abelian groups. This part can be skipped, until referred to. For integers m, n the notation $m \mid n$ means that m divides n, that is $n = km$ for some $k \in \mathbb{Z}$.

(a) In an abelian group A, the elements of finite order form a subgroup

$$\mathrm{t}A = \{x \in A \mid \exists\, n \in \mathbb{N}^* \text{ such that } nx = 0\}, \tag{1.28}$$

called the *torsion subgroup* of A. We say that A is a *torsion group* if $A = \mathrm{t}A$; we say that it is *torsion free* if $\mathrm{t}A = 0$.

The quotient $A/\mathrm{t}A$ is obviously torsion free.

(b) An abelian group A is *finitely generated* if there is a finite family $x_1, ...,$ x_n of elements that generates it: every $x \in A$ can be written as a linear combination $\Sigma_{i \in I}\, \lambda_i x_i$ of the generators, with integral coefficients.

We recall a classical theorem, that determines the structure of such groups. The proof can be found in any book on abelian groups (for instance [Fu]). Its extension to modules on principal ideal domains can be found in [Bou3] Section VII.4.4.

(c) (The Structure Theorem of Finitely Generated Abelian Groups) *A finitely generated abelian group A is (non-canonically) isomorphic to a direct sum*

$$A \cong \mathbb{Z}^n \oplus \mathbb{Z}_{\tau_1} \oplus ... \oplus \mathbb{Z}_{\tau_k}, \qquad \tau_1 \mid \tau_2 \mid ... \mid \tau_{k-1} \mid \tau_k, \tag{1.29}$$

that determines the following natural numbers:

- *the* rank $n = \mathrm{rk}A$ *of A,*

- *the* torsion coefficients $\tau_1, ..., \tau_k \in \mathbb{N}^*$, *ordered by divisibility $(k \geqslant 0)$.*

The sequence $(n, \tau_1, ..., \tau_k)$ classifies A, up to isomorphism.

(d) For a finitely generated abelian group A, the decomposition (1.28) can plainly be restricted to an isomorphism of the torsion subgroups

$$\mathrm{t}A \cong \mathbb{Z}_{\tau_1} \oplus ... \oplus \mathbb{Z}_{\tau_k}, \qquad \tau_1 \mid \tau_2 \mid ... \mid \tau_{k-1} \mid \tau_k. \tag{1.30}$$

(e) A can thus be decomposed in the direct sum $A \cong \mathbb{Z}^n \oplus tA$. But note that there is no canonical subgroup of A isomorphic to $\mathbb{Z}^n$.

(f) A finitely generated abelian group A which is torsion free is necessarily a free abelian group. All its bases have the same number of elements, namely $n = \mathrm{rk}A$.

(g) The additive group $\mathbb{Q}$ of rational numbers is not finitely generated. It is torsion free, and is not free: no element is a generator, but any two distinct elements are linearly dependent on $\mathbb{Z}$.

(h) The cyclic group $\mathbb{Z}_k$ has one torsion coefficient, namely k. The Klein four-element group $\mathbb{Z}_2 \oplus \mathbb{Z}_2$ has torsion coefficients $(2,2)$, which distinguishes it from the cyclic group $\mathbb{Z}_4$. The direct sum $\mathbb{Z}_2 \oplus \mathbb{Z}_3$ has torsion coefficient 6.

1.3 The terminology of categories and functors

Categories and functors were introduced by Eilenberg and Mac Lane [EM1] in 1945. We review here these issues, after the hints given in the previous section.

A difficulty of category theory, for a beginner, is the multi-layered structure of this discipline: three layers for the theory itself (categories, functors and their transformations), and two additional layers inside each category (objects and arrows). Starting from examples (as in the previous section) and working on exercises is the best way to master this feature.

Dealing with categories, one has to avoid the usual contradictions related to 'the set of all sets'. Here we make use of a particular set theory where there are *sets* and *classes*; every set is also a class, while a *proper* class is not a set: for instance the *class of all sets* and the *class of all groups* are proper (see [AHS] or the Appendix of [Ke]). This approach is followed in [Mi, Fr, AHS].

*When two levels, like sets and classes, are insufficient, one can introduce a third level of 'hyperclasses'. Alternatively, a more flexible setting widely used for categories is ordinary Set Theory (say ZFC) together with the assumption of the existence of a Grothendieck *universe*, or of a suitable hierarchy of universes (cf. [M2], Section I.6).*

1.3.1 Definition

A *category* C consists of the following data:

(a) a class $\mathrm{Ob}\,\mathsf{C}$, whose elements are called *objects* of C,

(b) for every pair X, Y of objects, a set $\mathsf{C}(X, Y)$ (called a *hom-set*) whose elements are called *morphisms* (or *arrows*) of C from X to Y and denoted as $f\colon X \to Y$,

(c) for every triple X, Y, Z of objects of C, a mapping of *composition*

$$\mathsf{C}(X,Y) \times \mathsf{C}(Y,Z) \to \mathsf{C}(X,Z), \qquad (f,g) \mapsto gf,$$

where gf is also written as $g.f$.

These data must satisfy the following axioms.

(i) (*Associativity*) Given three consecutive arrows $f\colon X \to Y$, $g\colon Y \to Z$ and $h\colon Z \to W$, one has: $h(gf) = (hg)f$.

(ii) (*Identities*) Every object X has an endomorphism $e\colon X \to X$ which acts as an identity whenever composition makes sense: if $f\colon X' \to X$ and $g\colon X \to X''$, then $ef = f$ and $ge = g$.

One shows, in the usual way, that e is determined by X; it is called the *identity* of X and written as $\mathrm{id}\, X$, or 1_X.

We generally assume that the following condition is also satisfied:

(iii) (*Separation*) For X, X', Y, Y' objects of C, the sets $\mathsf{C}(X,Y)$ and $\mathsf{C}(X',Y')$ are disjoint, unless $X = X'$ and $Y = Y'$.

Therefore a morphism $f\colon X \to Y$ has a well-determined *domain* and *codomain*

$$\mathrm{Dom}\,(f) = X, \qquad \mathrm{Cod}\,(f) = Y.$$

Concretely, when constructing a category, one can forget about condition (iii), since one can always satisfy it by redefining a morphism $\hat{f}\colon X \to Y$ as a triple (X, Y, f) where f is a morphism from X to Y in the original sense (possibly not satisfying the Separation axiom).

Mor C denotes the class of all the morphisms of C, i.e. the disjoint union of its hom-sets. Two morphisms f, g of C are said to be *parallel* when they have the same domain and the same codomain. For every object X, the set $\mathsf{C}(X) = \mathsf{C}(X,X)$ of endomorphisms $X \to X$ forms a *monoid* (i.e. a unital semigroup).

If C is a category, the *opposite* (or *dual*) category, written as C^{op}, has the same objects as C, reversed arrows and reversed composition

$$\mathsf{C}^{\mathrm{op}}(X,Y) = \mathsf{C}(Y,X), \qquad g * f = fg, \qquad \mathrm{id}^{\,\mathrm{op}}(X) = \mathrm{id}\, X. \qquad (1.31)$$

Every topic of category theory has a dual instance, which comes from the opposite category (or categories). A dual issue is generally distinguished by the prefix 'co-'.

The term 'map' is generally used as equivalent to morphism; here it will always mean 'continuous mapping' between topological spaces.

Complements. A category is *discrete* if the only arrows are its identities. A category is *indiscrete* if for any two objects X, Y there is precisely one arrow $X \to Y$.

A category is *connected* if it is not empty and any two objects X, Y are linked by some 'zig-zag sequence' of arrows

$$X \to X_1 \leftarrow X_2 \to \ldots \leftarrow X_{n-1} \to Y. \tag{1.32}$$

We follow here the usual convention, excluding the empty category from the connected ones. On the other hand, for connected and path-connected spaces, we are not excluding the empty space, although this might be preferable in the domain of Algebraic Topology (see 1.2.3(d), 1.4.6(i), etc.).

1.3.2 Small and large categories

We have assumed that a category C has a *class* $Ob\, C$ of objects (e.g. the class of all sets, or the class of all topological spaces) and, for every pair X, Y of objects, a *set* $C(X, Y)$ of morphisms from X to Y.

The categories of structured sets that we shall consider, like the examples of 1.2.1, are generally *large* categories, where the objects (or equivalently the morphisms) form a *proper class* (i.e. not a set). A category C is said to be *small* if the class $Ob\, C$ is a set. A *finite* category is a small category whose set of morphisms is finite; then the same is true of its set of objects, since an object is determined by its identity.

Examples and complements. (a) A set S will often be viewed as a discrete small category: its objects are the elements of S, and the only arrows are their (formal) identities; here $S^{\mathrm{op}} = S$.

(b) A preordered set X is equipped with a relation $x \prec x'$ which is assumed to be reflexive and transitive. It will be viewed as a small category, where the objects are the elements of X and the set $X(x, x')$ contains a unique arrow if $x \prec x'$ (which can be written as $(x, x')\colon x \to x'$), and no arrow otherwise.

Composition and identities are uniquely determined, as follows

$$(x', x'').(x, x') = (x, x''), \qquad \mathrm{id}\, x = (x, x). \tag{1.33}$$

In this sense, *categories generalise preordered sets*. Their dualities agree: the category associated to the opposite preordered set X^{op} (with reversed preordering) is dual to the category associated to X. Loosely speaking, the extension from preordered sets (or classes) to categories consists in allowing different arrows between two given objects.

In particular, each finite ordinal defines a category, which is often written as $\mathbf{0}, \mathbf{1}, \mathbf{2}\ldots$ Thus, $\mathbf{0}$ is the empty category; $\mathbf{1}$ is the *singleton category*, i.e. the discrete category on one object; **2** is the *arrow category*, with two objects (0 and 1), and one non-identity arrow, $0 \to 1$.

A (partial) *order* relation is a preordering which is anti-symmetric: $x \prec x'$ and $x' \prec x'$ implies $x = x'$.

(c) A monoid M can be viewed as a small category with one object, say $*$, and one set of morphisms M, from $*$ to $*$; composition is the multiplication of M. In this sense, *categories generalise monoids*. Again, their dualities coincide: the opposite monoid M^{op} corresponds to the opposite category. Loosely speaking, a category can be viewed as a 'many-object' extension of a monoid.

(d) (*Commutative diagrams*) Objects and morphisms of a category are often represented by vertices and arrows in a *diagram*, as in the examples below, to make evident their relationship and which compositions are legitimate

$$
\begin{array}{ccc}
\begin{array}{ccc}
X & \xrightarrow{f} & Y \\[-2pt]
& {}_{h}\searrow & \downarrow{\scriptstyle g} \\[-2pt]
& & Z
\end{array}
&
\begin{array}{ccc}
A & \xrightarrow{f} & B \\[-2pt]
{\scriptstyle h}\downarrow & \searrow^{d} & \downarrow{\scriptstyle k} \\[-2pt]
C & \xrightarrow{g} & D
\end{array}
&
\qquad X \mathrel{\substack{\xrightarrow{u} \\ \xleftarrow{v}}} Y
\end{array}
\qquad (1.34)
$$

We say that such a diagram is *commutative* if:

(i) whenever we have two 'paths' of consecutive arrows, from a certain object to another, the two composed morphisms are the same,

(ii) whenever we have a 'loop' of consecutive arrows, from an object to itself, the composed morphism is the identity of that object.

Thus, the first diagram above is commutative if and only if $gf = h$. For the second, commutativity means that $kf = d = gh$. For the third, it means that $vu = \mathrm{id}\, X$ and $uv = \mathrm{id}\, Y$ (so that these morphisms are inverse to each other).

(e) A small category is (associated to) a preordered set (as in point (b)) if and only if all diagrams in such a category commute.

1.3.3 Isomorphisms and retracts

As already said in 1.2.4, a morphism $f\colon X \to Y$ in a category C is said to be *invertible*, or an *isomorphism*, if it has an inverse, i.e. a morphism $g\colon Y \to X$ such that $gf = 1_X$ and $fg = 1_Y$. The latter is determined by f; it is called the *inverse* of f and written as f^{-1}. (Isomorphisms of topological spaces are called homeomorphisms.)

The *isomorphism relation* $X \cong Y$ between objects of C (meaning that there is an isomorphism $X \to Y$) is an equivalence relation: an obvious consequence of identity arrows, inverting isomorphisms and composing them.

A *groupoid* is a category where every map is invertible. Every discrete or indiscrete category is a groupoid.

The fundamental groupoid of a space X contains all the fundamental groups $\pi_1(X, x)$, for $x \in X$; it will be constructed in Section 6.1.

One-side inverses are also important. If, in a category C, we have two morphisms m and p such that

$$m\colon A \rightleftarrows X \colon p, \qquad pm = \operatorname{id} A, \qquad (1.35)$$

then m is called a *section* (of p), p is called a *retraction* (of m), and A is called a *retract* of X.

In the usual categories of structured sets and structure-preserving mappings, m is injective and p is surjective: we can replace A with its image $m(A) \subset X$, endowed with the restricted structure. (This is common practice, as in 1.1.3 and in the examples below.)

Exercises and complements. Retracts are easily characterised, below, in Set and Ab. Examples in Top have been examined in 1.1.3 and 1.1.6; later, this issue will be studied with the help of homology functors.

(a) In Set a retract of a set $X \neq \emptyset$ is any non-empty subset. The empty set is only a retract of itself. In $\mathsf{Set}_\bullet$ a retract of an object (X, x_0) is any pointed subset (A, x_0).

(b) In Ab, retracts coincide with 'direct summands'. More precisely, a subgroup A of an abelian group X is a retract if and only if there exists a subgroup $B \subset X$ such that $A \cap B = \{0\}$ and $A + B = X$. Then X is said to be the *internal direct sum* of A, B and is canonically isomorphic to $A \oplus B$.

Note that $\mathbb{Z}$ is isomorphic to $2\mathbb{Z} \oplus 0$, but $2\mathbb{Z}$ is not a direct summand of $\mathbb{Z}$: the previous directed sum is not an internal one.

(c) A cyclic group X of infinite or prime order has no other retracts than the trivial, unavoidable ones: $\{0\}$ and X.

(d) In Top, suppose that A is a retract of X, as defined in (1.35). Prove that the map m is a topological embedding, i.e. an injective map $m\colon A \to X$ that makes A homeomorphic to the subspace $m(A) \subset X$ (see 1.1.7(a)).

Identifying A with this subspace we recover the usual definition of retracts in Topology: a subspace $A \subset X$ that admits a retraction, i.e. a map $p\colon X \to A$ such that $p(x) = x$ for all $x \in A$.

(e) In the same hypothesis, prove that p is a topological projection, i.e. a surjective map $p\colon X \to A$ that makes A homeomorphic to the quotient X/R_p, modulo the equivalence relation associated to p (see 1.1.7(b)).

We can also view a retract of X as a quotient $p\colon X \to X/R$ that admits a (continuous) section, i.e. a map $m\colon X/R \to X$ such that $pm = \operatorname{id} X/R$.

1.3.4 Subcategories, quotients and products of categories

(a) Let C be a category. A *subcategory* D is defined by assigning:

- a subclass $\text{Ob}\,D \subset \text{Ob}\,C$, whose elements are called *objects of* D,

- for every pair of objects X, Y of D, a subset $D(X, Y) \subset C(X, Y)$, whose elements are called *morphisms of* D, from X to Y,

so that the following conditions hold:

(i) the composite in C of every pair of consecutive morphisms of D belongs to D,

(ii) the identity in C of every object of D belongs to D.

Then D, equipped with the restricted composition law, is a category.

We say that D is a *full* subcategory of C if, for every pair of objects X, Y of D, we have $D(X, Y) = C(X, Y)$, so that D is determined by assigning a subclass of objects. We say that D is a *wide* subcategory of C if it has the same objects.

For instance, Ab is a full subcategory of Gp. The open continuous mappings form a wide subcategory of Top.

(b) A *congruence* $R = (R_{XY})$ in a category C consists of a family of equivalence relations R_{XY} (also written as R) in each set of morphisms $C(X, Y)$, that is consistent with composition:

(iii) if $f R_{XY} f'$ and $g R_{YZ} g'$, then $gf \, R_{XZ} \, g'f'$.

Equivalently (by transitivity) it is sufficient to say that:

(iii$'$) if $f R f'$ then $fh \, R f'h$ and $kf R \, kf'$ (for all legitimate compositions).

Then one defines the *quotient category* $D = C/R$: the objects are those of C, and $D(X, Y) = C(X, Y)/R_{XY}$; in other words, a morphism $[f]\colon X \to Y$ in D is an equivalence class of morphisms $X \to Y$ in C. The composition is induced by that of C, which is legitimate because of condition (iii):

$$[g][f] = [gf]. \tag{1.36}$$

For instance, in Top the homotopy relation $f \simeq f'$ is a congruence of categories: see 1.4.4.

The relation of isomorphism is wider in a quotient category (but it may coincide with the original one, also in a non-trivial quotient).

(c) If C and D are categories, one defines the *product category* $C \times D$. An object is a pair (X, Y) where X is in C and Y in D; a morphism is a pair of morphisms

$$(f, g)\colon (X, Y) \to (X', Y'), \qquad f \in C(X, X'), \quad g \in D(Y, Y'). \tag{1.37}$$

The composition of (f, g) with a consecutive morphism (f', g') is (obviously) defined componentwise: $(f', g')(f, g) = (f'f, g'g)$.

More generally one defines the *cartesian product* $\mathsf{C} = \prod_{i \in I} \mathsf{C}_i$ of a family of categories $(\mathsf{C}_i)_{i \in I}$ indexed by a set I: an object of C is a family $(A_i)_{i \in I}$ where $A_i \in \mathrm{Ob}\,(\mathsf{C}_i)$ (for every i), and a morphism $f = (f_i)\colon (A_i) \to (B_i)$ is a family of morphisms $f_i \in \mathsf{C}_i(A_i, B_i)$; the composition is componentwise and $\mathrm{id}\,((A_i)_{i \in I}) = (\mathrm{id}\,A_i)_{i \in I}$. A product of groupoids is a groupoid.

Cartesian powers, like $\mathsf{C}^2 = \mathsf{C} \times \mathsf{C}$ and $\mathsf{C}^I = \prod_{i \in I} \mathsf{C}$, will also be used.

1.3.5 Functors

A (covariant) *functor* $F\colon \mathsf{C} \to \mathsf{D}$ consists of the following data:

(a) a mapping $F_0\colon \mathrm{Ob}\,\mathsf{C} \to \mathrm{Ob}\,\mathsf{D}$, whose action is generally written as $X \mapsto F(X)$,

(b) for every pair of objects X, X' in C, a mapping

$$F_{XX'}\colon \mathsf{C}(X, X') \longrightarrow \mathsf{D}(F(X), F(X')),$$

whose action is generally written as $f \mapsto F(f)$.

Composition and identities must be preserved:

(i) if f, g are consecutive maps in C, then $F(gf) = F(g).F(f)$,

(ii) if X is in C, then $F(\mathrm{id}\,X) = \mathrm{id}\,(F(X))$.

For a second functor $G\colon \mathsf{D} \to \mathsf{E}$, one defines in the obvious way the *composed functor* $GF\colon \mathsf{C} \to \mathsf{E}$. The composition of consecutive functors is associative and has identities: the *identity functor* of each category

$$\mathrm{id}\,\mathsf{C}\colon \mathsf{C} \to \mathsf{C}, \qquad X \mapsto X, \qquad f \mapsto f.$$

A functor preserves commutative diagrams, isomorphisms, retracts, sections and retractions.

A *contravariant functor* $F\colon \mathsf{C} \dashrightarrow \mathsf{D}$ can be defined as a covariant functor $\mathsf{C}^{\mathrm{op}} \to \mathsf{D}$, or equivalently $\mathsf{C} \to \mathsf{D}^{\mathrm{op}}$. Composing two contravariant functors one gets a covariant one.

A *functor in two variables* is an ordinary functor $F\colon \mathsf{C} \times \mathsf{D} \to \mathsf{E}$ defined on the product of two categories. Fixing an object X_0 in C, we have a functor $F(X_0, -)\colon \mathsf{D} \to \mathsf{E}$; and symmetrically.

Cat (resp. Gpd) will denote the category of small categories (resp. small groupoids) and their functors. (These categories also have a 2-dimensional structure, with higher morphisms $\varphi\colon F \to G$ between functors: see Section 1.5.)

The action of a functor F on objects and morphisms will often be written without parentheses: FX and Ff for $F(X)$ and $F(f)$.

A functor (resp. a contravariant functor) between two preordered sets, viewed as categories, is a monotone (resp. antitone) mapping. A functor between two monoids, viewed as categories, is a homomorphism of monoids.

1.3.6 Isomorphic categories

An *isomorphism of categories* is a functor $F\colon \mathsf{C} \to \mathsf{D}$ which is invertible: there is a functor $G\colon \mathsf{D} \to \mathsf{C}$ such that $GF = \mathrm{id}\,\mathsf{C}$ and $FG = \mathrm{id}\,\mathsf{D}$. The functor G is determined by F; it is called the *inverse* functor, and written as F^{-1}.

An easy verification proves that the functor F is an isomorphism if and only if all the mappings F_0 and $F_{XX'}$ considered above are bijective.

Being isomorphic categories is plainly an equivalence relation, written as $\mathsf{C} \cong \mathsf{D}$.

Categories linked by an obvious isomorphism are often viewed as 'the same thing'. For instance, everybody knows that an abelian group has a unique structure of module on the ring $\mathbb{Z}$ of integers (where $2x = x + x$, and so on), preserved by every homomorphism of abelian groups. This fact readily shows that the category Ab is canonically isomorphic to the category $\mathbb{Z}\,\mathsf{Mod}$ of modules on the ring of integers; one generally makes no distinction between these categories.

A topological space can be defined in many equivalent ways (based on open sets, closed sets, neighbourhoods, a closure operator, etc.); from a formal point of view, this yields many isomorphic categories which is rarely convenient to distinguish.

1.3.7 Forgetful and structural functors

(a) Forgetting the structure, or part of it, yields various examples of functors between categories of structured sets, like the following obvious instances

$$
\begin{array}{lll}
\mathsf{Top} \to \mathsf{Set}, & \mathsf{Top}_{\bullet} \to \mathsf{Top}, & \mathsf{Top}_{\bullet} \to \mathsf{Set}_{\bullet}, \\
\mathsf{Set}_{\bullet} \to \mathsf{Set}, & \mathsf{Rng} \to \mathsf{Ab}, & \mathsf{Ab} \to \mathsf{Set}.
\end{array}
\tag{1.38}
$$

The first, for instance, forgets the topology, taking a topological space X to its underlying set $|X|$, and a continuous mapping $X \to Y$ to the underlying mapping $|X| \to |Y|$ of sets. The second forgets the basepoint; and so on.

These *forgetful functors* are often denoted by the letter U, which refers to the *underlying* set, or *underlying* space, or *underlying* abelian group, etc.

(b) A subcategory D of C yields an *inclusion* functor D → C, which we also write as D ⊂ C. For instance, Ab ⊂ Gp. These functors *forget properties*, of objects or morphisms, rather than structure.

(c) A congruence R in a category C yields an obvious *projection functor* $P\colon$ C → C$/R$, which is the identity on objects and sends a morphism f to its equivalence class $[f]$. For instance, we are interested in the projection functor Top → hoTop = Top$/\simeq$ (see (1.62)).

(d) A product category C = $\Pi_{i \in I}$ C$_i$ has a family of projection functors $P_i\colon$ C → C$_i$.

1.3.8 Hom functors

(a) Every category C has a functor of morphisms, or *hom-functor*:

$$\mathrm{Mor}_C\colon C^{\mathrm{op}} \times C \to \mathsf{Set}, \quad (X, Y) \mapsto C(X, Y), \quad (h, k) \mapsto k.-.h, \qquad (1.39)$$

where, if $h\colon X' \to X$ and $k\colon Y \to Y'$ are in C, the mapping $k.-.h$

$$k.-.h\colon C(X, Y) \to C(X', Y'), \qquad f \mapsto kfh, \qquad (1.40)$$

means pre-composing with h and post-composing with k.

The hom-functor (1.39) is viewed as a functor on two variables of C, 'contravariant on the first and covariant on the second'. Note that it takes values in Set because of our assumption, in 1.3.1(b), that all classes $C(X, Y)$ be *sets*.

(b) Fixing the first variable we get a covariant functor on C

$$C(X_0, -)\colon C \to \mathsf{Set},$$
$$k \mapsto k_\sharp = k.-\colon C(X_0, Y) \to C(X_0, Y') \qquad (k\colon Y \to Y' \text{ in } C), \qquad (1.41)$$

that is said to be *representable*, and *represented by the object* X_0. This notion will be extended to any functor C → Set isomorphic to $C(X_0, -)$, as defined in 1.5.2.

Dually, fixing the second variable we get a functor

$$C(-, Y_0)\colon C^{\mathrm{op}} \to \mathsf{Set},$$
$$h \mapsto h^\sharp = -.h\colon C(X, Y_0) \to C(X', Y_0) \qquad (h\colon X' \to X \text{ in } C), \qquad (1.42)$$

that is contravariant on C, and *representable* as such. Again, we say the same of a functor $C^{\mathrm{op}} \to$ Set isomorphic to the former. Contravariant functors are often constructed in this way.

(c) In the category Ab of abelian groups, the set of homomorphisms $A \to B$ has a canonical structure of abelian group, by the pointwise addition $(f+g)(x) = f(x) + g(x)$. This abelian group is denoted as $\mathrm{Hom}(A, B)$.

The category Ab has thus an *enriched version* of the hom-functor, with values in the category itself (instead of Set)

$$\mathrm{Hom} \colon \mathsf{Ab}^{\mathrm{op}} \times \mathsf{Ab} \to \mathsf{Ab},$$
$$(A, B) \mapsto \mathrm{Hom}(A, B), \qquad (h, k) \mapsto k. - .h. \tag{1.43}$$

In fact one can easily verify that the mapping

$$k. - .h \colon \mathrm{Hom}(A, B) \to \mathrm{Hom}(A', B')$$

preserves the addition of homomorphisms: $k(f+g)h = kfh + kgh$.

(d) We have already used representable contravariant functors in the previous section, in an enriched form (with values in rings): the functors $P = \mathsf{Set}(-, \mathbb{Z})$ in (1.27), and $\mathcal{A} = \mathsf{Top}(-, \mathbb{Z})$ in (1.15). This comes out of the fact that $\mathbb{Z}$ is a ring, and a topological ring (with the discrete topology).

1.4 Paths and homotopy

We review here the elementary issues of homotopy theory, also to fix the terminology we are using.

Typically, the 'invariants' of Algebraic Topology are invariant up to homotopy — a stronger property than invariance up to homeomorphism; a homology functor H_k extracts from the complexity of a space, or a map, a limited information about k-dimensional connectedness, which we can hope to compute; choosing k, or collecting all the information provided by these functors (or others), we may be able to solve the problem we are studying.

The two-valued index α (or β) varies in the set $\{0, 1\}$, also written as $\{-, +\}$.

1.4.1 Euclidean spaces

We use the following notation, for spaces which play an important role in Topology and Algebraic Topology.

The space $\mathbb{R}^n$ of n-tuples of real numbers $x = (x_1, ..., x_n)$ is equipped with the euclidean topology, determined by the euclidean norm and the euclidean distance

$$||x|| = (x_1^2 + ... + x_n^2)^{1/2}, \qquad d(x, y) = ||x - y||, \tag{1.44}$$

or equivalently as a cartesian power of the euclidean line $\mathbb{R}$.

Identifying $\mathbb{R}^n$ with the hyperplane $\mathbb{R}^n \times \{0\} \subset \mathbb{R}^{n+1}$, these spaces form a nested family, and the singleton $\mathbb{R}^0 = \{0\}$ is included in all of them

$$\{0\} = \mathbb{R}^0 \subset \mathbb{R}^1 \subset \mathbb{R}^2 \subset \ldots \ \mathbb{R}^n \ \subset \ldots \tag{1.45}$$

A *euclidean space* is any subspace of some $\mathbb{R}^n$, and an *open euclidean space* is any open subspace of some $\mathbb{R}^n$.

$\mathbb{I}$ denotes the standard compact interval $[0,1]$ of $\mathbb{R}$, on which paths and homotopies are parametrised. Its cartesian power $\mathbb{I}^n$ is the *standard n-cube*, a subspace of $\mathbb{R}^n$. We are also interested in the standard n-disc (or closed ball)

$$\mathbb{D}^n = \{x \in \mathbb{R}^n \mid ||x|| \leqslant 1\}. \tag{1.46}$$

Inside $\mathbb{R}^{n+1}$, we also have the *standard n-sphere*

$$\mathbb{S}^n = \{x \in \mathbb{R}^{n+1} \mid ||x|| = 1\} \qquad (n \geqslant 0), \tag{1.47}$$

with basepoint at $e_1 = (1, 0, \ldots, 0)$, when useful.

(We can specify $e_1^n \in \mathbb{S}^n \subset \mathbb{R}^{n+1}$, but in fact all these points are identified with $1 \in \mathbb{R}$, in the nesting (1.45); and also with the unit of $\mathbb{C}$.)

In particular, $\mathbb{S}^0 = \{-1, 1\} \subset \mathbb{R}$ is a discrete space on two points, pointed at 1. Adding a lower sphere $\mathbb{S}^{-1} = \emptyset$ as the sphere of $\mathbb{R}^0 = \{0\}$, the discs and the spheres form a nested family of subspaces of the family (1.45)

$$\begin{aligned}\{0\} = \mathbb{D}^0 \ &\subset \ \mathbb{D}^1 \ \subset \ \mathbb{D}^2 \ \subset \ldots \ \mathbb{D}^n \ \subset \ldots \\ \emptyset = \mathbb{S}^{-1} \ &\subset \ \mathbb{S}^0 \ \subset \ \mathbb{S}^1 \ \subset \ldots \ \mathbb{S}^{n-1} \ \subset \ldots.\end{aligned} \tag{1.48}$$

where $\mathbb{S}^{n-1} = \underline{\partial}\mathbb{D}^n$ is the *border* of $\mathbb{D}^n$ in $\mathbb{R}^n$, for all $n \geqslant 0$. (The term 'boundary' will be used in other meanings.)

The convenience of adding $\mathbb{S}^{-1}$ will also appear elsewhere. Its topological dimension should be set at $-\infty$, to agree with the formula $\dim(X \times Y) = \dim X + \dim Y$.

The standard circle $\mathbb{S}^1$ can also be viewed as a subspace of the complex plane, which topologically is the same as $\mathbb{R}^2$

$$\mathbb{S}^1 = \{z \in \mathbb{C} \mid |z| = 1\}. \tag{1.49}$$

The ordinary parametrisation of the circle on the standard interval

$$p: \mathbb{I} \to \mathbb{S}^1, \qquad p(t) = (\cos 2\pi t, \sin 2\pi t) = e^{2\pi it}, \tag{1.50}$$

called the *exponential map*, covers the circle counterclockwise (in the oriented plane), identifying the two points of the border $\underline{\partial}\mathbb{I} = \{0, 1\}$ in $\mathbb{R}$.

Writing $\mathbb{I}/\underline{\partial}\mathbb{I}$ the quotient space, the map p induces a bijective continuous mapping

$$p': \mathbb{I}/\underline{\partial}\mathbb{I} \to \mathbb{S}^1, \qquad p'[t] = (\cos 2\pi t, \sin 2\pi t) = e^{2\pi it}, \tag{1.51}$$

which is a homeomorphism (as any bijective map from a compact to a Hausdorff space).

1.4.2 Paths and loops

A *path* in the space X is a map $a\colon \mathbb{I} \to X$ defined on the standard euclidean interval $\mathbb{I} = [0,1]$.

The basic structure of the interval $\mathbb{I}$ consists of four maps, linking the interval to the singleton $\mathbb{I}^0 = \{*\}$, its 0-th cartesian power:

$$\partial^\alpha\colon \{*\} \rightleftarrows \mathbb{I}, \qquad \partial^-(*) = 0, \ \ \partial^+(*) = 1 \qquad\qquad (faces),$$

$$e\colon \mathbb{I} \to \{*\}, \qquad e(t) = * \qquad\qquad\qquad\qquad (degeneracy), \qquad (1.52)$$

$$r\colon \mathbb{I} \to \mathbb{I}, \qquad r(t) = 1 - t \qquad\qquad\qquad\quad (reversion).$$

Identifying a point $x \in X$ with the corresponding map $x\colon \{*\} \to X$, this basic structure determines:

(i) the *endpoints* of a path $a\colon \mathbb{I} \to X$, namely the points $a\partial^- = a(0)$ and $a\partial^+ = a(1)$,

(ii) the *trivial path*, or constant path, at the point x, written as $e_x = xe$,

(iii) the *reversed* path of a, written as $a^\sharp = ar$.

We write as $P(X)$ the set of paths of the space X, and

$$P(X, x_0, x_1) = \{a \in P(X) \mid a(0) = x_0, \ a(1) = x_1\}, \qquad (1.53)$$

will be the set of paths of X, *from x_0 to x_1*. (These sets are canonically equipped with the compact-open topology, reviewed in Section 7.2.)

Two paths $a, b\colon \mathbb{I} \to X$ are *consecutive* if $a(1) = b(0)$, that is $a\partial^+ = b\partial^-$; then they have a *concatenated path* $a * b\colon \mathbb{I} \to X$ (see Exercise (a))

$$(a * b)(t) = \begin{cases} a(2t), & \text{for } 0 \leqslant t \leqslant 1/2, \\ b(2t - 1), & \text{for } 1/2 \leqslant t \leqslant 1. \end{cases} \qquad (1.54)$$

This *partial* operation is not associative, generally; we shall see that it works well up to homotopy with fixed endpoints.

A *loop* at $x_0 \in X$ is a path from x_0 to x_0, called the *basepoint* of the loop. The set of loops of X at x_0

$$\Omega(X, x_0) = P(X, x_0, x_0) = \{a \in \mathsf{Top}(\mathbb{I}, X) \mid a(0) = x_0 = a(1)\}, \qquad (1.55)$$

inherits a binary operation of concatenation, which again is not associative (except trivial cases).

Exercises and complements. (a) Verify that the mapping $a * b \colon \mathbb{I} \to X$ is well defined in (1.54), and continuous.

Although this fact is quite elementary, the *Finite Closed Cover Lemma*, in 1.1.8, is the standard tool for concatenations of paths, homotopies, etc.

(b) For two points $x, x' \in X$, we write $x \simeq x'$ if there is a path in X from x to x'. It is an equivalence relation, because of trivial paths, reversed paths and path-concatenation.

(c) The *path component* $[x]_P$ of the point $x \in X$ is its equivalence class for this relation. A space X is *path connected* if any two points are linked by a path; this implies that X is connected.

The empty space has no path component, and no connected component.

(d) A loop $a \colon \mathbb{I} \to X$ amounts to map $\hat{a} \colon \mathbb{S}^1 \to X$, where $\hat{a}(e^{2\pi i t}) = a(t)$. We say that a is a *simple loop* if $\hat{a}$ is injective; then, if X is Hausdorff, $\hat{a}$ is a topological embedding (see 1.1.7) and gives a homeomorphism of the circle onto its image, a subspace of X.

A loop at x_0 amounts to a pointed map $\hat{a} \colon (\mathbb{S}^1, e_1) \to (X, x_0)$.

1.4.3 Cylinder and homotopies

The *cylinder* $IX = X \times \mathbb{I}$ on the space X has the product topology.

Given two maps $f, g \colon X \to Y$ (in Top), a homotopy $\varphi \colon f \simeq g$ 'is' a map $\varphi \colon X \times \mathbb{I} \to Y$ defined on the cylinder IX, that coincides with f on the lower basis and with g on the upper one

$$\varphi \colon X \times \mathbb{I} \to Y,$$
$$\varphi(x, 0) = f(x), \quad \varphi(x, 1) = g(x) \qquad (\text{for } x \in X), \tag{1.56}$$

forming a continuous deformation of f into g.

The homotopy will also be written as $\varphi \colon f \simeq g \colon X \to Y$. The map (1.56) will be written as $\hat{\varphi}$ when we want to distinguish it from the homotopy which it represents.

All this is based on the basic structure of the cylinder $IX = X \times \mathbb{I}$, inherited from the structural maps (1.52) of the standard interval, and written as the latter

$$\partial^\alpha \colon X \to IX, \quad \partial^-(x) = (x, 0), \ \partial^+(x) = (x, 1) \qquad (faces),$$
$$e \colon IX \to X, \quad e(x, t) = x \qquad (degeneracy), \tag{1.57}$$
$$r \colon IX \to I), \quad r(x, t) = (x, 1 - t) \qquad (reversion).$$

This structure gives:

(i) the *faces* $\varphi \partial^\alpha \colon X \to Y$ of the homotopy, namely the maps $\varphi \partial^- = f$ and $\varphi \partial^+ = g$,

(ii) the *trivial homotopy* $e_f = fe$ of a map f,

(iii) the *reversed homotopy* $\varphi^\sharp = \varphi r$.

More precisely, the homotopy $\varphi^\sharp \colon g \simeq f$ is represented by the map $\hat{\varphi}r \colon IX \to Y$.

Identifying $I\{*\} = \mathbb{I}$, a path $a \colon \mathbb{I} \to X$ is the same as a homotopy $a \colon x \simeq x' \colon \{*\} \to X$ between its endpoints (consistently with the notation $x \simeq x'$ we are using). Moreover the structural maps of the standard interval coincide with the previous maps (1.57), when X is the singleton.

Two consecutive homotopies $\varphi \colon f \simeq g$ and $\psi \colon g \simeq h$ of maps $X \to Y$ have a *concatenation* $\varphi * \psi \colon f \simeq h$

$$(\varphi * \psi)(x,t) = \begin{cases} \varphi(x, 2t), & \text{for } (x,t) \in X \times [0, 1/2], \\ \psi(x, 2t - 1), & \text{for } (x,t) \in X \times [1/2, 1], \end{cases} \qquad (1.58)$$

which extends the concatenation of paths (see Exercise (a)).

Homotopy equivalences and deformation retracts have already been dealt with, at a basic level, in 1.1.3.

Exercises and complements. (a) Verify that the mapping $\varphi * \psi \colon X \times \mathbb{I} \to Y$ is well defined in (1.58), and continuous.

(b) Let X be a contractible space, i.e. homotopy equivalent to the singleton (see 1.1.3). Then X is path-connected and non-empty, and any singleton $\{x\}$ of X is a deformation retract.

(c) The following conditions on a space Y are equivalent:
- Y is *totally path disconnected*, i.e. every path in Y is trivial,
- for every space X, every homotopy $\varphi \colon f \simeq g \colon X \to Y$ is trivial.

*(d) The interval $\mathbb{I}$ (and therefore the cylinder $I(X)$) has a higher structure

$$\begin{aligned} g^- &\colon \mathbb{I}^2 \to \mathbb{I}, & g^-(t, t') &= \max(t, t'), \\ g^+ &\colon \mathbb{I}^2 \to \mathbb{I}, & g^+(t, t') &= \min(t, t') & (\textit{connections}), & \qquad (1.59) \\ s &\colon \mathbb{I}^2 \to \mathbb{I}^2, & s(t, t') &= (t', t) & (\textit{transposition}), \end{aligned}$$

which governs double homotopies (i.e. homotopies of homotopies) through the double cylinder $I^2(X) = X \times \mathbb{I}^2$: see [G5], Subsection 5.2.2.

The whole structure satisfies the axioms of a 'diad' — a cubical version of a monad [G1]. Double homotopies will be frequently used in Chapter 6. The formal structure of the cylinder is also investigated in [KP].

1.4.4 The homotopy category

Two parallel maps $f, g \colon X \to Y$ in Top are said to be *homotopic*, written as $f \simeq g$, if there exists a homotopy $\varphi \colon f \simeq g$. This is an equivalence relation in $\mathsf{Top}(X, Y)$, because of trivial homotopies, reversed homotopies and homotopy concatenation.

Furthermore, homotopies have a *whisker composition* with maps

$$X' \xrightarrow{\ h\ } X \underset{g}{\overset{f}{\rightrightarrows}} {\scriptstyle\downarrow\varphi}\ Y \xrightarrow{\ k\ } Y' \tag{1.60}$$

$$k\varphi h \colon kfh \to kgh \colon X' \to Y'.$$

where $k\varphi h$ is represented on the cylinder by the map:

$$k.\hat{\varphi}.I(h) \colon IX' \to Y', \qquad (x', t) \mapsto k(\varphi(h(x'), t), \tag{1.61}$$

for $(x', t) \in X' \times \mathbb{I}$. The homotopy relation is thus a congruence of categories in Top.

The quotient category $\mathrm{ho}\mathsf{Top} = \mathsf{Top}/\simeq$ is called the *homotopy category of topological spaces*: its objects are the topological spaces, and a morphism $[f] \colon X \to Y$ is a homotopy class of continuous mappings $X \to Y$. We have a projection functor

$$P \colon \mathsf{Top} \to \mathrm{ho}\mathsf{Top}, \qquad P(X) = X, \quad P(f) = [f]. \tag{1.62}$$

The classical notation $[X, Y]$ stands for the set of these homotopy classes $[f] \colon X \to Y$

$$[X, Y] = \mathrm{ho}\mathsf{Top}(X, Y) = \mathsf{Top}(X, Y)/\simeq. \tag{1.63}$$

An isomorphism in $\mathrm{ho}\mathsf{Top}$ is the same as a homotopy equivalence in Top, namely a pair of maps $f \colon X \rightleftarrows Y : g$ such that $gf \simeq \mathrm{id}\, X$ and $fg \simeq \mathrm{id}\, Y$.

A functor $F \colon \mathsf{Top} \to \mathsf{D}$ with values in an arbitrary category is said to be *homotopy invariant* if $F(f) = F(g)$ whenever $f \simeq g$ in Top, so that F induces a functor

$$\overline{F} \colon \mathrm{ho}\mathsf{Top} \to \mathsf{D}, \qquad \overline{F}(X) = F(X), \quad \overline{F}(f) = F[f], \tag{1.64}$$

determined by the relation $F = \overline{F}P$. Typically, this will be the case of all homology and homotopy functors. For simplicity, we may use the same symbol for F and $\overline{F}$.

A homotopy invariant functor $\mathsf{Top} \to \mathsf{D}$ takes a homotopy equivalence $f \colon X \to Y$ in Top to an isomorphism in D.

*The *weak homotopy category* $\mathrm{Ho}\mathsf{Top}$ is obtained by a more complex procedure, formally inverting all the continuous mappings that — loosely speaking — induce isomorphisms in all homotopy sets and groups [GaZ, Bo1]. It can also be obtained as the full subcategory of $\mathrm{ho}\mathsf{Top}$ whose objects are the spaces homotopically equivalent to CW-complexes.*

1.4.5 Pointed homotopy

The category $\mathsf{Top}_\bullet$ of pointed spaces has analogous issues. Given two parallel pointed maps $f, g \colon (X, x_0) \to (Y, y_0)$, a *pointed homotopy* $\varphi \colon f \simeq g$, or *homotopy of pointed maps*, is an ordinary homotopy that keeps the basepoint fixed

$$\varphi \colon X \times \mathbb{I} \to Y, \qquad \varphi(x, 0) = f(x), \quad \varphi(x, 1) = g(x),$$
$$\varphi(x_0, t) = y_0 \qquad\qquad (\text{for } x \in X,\ t \in \mathbb{I}). \tag{1.65}$$

Also here, pointed homotopies present trivial instances, reversion, concatenation and whisker composition with pointed maps. The relation $f \simeq g$ of pointed homotopy, or *homotopy of pointed maps*, is a congruence of categories in $\mathsf{Top}_\bullet$, and the quotient category will be written as $\mathsf{hoTop}_\bullet$.

A functor $F \colon \mathsf{Top}_\bullet \to \mathsf{D}$ which is *homotopy invariant* (up to pointed homotopy), induces a functor $\overline{F}$ (also written as F)

$$\overline{F} \colon \mathsf{hoTop}_\bullet \to \mathsf{D}, \qquad \overline{F}(X) = F(X), \quad \overline{F}(f) = F[f]. \tag{1.66}$$

A pointed map $f \colon (X, x_0) \to (Y, y_0)$ is a (pointed) *homotopy equivalence* if it has a weak inverse g

$$f \colon (X, x_0) \rightleftarrows (Y, y_0) : g, \qquad gf \simeq \mathrm{id}\,(X, x_0), \quad fg \simeq \mathrm{id}\,(Y, y_0). \tag{1.67}$$

Then we say that (X, x_0) and (Y, y_0) are *homotopically equivalent* in $\mathsf{Top}_\bullet$, or as pointed spaces, and we write $(X, x_0) \simeq (Y, y_0)$. This amounts to isomorphism in $\mathsf{hoTop}_\bullet$, and is an equivalence relation.

The notation $[(X, x_0), (Y, y_0)]_\bullet$ will stand for the set of homotopy classes of pointed maps $[f] \colon (X, x_0) \to (Y, y_0)$

$$[(X, x_0), (Y, y_0)]_\bullet = \mathsf{hoTop}_\bullet((X, x_0), (Y, y_0)). \tag{1.68}$$

Working in $\mathsf{Set}_\bullet$ and $\mathsf{Top}_\bullet$, any mapping or homotopy is understood to be pointed, by default. It may be convenient to write a pointed set (or space) as $(X, 0)$, viewed as a set (or space) equipped with a nullary operation $0 \colon \{*\} \to X$ (a constant). Or just as X, leaving the nullary operation understood (as is generally done with algebraic structures).

Accordingly, the homotopy hom-set (1.68) can be written as $[X, Y]_\bullet$, keeping the dot for clarity.

1.4.6 Basic homotopy functors

We define various homotopy invariant functors, covariant or contravariant, on Top or $\mathsf{Top}_\bullet$. The variable space, or pointed space, is denoted as X; the variable map, or pointed map, as $f \colon X \to Y$.

(i) The homotopy invariant functor $\Pi_0 \colon \mathsf{Top} \to \mathsf{Set}$ of path components (introduced in (1.14)) can be rewritten in the form

$$\Pi_0 \colon \mathsf{Top} \to \mathsf{Set}, \qquad \Pi_0(X) = |X|/\simeq,$$

$$f_* = \Pi_0(f) \colon \Pi_0(X) \to \Pi_0(Y), \qquad f_*([x]_P) = [f(x)]_P, \tag{1.69}$$

where $f \colon X \to Y$ is a map. In fact the quotient $|X|/\simeq$ is precisely the set of path components $[x]_P$ of X (see 1.4.2(c)), and the map $\Pi_0(f)$ defined in 1.2.2 acts as above.

We say that the space X is *0-connected* if $\Pi_0(X)$ is a singleton, which means that X is path connected and non-empty. Note that Π_0 sharply distinguishes the empty space from all other path connected spaces — as is usually the case in Algebraic Topology.

(ii) We have a similar functor from pointed spaces to pointed sets:

$$\pi_0 \colon \mathsf{Top}_\bullet \to \mathsf{Set}_\bullet, \qquad \pi_0(X) = (\Pi_0(X), [0_X]_P), \tag{1.70}$$

that acts as above on pointed maps. It will also be used in Chapter 6.

Note. We use the letter Π for action on spaces, and π on pointed spaces.

(iii) Fixing a pointed space S, there is a *homotopy functor*

$$\pi_S = [S, -]_\bullet \colon \mathsf{Top}_\bullet \to \mathsf{Set}_\bullet,$$

$$f_* \colon [S, X]_\bullet \to [S, Y]_\bullet, \qquad f_*([a]) = [fa], \tag{1.71}$$

which corresponds to a hom-functor $\mathsf{hoTop}_\bullet \to \mathsf{Set}_\bullet$ represented by the pointed space S, and is thus homotopy invariant.

(iv) The space S also determines a *cohomotopy functor*

$$\pi^S = [-, S]_\bullet \colon \mathsf{Top}_\bullet{}^{\mathrm{op}} \to \mathsf{Set}_\bullet,$$

$$f^* \colon [S, Y]_\bullet \to [S, X]_\bullet, \qquad f^*([b]) = [bf], \tag{1.72}$$

which 'is' a hom-functor $\mathsf{hoTop}_\bullet{}^{\mathrm{op}} \to \mathsf{Set}_\bullet$ represented by S, and is again homotopy invariant.

Exercises and complements. (a) The functor $\Pi_0 \colon \mathsf{hoTop} \to \mathsf{Set}$ is represented by the singleton $S = \{*\}$.

(b) The functor $\pi_0 \colon \mathsf{hoTop}_\bullet \to \mathsf{Set}_\bullet$ is represented by the sphere $\mathbb{S}^0 = \{-1, 1\}$, pointed at 1.

1.4.7 Homotopy of paths

For paths a, b in X, the general homotopy relation $a \simeq b$ of parallel maps is of little relevance (see Exercise (a)). We are interested in a stronger property, that will be useful to compute the homology group $H_1(X)$, in

Chapter 2, and will be the starting point for the definition of the fundamental groupoid $\Pi_1 X$, in Chapter 6.

We say that the paths $a, b \colon \mathbb{I} \to X$ are *homotopic with fixed endpoints,* or *2-homotopic,* if there is a homotopy h from a to b which does not move the endpoints, as shown in the figure below

$$h \colon \mathbb{I} \times \mathbb{I} \to X, \qquad h(s,0) = a(s), \qquad h(s,1) = b(s),$$
$$h(0,t) = x, \qquad h(1,t) = y \qquad (\text{for } s,t \in \mathbb{I}), \tag{1.73}$$

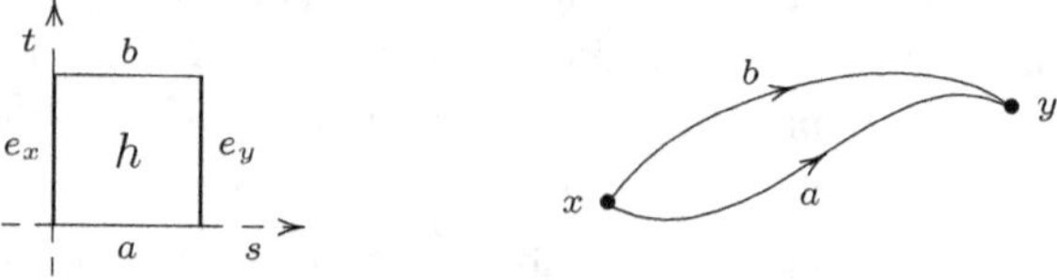

This requires that a, b have the same endpoints, $x = a(0) = b(0)$ and $y = a(1) = b(1)$; e_x and e_y are the constant loops at them. The 2-homotopy h will be denoted as $h \colon a \simeq_2 b \colon x \simeq y$.

In this context, we shall use s as the variable of the given paths, and t as the deformation variable. *The name of 2-homotopy and the notation $h \colon a \simeq_2 b \colon x \simeq y$ is suggested by the 2-cells in a 2-category, see 1.5.2(e), although here we have a lax version of such a structure.*

Loosely speaking, the property $a \simeq_2 b$ means that in X *there is no 1-dimensional hole between the given paths.* The space X is said to be *simply connected* if it is path connected and, for every pair $x, y \in X$, all paths in X from x to y are 2-homotopic; it is said to be *1-connected* if it is simply connected and non-empty. We shall see that these properties are invariant up to homotopy equivalence, in 6.1.5(d); in particular, every contractible space is 1-connected.

A finite sequence $a_1, a_2, ..., a_n$ of consecutive paths in X can be concatenated in various ways, by iterated binary concatenations, and it is important to show that all the paths we can obtain in this way are homotopic with fixed endpoints.

We write as $a_1 * a_2 * ... * a_n$ the *regular concatenation* of the sequence, based on the regular decomposition of the standard interval $\mathbb{I}$ in n subintervals of length $1/n$, indexed by $k = 1, ..., n$

$$(a_1 * a_2 * ... * a_n)(t) = a_k(nt - k + 1), \qquad \text{for } (k-1)/n \leqslant t \leqslant k/n. \tag{1.74}$$

In general, any decomposition of $\mathbb{I}$ in n closed subintervals meeting at the endpoints

$$\mathbb{I} = \bigcup_{k \geqslant 1} [t_{k-1}, t_k], \qquad 0 = t_0 < t_1 < ... < t_n = 1, \tag{1.75}$$

gives raise to *a* concatenation $c\colon \mathbb{I} \to X$ of the paths a_k $(k = 1, ..., n)$

$$c(t) = a_k((t - t_{k-1})/(t_k - t_{k-1})), \qquad \text{for } t \in [t_{k-1}, t_k]. \tag{1.76}$$

Exercise (d) shows that all of them are homotopic with fixed endpoints.

All iterated binary concatenations of $a_1, a_2, ..., a_n$ can be obtained in this way: for instance, $a_1 * (a_2 * a_3)$ comes out of the decomposition $0 < 1/2 < 3/4 < 1$.

Exercises and complements. (a) Two paths $a, b\colon \mathbb{I} \to X$ are homotopic (in the general sense, as maps) if and only if their images belong to the same path component of X. This says nothing more on X than the homotopy relation $x \simeq y$ of its points.

(b) Homotopy with fixed endpoints is an equivalence relation between paths (in a fixed space). Moreover, $a \simeq_2 b$ is equivalent to $a^\sharp \simeq_2 b^\sharp$.

(c) (*Reparametrised paths*) A *reparametrisation* of the standard interval will be a continuous function

$$f\colon \mathbb{I} \to \mathbb{I}, \qquad f(0) = 0, \quad f(1) = 1, \tag{1.77}$$

which we are not assuming to be invertible.

For every path $a\colon \mathbb{I} \to X$, the *reparametrised path* $af\colon \mathbb{I} \to X$ is 2-homotopic to the former.

Hints: use a 2-homotopy $h\colon \mathrm{id}\,\mathbb{I} \simeq_2 f\colon \mathbb{I} \to \mathbb{I}$ 'of reparametrisation'.

(d) For a finite sequence $a_1, a_2, ..., a_n$ of consecutive paths in X, every concatenation c (defined as in (1.76)) is 2-homotopic to the regular concatenation $a_1 * a_2 * ... * a_n$ defined in (1.74).

1.4.8 Homotopy of loops

For loops in X, we are interested in homotopy with fixed basepoint (still denoted as $a \simeq_2 b$), but also in a more general relation.

Two loops $a\colon x \simeq x$ and $b\colon y \simeq y$ in X are *homotopic as (free) loops*, or *loop homotopic*, if there is a homotopy $h\colon a \simeq b$ such that each intermediate path $h_t = h(-, t)\colon \mathbb{I} \to X$ of the deformation is a loop, as shown in the figure below

$$h\colon \mathbb{I} \times \mathbb{I} \to X, \qquad h(s, 0) = a(s), \qquad h(s, 1) = b(s),$$
$$h(0, t) = h(1, t) \qquad\qquad (\text{for } s, t \in \mathbb{I}), \tag{1.78}$$

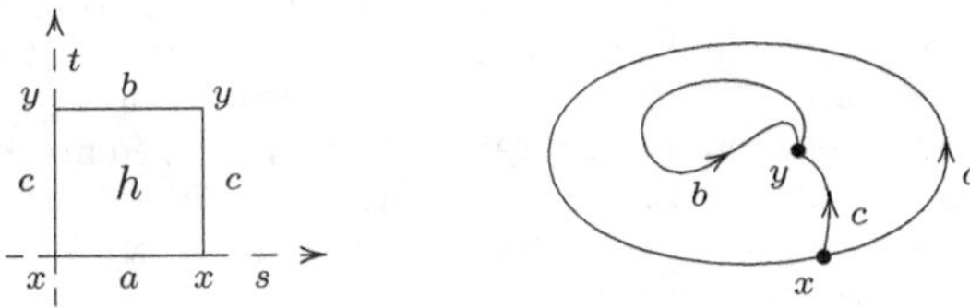

The figure represents a loop homotopy: each intermediate step h_t between a and b is a loop at the point $c(t) = h(0,t) = h(1,t)$; these points describe the path $c(t) = h(0,t)$. In a 2-homotopy $h\colon a \simeq_2 b$ (between two loops at the same basepoint), the path c is constant.

We shall see that loop homotopy is relevant in homology (in 2.2.8(c)), while the stronger condition of 2-homotopy of loops is relevant for the fundamental group (in Section 6.2). Two loops at the same basepoint can be homotopic as free loops, and not 2-homotopic: see Exercise 6.4.4(c).

Remarks and complements. (a) Again, loop homotopy in the space X is an equivalence relation, with the same proof as for 2-homotopy.

(b) The quotient of $\mathbb{I} \times \mathbb{I}$ modulo the equivalence relation that identifies $(0,t)$ and $(1,t)$ (for every $t \in \mathbb{I}$) is the cylinder $\mathbb{S}^1 \times \mathbb{I}$. The loop homotopy (1.78) amounts to an ordinary homotopy $\hat{a} \simeq \hat{b}\colon \mathbb{S}^1 \to X$, between the maps associated to a and b, in 1.4.2(d).

1.4.9 The Lebesgue Number Lemma

Studying maps $a\colon \mathbb{I}^n \to X$ defined on the standard cube (and paths, in particular), we shall have to decompose a on small cubes $A \subset \mathbb{I}^n$, where A is measured by its *diameter*, computed by the euclidean distance

$$\mathrm{diam}(A) = \sup\{d(x,y) \mid x,y \in A\} \leqslant \sqrt{n}. \tag{1.79}$$

The decomposition will use the following classical result, proved in many books on general topology: for instance in [Mu], Chapter 3, Lemma 2.7.5.

Lebesgue Number Lemma. *Let M be a compact metric space and $\mathcal{U}$ an open cover of M. There exists a number $\varepsilon > 0$ such that every subset $A \subset M$ with $\mathrm{diam}(A) < \varepsilon$ is contained in some element U of the cover.*

Such a number is called a *Lebesgue number of $\mathcal{U}$*.

1.5 Natural transformations and equivalence of categories

Between two functors $F, G\colon \mathsf{C} \to \mathsf{D}$ there can be 'higher arrows' $\varphi\colon F \to G$, called 'natural transformations' (and functorial isomorphisms, or natural isomorphisms, when they are invertible).

'Category', 'functor' and 'natural transformation' are the three basic terms of category theory, since the very beginning of the theory in [EM1].

It is interesting to note that only the last term is taken from the common language: one can say that Eilenberg and Mac Lane introduced categories and functors because they wanted to formalise the natural transformations that they were encountering in Algebra and Algebraic Topology (as remarked in [M2], at the end of Section I.4). Much in the same way as a general theory of *continuity* (a familiar term for a familiar notion) requires the introduction of *topological spaces* (denoted by a theoretical term).

1.5.1 Natural transformations

Given two *parallel* functors $F, G\colon \mathsf{C} \to \mathsf{D}$ (between the same categories), a *natural transformation* $\varphi\colon F \to G$ (also written as $\varphi\colon F \to G\colon \mathsf{C} \to \mathsf{D}$) consists of the following data:

- for each object X of C, a morphism $\varphi X\colon FX \to GX$ in D, called the *component* of φ on X,

so that, for every arrow $f\colon X \to X'$ in C, we have a commutative square in D

$$
\begin{array}{ccc}
FX & \xrightarrow{\ \varphi X\ } & GX \\
{\scriptstyle Ff}\downarrow & & \downarrow{\scriptstyle Gf} \\
FX' & \xrightarrow[\ \varphi X'\ ]{} & GX'
\end{array}
\qquad\qquad \varphi X'.Ff = Gf.\varphi X \ \ (= \varphi(f)). \qquad (1.80)
$$

This is called the *naturality condition* of φ on the morphism f. The component of φ on the object X is also written as φ_X, or as $\varphi\colon FX \to GX$. As above, the diagonal of the square can be written as $\varphi(f)$.

A natural transformation $\varphi\colon F \to G$ often arises from a 'canonical choice' of a family $(\varphi X)_X$ of morphisms, but these two aspects are not the same. There are canonical choices of such families that are not natural, and natural transformations 'defined' by means of the axiom of choice (see Exercise 6.1.9(b)).

The *identity* of a functor $F\colon \mathsf{C} \to \mathsf{D}$ is the natural transformation $\operatorname{id} F\colon F \to F$, with components $(\operatorname{id} F)X = \operatorname{id}(FX)$.

Exercises and complements. (a) Let $F\colon R\,\mathsf{Mod} \to R\,\mathsf{Mod}$ be the identity functor of the category of modules on a commutative (unital) ring R. Every scalar $\lambda \in R$ gives a natural transformation $\lambda\colon F \to F$ whose component $\lambda_A\colon A \to A$ on a module A is the multiplication $\lambda_A(x) = \lambda x$, by λ.

(b) Every natural transformation $\varphi\colon F \to F$ is of this form, determined by a unique $\lambda \in R$. *Hints:* use the component of φ on R, as a module on itself.

1.5.2 Two operations

Natural transformations $\varphi\colon F \to G$ and $\psi\colon G \to H$ between the same categories have a *vertical composition*, written as $\psi\varphi$ or $\psi.\varphi$

$$
\mathsf{C} \substack{\xrightarrow{\ F\ } \\ \downarrow\varphi \\ \xrightarrow{\quad} \\ \downarrow\psi \\ \xrightarrow[\ H\ ]{}} \mathsf{D}
\qquad\qquad (\psi\varphi)(X) = \psi X.\varphi X\colon FX \to HX, \qquad (1.81)
$$

which is associative; the identities of functors act as units.

There is also a *whisker composition*, or *reduced horizontal composition*, of natural transformations with functors, written as $K\varphi H$ (or $K{\circ}\varphi{\circ}H$, when useful to distinguish compositions)

$$C' \xrightarrow{\ H\ } C \underset{G}{\overset{F}{\Rightarrow}}{\downarrow\varphi} D \xrightarrow{\ K\ } D' \qquad (1.82)$$

$$K\varphi H : KFH \to KGH, \qquad (K\varphi H)(X') = K(\varphi(HX')).$$

An *isomorphism of functors*, or *functorial isomorphism*, or *natural isomorphism*, is a natural transformation $\varphi: F \to G$ which is invertible for the vertical composition. Then we write $F \cong G$; it is obviously an equivalence relation between parallel functors. Two examples can be found in the solution to Exercises 1.4.6(a), (b).

It is easy to verify that the natural transformation $\varphi: F \to G: C \to D$ is invertible if and only if all its components $\varphi X: FX \to GX$ are invertible in D: if this is the case, the inverse components $\psi X = (\varphi X)^{-1}: GX \to FX$ form a natural transformation $\psi: G \to F$ which is inverse to φ.

As a consequence, every natural transformation $\varphi: F \to G: C \to D$ with values in a groupoid D is invertible.

Exercises and complements. (The calculus of natural transformations.) The reader should prove the following properties of the compositions of natural transformations, which will often be used in computations. The (easy) solution to point (d) can be found in Chapter 8; the others are straightforward.

(a) $$\chi(\psi\varphi) = (\chi\psi)\varphi,$$
$$K'(K\varphi H)H' = (K'K)\varphi(HH') \qquad (\textit{associativities}),$$

(b) $$\varphi.\mathrm{id}\, F = \varphi = \mathrm{id}\, G.\varphi,$$
$$1_Y{\circ}\varphi{\circ}1_X = \varphi, \qquad K{\circ}\mathrm{id}\, F{\circ}H = \mathrm{id}\,(KFH) \qquad (\textit{identities}),$$

(c) $$K(\psi\varphi)H = (K\psi H).(K\varphi H) \qquad (\textit{distributivity}),$$

(d) $$(\psi G).(H\varphi) = (K\varphi).(\psi F) \qquad (\textit{interchange}).$$

The last formula is about two 'horizontally consecutive' natural transformations, as in the following diagram

$$C \underset{G}{\overset{F}{\Rightarrow}}{\downarrow\varphi} D \underset{K}{\overset{H}{\Rightarrow}}{\downarrow\psi} E. \qquad (1.83)$$

*(e) The category Cat of small categories and functors, enriched with the natural transformations as second order morphisms $\varphi \colon F \to_2 G$ between ordinary arrows, and the previous operations, becomes a two-dimensional structure called a 2-*category*. This is the beginning of higher dimensional Category Theory; one can find an introduction to this domain in [M3], Chapter XII, or [G5], Chapter 7.

1.5.3 The cylinder functor

As we have seen, homotopies are based on the cylinder $X \times \mathbb{I}$. This transformation $I = - \times \mathbb{I}$ acts on a general map $f \colon X \to Y$ in the obvious way (already used above), forming the *cylinder functor* of topological spaces

$$I \colon \mathsf{Top} \to \mathsf{Top}, \qquad IX = X \times \mathbb{I},$$
$$If = f \times \mathrm{id}\,\mathbb{I} \colon X \times \mathbb{I} \to Y \times \mathbb{I}, \qquad (If)(x,t) = (f(x),t), \tag{1.84}$$

where $x \in X$ and $t \in \mathbb{I}$. The map $f \times \mathrm{id}\,\mathbb{I}$ is also written as $f \times \mathbb{I}$.

Now, the faces $(\partial^\alpha \colon X \to IX)$, degeneracy $(e \colon IX \to X)$ and reversion $(r \colon IX \to IX)$ described in (1.57) are the components on the space X of four natural transformations of endofunctors

$$\partial^\alpha \colon \mathrm{id} \to I \colon \mathsf{Top} \to \mathsf{Top} \qquad (faces),$$
$$e \colon I \to \mathrm{id} \colon \mathsf{Top} \to \mathsf{Top} \qquad (degeneracy), \tag{1.85}$$
$$r \colon I \to I \colon \mathsf{Top} \to \mathsf{Top} \qquad (reversion).$$

As in 1.4.3, the object X is often left understood in the notation of the component: the notation $\partial^\alpha \colon X \to IX$ stays for $\partial^\alpha X \colon X \to IX$, and so on.

We have already mentioned in 1.4.3(d) the higher structure related to double homotopies.

Exercises and complements. (a) Prove the naturality of the transformations in (1.85).

(b) If $F, G \colon \mathsf{Top} \to \mathsf{D}$ are isomorphic functors and F is homotopy invariant, so is G.

(c) Homotopies can be equivalently based on the path endofunctor $P \colon \mathsf{Top} \to \mathsf{Top}$: see 7.2.5.

1.5.4 Categories of functors

Let S be a small category. A *diagram* in the category C, *of type* S (or *based on* S) will be any functor $X \colon \mathsf{S} \to \mathsf{C}$.

It can be written in *index notation*, with indices $i \in \mathsf{Ob\,S}$ and $a \colon i \to j$ in $\mathsf{Mor\,S}$:

$$X \colon \mathsf{S} \to \mathsf{C}, \qquad i \mapsto X_i, \quad a \mapsto (X_a \colon X_i \to X_j), \tag{1.86}$$

or as a system $((X_i), (X_a))$.

We write as C^S the category whose objects are the functors $X \colon \mathsf{S} \to \mathsf{C}$, and whose morphisms are the natural transformations $X \to Y \colon \mathsf{S} \to \mathsf{C}$, with vertical composition. It is also written as $\mathsf{Cat(S,C)}$ (even if C is not assumed to be small).

In particular the arrow category $\mathbf{2}$, with two objects (0 and 1) and one non-identity arrow, $0 \to 1$ (see 1.3.2(b)) gives the category $\mathsf{C}^\mathbf{2}$ *of morphisms* of C, where an object is a morphism $x \colon X_0 \to X_1$ of C, and a map $f = (f_0, f_1) \colon x \to y$ is a commutative square of C, as in the left diagram below

$$
\begin{array}{ccccc}
X_0 \xrightarrow{\ f_0\ } Y_0 & \qquad & X_0 \xrightarrow{\ f_0\ } Y_0 \xrightarrow{\ g_0\ } Z_0 \\
\ \downarrow{\scriptstyle x} \qquad \downarrow{\scriptstyle y} & \qquad & \ \downarrow{\scriptstyle x} \qquad \downarrow{\scriptstyle y} \qquad \downarrow{\scriptstyle z} \\
X_1 \xrightarrow[\ f_1\ ]{} Y_1 & \qquad & X_1 \xrightarrow[\ f_1\ ]{} Y_1 \xrightarrow[\ g_1\ ]{} Z_1
\end{array}
\tag{1.87}
$$

These maps are composed *by pasting* commutative squares, as in the right diagram above. The identity of x is: $\mathrm{id}\,(x) = (\mathrm{id}\,X_0, \mathrm{id}\,X_1) \colon x \to x$.

Writing an object of C^S in index notation, as above, the *diagonal functor*

$$D \colon \mathsf{C} \to \mathsf{C}^\mathsf{S},$$
$$(DA)_i = A, \qquad (DA)_a = \mathrm{id}\,A \qquad (i \in \mathsf{Ob\,S},\ a \in \mathsf{Mor\,S}), \tag{1.88}$$

sends an object A to the constant functor at A, and a morphism $f \colon A \to B$ to the natural transformation $Df \colon DA \to DB \colon \mathsf{S} \to \mathsf{C}$ with constant components $(Df)_i = f$.

Exercises and complements. (a) $\mathsf{C}^\mathbf{0}$ is the singleton category $\mathbf{1}$ (also when $\mathsf{C} = \mathbf{0}$). $\mathsf{C}^\mathbf{1}$ is isomorphic to C. $\mathsf{C}^\mathbf{2}$ is described above. $\mathsf{C}^\mathbf{3}$ is the *category of pairs of consecutive arrows* of C. $\mathsf{C}^{\mathbf{2} \times \mathbf{2}}$ is the *category of commutative squares* of C. For a set I (a discrete category), C^I is the power category $\prod_{i \in I} \mathsf{C}$.

Interesting categories of diagrams in Top are obtained in this way.

(b) The diagonal functor $D \colon \mathsf{C} \to \mathsf{C}^\mathsf{S}$ is an embedding, provided that S is not empty.

(c) A natural transformation $\varphi \colon F \to G \colon \mathsf{C} \to \mathsf{D}$ can be viewed as a functor $\mathsf{C} \times \mathbf{2} \to \mathsf{D}$, or equivalently as a functor $\mathsf{C} \to \mathsf{D}^\mathbf{2}$.

The cylinder functor $I = - \times \mathbf{2} \colon \mathsf{Cat} \to \mathsf{Cat}$ represents natural transformations as directed homotopies between functors: cf. [G1], Section 1.1.6.

1.5.5 *Equivalence of categories*

Categories have a notion of equivalence, more general than isomorphism, that plays an important role in the theory of categories, but will play a restricted role in this book (mostly with respect to fundamental groupoids, in 6.1.4).

An *equivalence of categories* is a functor $F: C \to D$ which is invertible up to functorial isomorphism: there is a functor $G: D \to C$ such that $GF \cong \mathrm{id}\, C$ and $FG \cong \mathrm{id}\, D$. The functor G can be called a *weak inverse* of F.

One says that the categories C, D are *equivalent*, written as $C \simeq D$, if there exists an equivalence of categories between them. This is indeed an equivalence relation, as can be easily proved.

The reader will note a formal analogy, where

(i) topological spaces, maps, homotopies $\varphi: f \simeq g: X \to Y$, their operations, and homotopy equivalence,

correspond to:

(ii) categories, functors, natural isomorphisms $\varphi: F \to G: C \to D$, their operations, and categorical equivalence.

*This analogy is even deeper in the domain of Directed Algebraic Topology, where *directed homotopies* need not be reversible and correspond to general *natural transformations*: see [G1].*

Exercises and complements. (a) A category C is equivalent to the singleton category **1** if and only if it is non-empty and indiscrete, i.e. each hom-set $C(X, Y)$ has precisely one element (see 1.3.1).

*(b) The calculus of homotopies is more complex than the calculus of natural transformations: homotopies only satisfy the properties 1.5.2(a)–(d) up to higher homotopies.

1.5.6 *Presheaves

A functor $S^{\mathrm{op}} \to C$, defined on the opposite category S^{op}, is also called a *presheaf* of C on the (small) category S. They form the *presheaf category* $\mathrm{Psh}(S, C) = C^{S^{\mathrm{op}}}$.

In particular, we have the category $\mathrm{Psh}(S, \mathrm{Set}) = \mathrm{Set}^{S^{\mathrm{op}}}$ *of presheaves of sets*, on S. The category S is canonically embedded in the latter, by the *Yoneda embedding*

$$y: S \to \mathrm{Set}^{S^{\mathrm{op}}}, \qquad y(i) = S(-, i): S^{\mathrm{op}} \to \mathrm{Set}, \qquad (1.89)$$

which sends every object $i \in \mathrm{Ob}\, S$ to the presheaf $y(i)$ *represented by i*.

Various categories of presheaves are used in homology and homotopy theory: cubical sets will be introduced in 2.2.2, simplicial sets in 2.8.5.

A reader interested in categories of presheaves and sheaves is referred to [MaM, Bo3].

1.6 Products, sums and universal properties

Cartesian products are a basic, elementary issue of topological spaces and algebraic structures. Yet, it will be useful to unify these issues in a single categorical concept, based on a universal property.

This is even more evident for the dual notion, a sum of objects, which has dissimilar realisations in the usual categories of structured sets: disjoint unions of sets and spaces, direct sums of abelian groups, tensor products of commutative rings, etc.

Moving from one category to another, for instance from topological spaces to abelian groups, every homology functor $H_n \colon \mathsf{Top} \to \mathsf{Ab}$ 'preserves sums', transforming disjoint unions of spaces into direct sums of abelian groups; this statement makes a precise sense with respect to the universal properties of these structures.

Universal properties have a general formalisation in category theory, that unifies diverse topics, from free algebraic structures to cartesian products and sums, from universal compactification to metric completion. This is briefly examined in 1.6.6 and 1.6.7; the global aspect of this issue — adjoint functors — will only be sketched in Section 7.1.

1.6.1 Products

In the categories we are considering (and many other categories of structured sets) we have cartesian products, constructed with the cartesian product of the underlying sets, on which we put the 'natural' structure of the kind we are considering, be it of algebraic character, or a topology, or something else.

All these procedures can be unified: we have a family $(X_i)_{i \in I}$ of objects of the category C, indexed by a set I, and we want to find an object X equipped with a family of morphisms $p_i \colon X \to X_i$ $(i \in I)$, called *cartesian projections*, which satisfies the following *universal property*:

$$
\begin{array}{ccc}
X & \xrightarrow{\ p_i\ } & X_i \\
\uparrow{\scriptstyle f} & \nearrow{\scriptstyle f_i} & \\
Y & &
\end{array}
\qquad\qquad (1.90)
$$

(i) for every object Y and every family of morphisms $f_i \colon Y \to X_i$ there is a unique morphism $f \colon Y \to X$ such that, for all $i \in I$, $p_i f = f$.

The morphisms f_i are called the *components* of f; the latter is often written as (f_i) — although it is not the same as the family of its components, of course.

The product of a family of objects need not exist. If it does, it is determined up to a unique *coherent* isomorphism, in the sense that if also Y is a product of the family $(X_i)_{i \in I}$ with projections $q_i \colon Y \to X_i$, then the unique morphism $f \colon X \to Y$ which commutes with all projections (i.e. $q_i f = p_i$, for all i) is an isomorphism.

In fact there is also a unique morphism $g \colon Y \to X$ such that $p_i g = q_i$; moreover $gf = \operatorname{id} X$ (because $p_i(gf) = p_i(\operatorname{id} X)$, for all i) and $fg = \operatorname{id} Y$.

We speak thus of *the* product of the family (X_i), denoted as $\Pi_i X_i$ or $X_1 \times ... \times X_n$ in a finite case. We say that a category C *has products* (resp. *finite products*) if every family (resp. finite family) of objects has a product in C.

In particular the product of the empty family of objects $\emptyset \to \operatorname{Ob} C$ means an object X (equipped with no projections) such that for every object Y (equipped with no arrows) there is a unique morphism $f \colon Y \to X$ (satisfying no condition). The solution is called the *terminal* object of C; again, it need not exist, but is determined up to a unique isomorphism.

Examples and complements. (a) In Set, Top, Ab, Gp, Rng all products exist and are the usual cartesian ones (built on the product of the underlying sets). The terminal object is the singleton, with the fitting structure.

We recall that a topological product X with projections $p_i \colon X \to X_i$ has the coarsest topology making all projections continuous. This structure is less obvious than the other product structures we have mentioned; the universal property confirms that it is indeed the 'right' choice, in Top.

The reader will also note that the projections $p_i \colon X \to X_i$ of a product of sets, or topological spaces, need not be surjective: this fails if (and only if) some factors are empty and others are not.

(b) Products in Set. and Top. are readily constructed with the unpointed analogues

$$\Pi_i \, (X_i, \overline{x}_i) = (\Pi_i \, X_i, (\overline{x}_i)). \tag{1.91}$$

(c) In the category X associated to a preordered set, the categorical product of a family of points $x_i \in X$ amounts to $\inf x_i$ (the greatest lower bound), and the terminal object amounts to the greatest element of X. These elements need not exist; they are determined up to the equivalence relation associated to our preorder, and uniquely determined in an ordered set.

(d) Products in Cat have been considered in 1.3.4; the terminal object is the singleton category **1**. Category theory can 'operate on itself', so to say.

(e) If the category C has (all) binary products, there is a functor

$$- \times - : \mathsf{C} \times \mathsf{C} \to \mathsf{C}, \qquad (X, Y) \mapsto X \times Y,$$
$$(f, g) \mapsto f \times g : X \times Y \to X' \times Y', \tag{1.92}$$

where $f : X \to X'$, $g : Y \to Y'$ and the morphism $f \times g$ has the following components, with respect to the projections p', q' of $X' \times Y'$

$$p'(f \times g) = fp : X \times Y \to X', \qquad q'(f \times g) = fq : X \times Y \to Y'. \tag{1.93}$$

(f) Similarly, if the category C has products, for every set I there is a functor

$$\Pi_{i \in I} : \mathsf{C}^I \to \mathsf{C}, \qquad (X_i) \mapsto \Pi_i X_i, \qquad (f_i) \mapsto \Pi_i f_i, \tag{1.94}$$

where $f = \Pi_i f_i : \Pi_i X_i \to \Pi_i Y_i$ is characterised by: $p_i f = f_i p_i$ (writing the projections of X and Y by the same symbols).

1.6.2 Sums

The dual notion, the sum of a family $(X_i)_{i \in I}$ of objects, has dissimilar realisations in the usual categories of structured sets, and was recognised as a single issue in category theory.

The *sum*, or *coproduct*, of a family $(X_i)_{i \in I}$ of objects of C is an object X equipped with a family of morphisms $u_i : X_i \to X$ $(i \in I)$, called *injections*, which satisfy the following universal property:

$$\begin{array}{ccc} X_i & \xrightarrow{\ u_i\ } & X \\ & f_i \searrow & \downarrow f \\ & & Y \end{array} \tag{1.95}$$

(i*) for every object Y and every family of morphisms $f_i : X_i \to Y$, there is a unique morphism $f : X \to Y$ such that, for all $i \in I$, $fu_i = f_i$.

The map f will be written as $[f_i]$, by its *co-components*.

Again, if the sum of the family (X_i) exists, it is determined up to a unique coherent isomorphism, and denoted as $\Sigma_i X_i$, or $X_1 + \ldots + X_n$ in a finite case.

The sum of the empty family is the *initial* object X: this means that every object Y has precisely one morphism $X \to Y$.

Examples and complements. (a) Sums in Set are realised as 'disjoint unions'. There is a standard way of doing this, replacing each summand X_i by an isomorphic copy $X_i \times \{i\}$

$$\Sigma_i X_i = \bigcup_i X_i \times \{i\}, \qquad u_i(x) = (x, i) \qquad (\text{for } x \in X_i). \tag{1.96}$$

The initial object is the empty set.

(b) In **Top** we do the same, equipping the disjoint union X with the finest topology that makes all the injections continuous; in other words, the open subsets of X are arbitrary unions of open subsets of the summands. The initial object is the empty space.

(c) In **Ab** and $R\,$**Mod** categorical sums are realised as direct sums

$$\oplus A_i = \{(a_i) \in \Pi A_i \mid a_i = 0 \text{ except for a finite set of indices}\}, \qquad (1.97)$$

where the family (a_i) is essentially finite. Now, for every family of homomorphisms $f_i \colon A_i \to B$ we let $f((a_i)_{i \in I}) = \Sigma f_i(a_i)$, an essentially finite sum.

The initial object is the singleton.

(d) Sums in **Set.** and **Top.** will be described in 2.5.4.

(e) In the category X associated to a preordered set, the categorical sum of a family of points $x_i \in X$ amounts to $\sup x_i$ (the least upper bound), while the initial object amounts to the least element of X.

(f) In **Cat** a sum $\Sigma\, C_i$ of categories is their obvious disjoint union. The initial object is the empty category **0**.

(g) If the category **C** has binary sums, or arbitrary sums, we can build functors corresponding to those of 1.6.1(e) and 1.6.1(f).

(h) In **Gp** categorical sums are classically known as 'free products' (see 6.4.1). *In the category of commutative rings, categorical sums are realised as tensor products on $\mathbb{Z}$.*

1.6.3 Comments

(a) The categorical definitions of product and coproduct are dual to each other: each of them is obtained from the other by reversing all arrows and compositions.

This only makes sense within category theory, because the dual of a given category, formed by reversing its arrows and partial composition law, is a formal construction: the dual of a category of structured sets and mappings is *not* a category of the same kind — although, in certain cases, it may essentially be, as a result of some duality theorem (an example will be cited in 5.5.9(a)).

Similarly, the main topics of Homological Algebra that we shall build in categories of modules have 'dual versions', reversing arrows. This vague duality becomes a theoretical issue in the selfdual framework of abelian categories (see [M2], Chapter VIII, or [G5], Section 6.4), where proving a statement automatically proves its dual.

(b) We shall see that the singular homology functor $H_n \colon \mathsf{Top} \to \mathsf{Ab}$ *preserves sums*, in the sense that it transforms a sum $X = \sum X_i$ of topological spaces into the direct sum of their homology groups.

This means that the group $H_n(X)$ is canonically isomorphic to the direct sum $\oplus H_n(X_i)$: we should not expect these groups to coincide, in a literal sense.

The precise meaning of the property we are considering is that the family of structural injections $u_i \colon X_i \to X$ is transformed into a family

$$(H_n(X_i) \to H_n(X))_{i \in I}$$

that makes $H_n(X)$ into the sum of the groups $H_n(X_i)$. The concrete realisation of sums is of a marginal interest.

(c) It is easy to see that the contravariant functor $\mathcal{A}(X) = \mathsf{Top}(X, \mathbb{Z})$ introduced in (1.15) takes a sum of spaces (a product in $\mathsf{Top}^{\mathrm{op}}$) to a product of rings. This is a typical property of cohomology theories.

(d) A 'categorical definition' *is based on morphisms and their composition,* while the objects only step in as domains and codomains of arrows.

If we want to understand what unifies the product of a family (X_i) of sets, or topological spaces, or groups, we should forget the nature of the objects, and think of the family of cartesian projections $p_i \colon X \to X_i$, together with the previous property. Then — in each category we are interested in — we come back to the objects in order to prove that a solution exists. Fixing a particular one (typically, with the help of set theory) can be useful; but it can also be a hindrance, when functors are involved.

From a structural point of view, a category only 'knows' its objects by their morphisms and composition.

1.6.4 Finite direct sums

The reader may have remarked that, in $R\,\mathsf{Mod}$, a *finite* product ΠA_i and the corresponding finite sum ΣA_i are realised as the same object, namely $A = \oplus A_i$.

Essentially, we are saying two things. First, this category has a *zero object*, which is both terminal (the empty product) and initial (the empty sum): this is the trivial module 0.

Secondly, for a pair of modules A, B, the direct sum $C = A \oplus B$ is determined by a *butterfly diagram* $(C; u, v, p, q)$ satisfying the following equivalent properties — where we use the notation by components and co-components of 1.6.1–1.6.2, and the pointwise addition of parallel homomorphisms

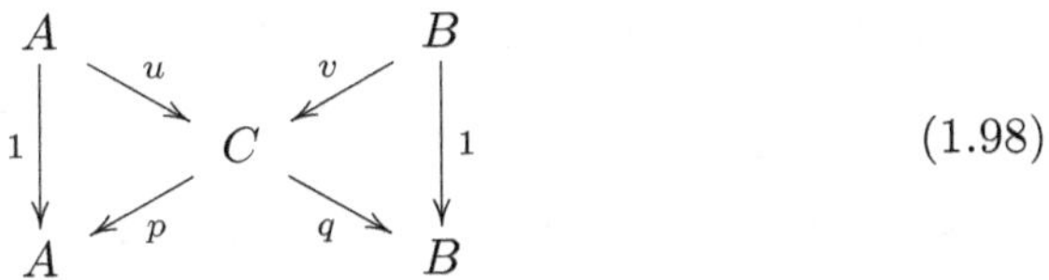

$$\tag{1.98}$$

(i) $(C; p, q)$ is the product of A, B and $u = (\operatorname{id} A, 0)$, $v = (0, \operatorname{id} B)$,

(i*) $(C; u, v)$ is the sum of A, B and $p = [\operatorname{id} A, 0]$, $q = [0, \operatorname{id} B]$,

(ii) $pu = \operatorname{id} A$, $qv = \operatorname{id} B$, $up + vq = \operatorname{id} C$.

The equivalence is (easily) proved in Exercise (a). The object C is thus a product and a coproduct, by distinct structural arrows (of course): the projections p, q and the injections u, v. It is also called the *biproduct* of A and B.

A finite biproduct $A = \oplus A_i$ of modules is similarly analysed: it has projections $p_i \colon A \to A_i$ and injections $u_i \colon A_i \to A$ satisfying respectively the universal properties of product and sum, and the equations:

$$p_i u_i = \operatorname{id} A_i, \qquad \Sigma_i \, u_i p_i = \operatorname{id} A. \tag{1.99}$$

For the empty biproduct, condition (1.99) reduces to $\operatorname{id} A = 0$ (the zero endomorphism), which indeed characterises the trivial module.

Exercises and complements. (a) Prove the equivalence of properties (i), (i*) and (ii).

(b) The binary product and sum also coincide on morphisms, giving the *biproduct functor*

$$- \oplus - \colon R\,\mathsf{Mod}^2 \to R\,\mathsf{Mod}, \qquad (A, B) \mapsto A \oplus B,$$
$$(f, g) \mapsto f \oplus g, \qquad (f \oplus g)(a, b) = (f(a), g(b)). \tag{1.100}$$

1.6.5 Biproducts and additive categories

Biproducts and the addition $f + g$ of parallel morphisms are analysed in a general way, in category theory. In order to keep the present exposition elementary and concrete, we only hint here at this topic; an interested reader can see [M2], Section VIII.2, or [G5], Section 6.4.

First, a *zero object* in a category C, often written as 0, is both initial and terminal. This exists in Ab, $R\,$Mod, Gp, Set$_\bullet$ and Top$_\bullet$, where the zero object is the (appropriate) singleton; it does not exist in Set, Top and Cat where the initial and terminal object are distinct.

A category with zero object is said to be *pointed*; then each pair of objects A, B has a *zero-morphism* $0_{AB} \colon A \to B$ (also written as 0), which is given by the composite $A \to 0 \to B$.

If C is a pointed category, the *biproduct* of a family of objects (A_i) is an object A satisfying:

- the universal property of the product, by a family of projections $p_i \colon A \to A_i$,
- the universal property of the sum, by a family of injections $u_i \colon A_i \to A$,
- the equations $p_i u_i = \mathrm{id}\, A_i$ and $p_j u_i = 0 \colon A_i \to A_j$ (for $i \neq j$).

The solution can be written as $\oplus A_i$. The empty biproduct is the zero object.

As a related notion, a *preadditive*, or $\mathbb{Z}$-*linear*, category is a category C where every hom-set $C(A, B)$ is equipped with a structure of abelian group, generally written as $f + g$, so that composition is additive in each variable (or bilinear over $\mathbb{Z}$)

$$(f + g)h = fh + gh, \qquad k(f + g) = kf + kg \qquad (\textit{distributivity}), \quad (1.101)$$

where $h \colon A' \to A$ and $k \colon B \to B'$.

Finally, an *additive category* C is a preadditive category with finite products, or equivalently with finite sums, which are then biproducts. The equivalence is proved in the references cited above, where we also see that, *in this case*, the addition of parallel maps is determined by the categorical structure.

Typical examples are again Ab and R Mod, where $f + g$ is the pointwise addition of homomorphisms (see 1.3.8(c)). The categories of chain complexes introduced in the following chapter are also additive categories.

Exercises and complements. (a) In a preadditive category C, an object Z is initial if and only if $C(Z, Z)$ is the trivial group, if and only if Z is terminal (and a zero object). If the zero object exists, the zero-morphism $0_{AB} \colon A \to B$ is the same as the additive identity of the abelian group $C(A, B)$.

(b) A preadditive category need not have a zero object. *Hints*: use a subcategory of Ab.

(c) The category Rng of (unital) rings does not have a zero object. The larger category Rng$'$ *of possibly non-unital rings* has a zero object.

(d) Let C be pointed category with binary sums and products; these can be distinct, but there always is a canonical morphism $f \colon X + Y \to X \times Y$, which is invertible when X or Y is 0.

(e) A preadditive category on a single object is the same as a (unital) ring.

*(f) Arbitrary biproducts exist in rather peculiar cases: for instance in the (selfdual) category of relations of sets, where they are realised as disjoint unions.

1.6.6 Universal arrows

We have encountered diverse universal properties, and many more will be seen in the sequel. There is a general way of formalising them, based on a functor $U\colon \mathsf{A} \to \mathsf{X}$ and an object X of the codomain category X. There are two forms, dual to each other.

(a) A *universal arrow from the object X to the functor U* is a pair $(A, \eta\colon X \to UA)$ consisting of an object A of A and an arrow η of X which is universal, in the sense that every similar pair $(B, f\colon X \to UB)$ factorises uniquely through (A, η). In other words, there exists a unique morphism $g\colon A \to B$ in A such that the following triangle commutes in X

$$Ug.\eta = f. \tag{1.102}$$

The pair (A, η) may exist or not, but is determined up to a unique isomorphism: if $(A', \eta'\colon X \to UA')$ is also a solution, the unique map $g\colon A \to A'$ such that $Ug.\eta = \eta'$ is an isomorphism of A.

In fact, by the usual argument, there is also a unique map $h\colon A' \to A$ in A such that $(Uh)\eta' = \eta$. The composite $hg\colon A \to A$ coincides with $\mathrm{id}\, A$, because $U(hg)\eta = \eta = U(\mathrm{id}\, A)\eta$. In the same way $gh = \mathrm{id}\, A'$, so that g and h are inverse to each other, in A.

(b) Dually, a *universal arrow from the functor U to the object X* is a pair $(A, \varepsilon\colon UA \to X)$ consisting of an object A of A and an arrow ε of X such that every similar pair $(B, f\colon UB \to X)$ factorises uniquely through (A, ε): there exists a unique $g\colon B \to A$ in A such that the following triangle commutes in X

$$\varepsilon.Ug = f. \tag{1.103}$$

(c) Let us point out that, from a formal point of view, *existence* and *uniqueness* of solutions are independent facts: existence means that some solution exists, while uniqueness means that, *if* there are two solutions, *then* they coincide (maybe up to some equivalence relation, as here).

Nevertheless, we often express uniqueness in a 'psychologically prudent' way, saying that the solution is unique *provided it exists* — a formally useless condition.

*(d) Universal arrows compose: for a composed functor $UV\colon \mathsf{B} \to \mathsf{A} \to \mathsf{X}$, given

- a universal arrow $(A, \eta\colon X \to UA)$ from an object X to U,
- a universal arrow $(B, \zeta\colon A \to VB)$ from the previous object A to V,

it is easy to verify that we have a universal arrow from X to UV, constructed as follows

$$(B, U\zeta.\eta\colon X \to UA \to UVB). \tag{1.104}$$

Dually, universal arrows from functors to objects can also be composed.

1.6.7 Exercises and complements

(a) (*Free abelian groups*) The free abelian group $\mathbb{Z}X$ on the set X comes equipped with a mapping $\eta\colon X \to |\mathbb{Z}X|$ (the insertion of the basis) satisfying the universal property stated in (1.22). This means that the pair $(\mathbb{Z}X, \eta)$ is a universal arrow from the object X (a set) to the forgetful functor $U\colon \mathsf{Ab} \to \mathsf{Set}$.

(b) (*Free objects*) In general, for any category A equipped with a functor $U\colon \mathsf{A} \to \mathsf{Set}$, the *free A-object over the set X* (with respect to the functor U) is defined as a universal arrow $(A, \eta\colon X \to UA)$, from the set X to the forgetful functor U.

This terminology is typically used for a category A of algebraic structures, with respect to its forgetful functor $UA = |A|$ (see 1.3.7). Other instances will be considered in the sequel:

- the free R-module RX in (4.5),
- the free group $G(X)$ in 6.4.1(c).

    The solution always exists if A is a category of 'equational algebras': see [G5], Chapter 4. Free commutative rings are constructed with polynomial rings; free rings with 'non-commutative polynomials'.

(c) (*Free monoids*) Prove that the free monoid $M(X)$ on the set X can be constructed as the disjoint union $\sum_{n \in \mathbb{N}} X^n$ of the cartesian powers of the set X. This will be useful further on.

(d) (*Products and sums*) Products and sums in a category C can be viewed as universal arrows. For a set I, a family $(X_i)_{i \in I}$ of objects of C is an object of the cartesian-power category $\mathsf{C}^I = \prod_{i \in I} \mathsf{C}$ (see 1.3.4), and we have a *diagonal functor*

$$D\colon \mathsf{C} \to \mathsf{C}^I, \qquad D(X) = (X)_{i \in I}, \quad D(f) = (f)_{i \in I}, \tag{1.105}$$

taking objects and arrows of C to constant families.

Now, the product of the family $(X_i)_{i \in I}$, defined by the universal property 1.6.1(i), is a universal arrow $(X, \varepsilon \colon DX \to (X_i)_{i \in I})$ from the functor D to the object $(X_i)_{i \in I}$. The morphism $\varepsilon = (p_i \colon X \to X_i)_{i \in I}$ gives the family of projections.

Dually, the sum of the family $(X_i)_{i \in I}$, defined by the universal property 1.6.2(i*), is a universal arrow $(X, \eta \colon (X_i)_{i \in I} \to DX)$ from the object $(X_i)_{i \in I}$ to the functor D. The morphism $\eta = (u_i \colon X_i \to X)_{i \in I}$ gives the family of injections.

2
Singular homology

The modern approach to homology theories associates to a topological space a canonical combinatorial structure, without looking for economical presentations, which cannot be natural. This combinatorial structure is usually enormous, but we are interested in its homology groups, which can be computed by appropriate means, and are finitely generated for the spaces we want to study.

In this chapter we present singular homology, the simplest homology theory of general topological spaces. In this theory, a space X is explored by studying all the maps $a\colon M^n \to X$, defined on a standard model of dimension n; these maps are not supposed to be topological embeddings (as in the initial approaches); allowing 'singularities' was the origin of the name.

We adopt the *cubical* construction (where M^n is the standard euclidean cube $\mathbb{I}^n$), rather than the *simplicial* one (where M^n is an n-dimensional tetrahedron): in various cases, this allows simpler geometrical constructions, essentially because cubes are closed under cartesian products, while the products of tetrahedra should be covered with tetrahedra. (This aspect will be evident in the proof of the homotopy invariance of singular homology, in Theorem 2.2.6.) The simplicial form is briefly presented in Section 2.8, and will be proved to be equivalent to the cubical one in Section 5.6.

Relative singular homology will be introduced in the next chapter; singular homology and cohomology with coefficients in a group is deferred to Chapter 4.

The two-valued index α (or β) varies in the cardinal set $2 = \{0,1\} = \{-,+\}$. The trivial group is always written as 0.

Literature. Two elementary textbooks, by Vick [Vi] and Massey [Mas3], follow respectively the simplicial and the cubical construction. More advanced texts on Algebraic Topology, cited in the general Introduction, generally use the simplicial form. Hilton–Wylie [HiW] uses both forms.

2.1 Chain complexes and their homology

Before dealing with topological spaces, we have to fix some preliminary algebraic tools: exact sequences and chain complexes of abelian groups.

We are here in the domain of Homological Algebra, at its beginning. This subject arose in Algebraic Topology, when the importance of 'exact sequences' of abelian groups was remarked (by Hurewicz [Hur], in 1941) and soon acquired a central role, registered in Eilenberg–Steenrod's axioms of homology theory ([ES], in 1952).

An exact sequence is a quite concrete notion, defined by a simple condition: $\operatorname{Im} f = \operatorname{Ker} g$, for a pair (f, g) of consecutive homomorphisms of abelian groups $A \to B \to C$. But it will pay to examine this issue in a slightly more formal way, where subgroups are replaced by monomorphisms, quotient groups are replaced by epimorphisms, and the image $\operatorname{Im} f \subset B$ of $f \colon A \to B$ is often replaced by the 'cokernel' $\operatorname{Cok} f = B/f(A)$, a quotient of B that plays a role symmetrical to $\operatorname{Ker} f$.

The reader is invited to follow this formalisation, without loosing contact with its concrete interpretation, although it may initially seem that we are expressing simple things in a complicated way.

Exact sequences and chain complexes of R-modules are dealt with in the same way, but will not be used before Section 3.6.

2.1.1 Kernels and cokernels

Let $f \colon A \to B$ be a homomorphism of abelian groups, also called a morphism in **Ab**. We recall that f is said to be a *monomorphism* (or *mono*) if it is injective, and an *epimorphism* (or *epi*) if it is surjective.

An arrow $\rightarrowtail$ will always denote a monomorphism, while $\twoheadrightarrow$ stands for an epimorphism.

A symmetric role will be played, on the one hand by the kernel of f (a subgroup of the domain A), and on the other hand by the cokernel of f (a quotient of the codomain B, less used in Algebra)

$$\operatorname{Ker} f = \{x \in A \mid f(x) = 0\}, \qquad \operatorname{Cok} f = B/f(A). \tag{2.1}$$

More precisely, we shall use the embedding of $\operatorname{Ker} f$ in A, and the projection of B onto $\operatorname{Cok} f$, written as

$$\ker f \colon \operatorname{Ker} f \rightarrowtail A, \qquad \operatorname{cok} f \colon B \twoheadrightarrow \operatorname{Cok} f, \tag{2.2}$$

and characterised in Exercises 2.1.2 by universal properties that will be frequently used.

Concretely, the image $\operatorname{Im} f = f(A) \subset B$ comes first, and the cokernel of f is a derived notion. Yet, in Homological Algebra, the cokernel has a primitive role, symmetric to the role of the kernel, and $\operatorname{Im} f$ tends to come out as $\operatorname{Ker}(\operatorname{cok} f)$, the kernel object of the cokernel morphism.

This is already evident in the *canonical factorisation* of the homomorphism $f = mgp$ (or first Noether isomorphism theorem), analysed in the following commutative diagram, where:

$$\operatorname{Ker} f \xrightarrow{\ \ker f\ } A \xrightarrow{\ f\ } B \xrightarrow{\ \operatorname{cok} f\ } \operatorname{Cok} f \tag{2.3}$$

- $\operatorname{Coim} f = A/\operatorname{Ker} f$ is the *coimage* of f and $p = \operatorname{cok}(\ker f)\colon A \twoheadrightarrow \operatorname{Coim} f$ is the projection,

- $\operatorname{Im} f = \operatorname{Ker}(\operatorname{cok} f)$ is the image of f and $m = \ker(\operatorname{cok} f)\colon \operatorname{Im} f \rightarrowtail B$ is the inclusion,

- $g\colon \operatorname{Coim} f \to \operatorname{Im} f$ is the induced Noether isomorphism.

2.1.2 Exercises and complements

The following elementary results on kernels and cokernels in Ab are important, and of frequent use.

(a) The kernel morphism $\ker f\colon \operatorname{Ker} f \rightarrowtail A$ of $f\colon A \to B$ is characterised up to a coherent isomorphism by the following *universal property*:

(i) $f(\ker f) = 0$, and every morphism h such that $fh = 0$ factorises uniquely through $\ker f$

$$\operatorname{Ker} f \xrightarrow{\ \ker f\ } A \xrightarrow{\ f\ } B \tag{2.4}$$

which means that there is a unique morphism u such that $h = (\ker f)u$.

Loosely speaking, the universal property says that $\ker f$ is the 'best morphism' which annihilates f on its domain; the property does not assume that $\ker f$ be injective: this is a consequence.

A morphism $k\colon K \to A$ that satisfies this property will be said to be *a kernel* of f; then there is a unique morphism $u\colon K \to \operatorname{Ker} f$ such that $(\ker f)u = k$, and it is an isomorphism.

Let us note that an 'exact functor' will preserve kernels in this generalised sense (see 4.2.2).

(b) The cokernel morphism $\operatorname{cok} f\colon B \twoheadrightarrow \operatorname{Cok} f$ is characterised up to a coherent isomorphism by the following universal property:

(i*) $(\operatorname{cok} f)f = 0$, and every morphism h such that $hf = 0$ factorises uniquely through $\operatorname{cok} f$,

$$ A \xrightarrow{\ f\ } B \xrightarrow{\ \operatorname{cok} f\ } \operatorname{Cok} f \tag{2.5}$$

i.e. there is a unique morphism u such that $h = u(\operatorname{cok} f)$.

A morphism $c\colon B \to C$ that satisfies this property will be said to be *a cokernel* of f; then there is a unique morphism $u\colon \operatorname{Cok} f \to C$ such that $u(\operatorname{cok} f) = c$, and it is an isomorphism.

The universal property says that $\operatorname{cok} f$ is the 'best morphism' which annihilates f on its codomain. Again, an 'exact functor' will preserve cokernels in this wider sense.

(c) A commutative square $vf = gu$ has a unique extension to the following commutative diagram

$$
\begin{array}{ccccccc}
\operatorname{Ker} f & \xrightarrow{\ \ker f\ } & A & \xrightarrow{\ f\ } & B & \xrightarrow{\ \operatorname{cok} f\ } & \operatorname{Cok} f \\
\downarrow{\scriptstyle u'} & & \downarrow{\scriptstyle u} & & \downarrow{\scriptstyle v} & & \downarrow{\scriptstyle v'} \\
\operatorname{Ker} g & \xrightarrow[\ \ker g\ ]{} & C & \xrightarrow[\ g\ ]{} & D & \xrightarrow[\ \operatorname{cok} g\ ]{} & \operatorname{Cok} g
\end{array}
\tag{2.6}
$$

We say that u' is *induced by* u (on the monomorphisms $\ker f$ and $\ker g$), and that v' is *induced by* v (on the epimorphisms $\operatorname{cok} f$ and $\operatorname{cok} g$). Both kinds of induction are consistent with composition. In the first case we also speak of a *restricted* morphism.

(d) Starting again from the commutative square $vf = gu$, the canonical factorisations of f and g gives a commutative diagram — the *canonical factorisation* of the given square

$$
\begin{array}{ccccccccccc}
\operatorname{Ker} f & \rightarrowtail & A & \twoheadrightarrow & \operatorname{Coim} f & \xrightarrow{\ f'\ } & \operatorname{Im} f & \rightarrowtail & B & \twoheadrightarrow & \operatorname{Cok} f \\
\downarrow{\scriptstyle u'} & & \downarrow{\scriptstyle u} & & \downarrow{\scriptstyle u''} & & \downarrow{\scriptstyle v''} & & \downarrow{\scriptstyle v} & & \downarrow{\scriptstyle v'} \\
\operatorname{Ker} g & \rightarrowtail & C & \twoheadrightarrow & \operatorname{Coim} g & \xrightarrow[\ g'\ ]{} & \operatorname{Im} g & \rightarrowtail & D & \twoheadrightarrow & \operatorname{Cok} g
\end{array}
\tag{2.7}
$$

where f' and g' are the isomorphisms associated to f and g, u'' is induced by u and v'' by v.

(e) Every monomorphism $f \colon A' \to A$ is *normal* in Ab, which means that it is a kernel of some morphism, and in particular of the projection $\operatorname{cok} f \colon A \twoheadrightarrow A/(\operatorname{Im} f)$, as shown in the left diagram below, a particular case of the canonical factorisation of f (since $\operatorname{Ker} f = 0$)

$$(2.8)$$

Symmetrically, every epimorphism $f \colon A \twoheadrightarrow A''$ is *normal* in Ab, which means that it is a cokernel of some morphism, and in particular of the embedding $\operatorname{ker} f \colon \operatorname{Ker} f \rightarrowtail A)$, as shown in the right diagram above.

> *There is a sharp distinction between the (abelian) category Ab and the (nonabelian) category Gp of all groups, where monomorphisms need not be normal (but all epimorphisms still are).*

(f) Kernels and cokernels can be defined in any category with zero object, by the universal properties (i) and (i*). Set$_\bullet$ and Top$_\bullet$ have kernels and cokernels.

2.1.3 Exact sequences of abelian groups

Exact sequences are a main algebraic tool for homology and homotopy theory.

A pair (f, g) of consecutive homomorphisms, in Ab

$$A \xrightarrow{\ f\ } B \xrightarrow{\ g\ } C \tag{2.9}$$

is said to be *exact* (at B) if $\operatorname{Im} f = \operatorname{Ker} g$ (or equivalently if $\operatorname{Cok} f = \operatorname{Coim} g$). A sequence of consecutive homomorphisms (possibly lower or upper unbounded) is said to be *exact* if it is in all locations where the condition makes sense, i.e. excluding the first and the last position (if any).

For instance, the upper row of the canonical factorisation diagram (2.3) is exact, at A and B.

Exercises and complements. (a) The sequence (2.9) is said to be *of order two* if $gf = 0$, or equivalently $\operatorname{Im} f \subset \operatorname{Ker} g$. The sequence is exact if and only if the other inclusion holds too.

(b) (*Detecting trivial objects*) The sequence $0 \to A \to 0$ is exact if and only if $A = 0$.

(c) (*Detecting monos, epis and isos*) The sequence $0 \to A \xrightarrow{f} B \to 0$

- is exact at A if and only if f is mono,
- is exact at B if and only if f is epi,
- is exact (at A and B) if and only if f is an isomorphism.

    The reader will note that this terminology can be ambiguous: the objects A, B could be the same! In such a case one refers to locations in the sequence.

(d) (*Detecting kernels and cokernels*) The sequence

$$0 \longrightarrow A \xrightarrow{\ f\ } B \xrightarrow{\ g\ } C \longrightarrow 0 \qquad (2.10)$$

- is exact at A, B if and only if f is a kernel of g,
- is exact at B, C if and only if g is a cokernel of f.

(e) *Short exact sequences* are of particular interest. By definition, this is a sequence of form (2.10), that is exact (at A, B, C).

    This means that the sequence satisfies the following equivalent conditions:

  (i) f is a monomorphism, g is an epimorphism and $\operatorname{Im} f = \operatorname{Ker} g$,

  (ii) f is a kernel of g and g is a cokernel of f.

    A short exact sequence will often be written in the form $A \rightarrowtail B \twoheadrightarrow C$.

(f) In the short exact sequence (2.10) we can replace A with the subobject $B' = m(A)$ of B, and replace C with the quotient $B'' = B/B'$. In this way any short exact sequence can be replaced, up to isomorphism, with a 'standard' one

$$B' \xrightarrowtail{\ m\ } B \xrightarrow{\ p\ } B'' \qquad (2.11)$$

where m is the inclusion of a subobject and p is the projection on a quotient.

(g) (*Short exact sequences detect exactness*) A sequence (f, g) of consecutive homomorphisms is exact if and only if it can be inserted in the commutative diagram below, with slanting short exact sequences

$$
\begin{array}{ccccccc}
\operatorname{Ker} p & & & \operatorname{Im} g & & & \\
& \searrow & \overset{q}{\nearrow} & & \overset{n}{\searrow} & & \\
& A \xrightarrow{\ f\ } & B & \xrightarrow{\ g\ } & C & & \\
& \overset{p}{\searrow} & \overset{m}{\nearrow} & & & \searrow & \\
& & \operatorname{Im} f & & & & \operatorname{Cok} n
\end{array}
\qquad (2.12)
$$

(h) In the following exact sequence of abelian groups:

$$A \xrightarrow{\ f\ } B \xrightarrow{\ g\ } C \xrightarrow{\ h\ } D \qquad (2.13)$$

- f is epi $\Leftrightarrow g = 0 \Leftrightarrow h$ is mono,
- g is iso $\Leftrightarrow f$ and h are zero morphisms.

2.1.4 Induction Lemma

We have a solid diagram in Ab, *with short exact rows*

$$
\begin{array}{ccccc}
A' & \xrightarrow{\ m\ } & A & \xrightarrow{\ p\ } & A'' \\
{\scriptstyle f'}\downarrow & & \downarrow{\scriptstyle f} & & \downarrow{\scriptstyle f''} \\
B' & \xrightarrow[\ n\ ]{} & B & \xrightarrow[\ q\ ]{} & B''
\end{array}
\tag{2.14}
$$

The following conditions are readily equivalent:

(i) there is a (unique) morphism $f' \colon A' \to B'$ making the left square commutative,

(ii) there is a (unique) morphism $f'' \colon A'' \to B''$ making the right square commutative.

When this holds true, the morphisms f' and f'' are induced by f (on the monomorphisms m, n, or respectively on the epimorphisms p, q).

If the rows are short exact sequences in standard form, condition (i) can be rewritten as

(i') $f(A') \subset B'$.

Moreover, f' and f'' are computed as: $f'(x) = f(x)$, $f''[x] = [f(x)]$.

Proof It is a straightforward application of Exercise 2.1.2(c), where the terminology of induced morphisms has been introduced. $\qquad\square$

2.1.5 The category of chain complexes

A (positive) *chain complex* of abelian groups is a diagram

$$
A = ((A_n)_{n \geqslant 0},\ (\partial_n)_{n \geqslant 1})
$$

in Ab, consisting of a sequence of consecutive homomorphisms

$$
\ldots A_{n+1} \xrightarrow{\ \partial_{n+1}\ } A_n \xrightarrow{\ \partial_n\ } A_{n-1} \quad \ldots \quad A_1 \xrightarrow{\ \partial_1\ } A_0 \dashrightarrow 0 \ldots
\tag{2.15}
$$

with $\partial_n \partial_{n+1} = 0$ for $n \geqslant 1$.

The homomorphisms ∂_n are called *boundaries* or *differentials*. When useful, all terms are extended at the right with trivial groups A_n (for $n < 0$) and zero homomorphisms ∂_n (for $n \leqslant 0$).

The name of 'boundary' is related to the singular chain complex of a space (see (2.49)); the name of 'differential' is related to the cochain complex of differential forms (see Section 3.7).

A *morphism of chain complexes* $f\colon A \to B$, or *chain morphism*, is a family $(f_n\colon A_n \to B_n)_{n\geqslant 0}$ of homomorphisms in Ab (its *components*), that commute with the boundaries:

$$\begin{array}{ccc} A_n & \xrightarrow{\ \partial_n\ } & A_{n-1} \\ {\scriptstyle f_n}\downarrow & & \downarrow{\scriptstyle f_{n-1}} \\ B_n & \xrightarrow[\partial_n]{} & B_{n-1} \end{array} \qquad \partial_n f_n = f_{n-1}\partial_n \qquad (n > 0). \tag{2.16}$$

Again, these components are extended with zero homomorphisms in negative degree, when useful; note also that we are using the same notation for the differentials of A and B.

With the componentwise composition we get the category $\mathrm{Ch}_+\mathsf{Ab}$ of *chain complexes of abelian groups*. For every $n \geqslant 0$ we have a forgetful functor

$$U_n\colon \mathrm{Ch}_+\mathsf{Ab} \to \mathsf{Ab}, \qquad A \mapsto A_n, \quad f \mapsto f_n, \tag{2.17}$$

associating to any complex (or morphism) its *component* in degree n. On the other hand, there is a canonical embedding

$$J\colon \mathsf{Ab} \to \mathrm{Ch}_+\mathsf{Ab}, \qquad (JA)_0 = A, \qquad (JA)_n = 0,\ \text{for } n > 0. \tag{2.18}$$

The category $\mathrm{Ch}_+\mathsf{Ab}$ has a zero object, the trivial chain complex 0 whose components are trivial groups (see 1.6.5). It also has a componentwise addition of parallel morphisms $f, g\colon A \to B$

$$f + g\colon A \to B, \qquad (f + g)_n = f_n + g_n, \tag{2.19}$$

making $\mathrm{Ch}_+\mathsf{Ab}$ a preadditive category (see 1.6.5). Actually, it is an additive category, as (nearly obvious and) written down in 2.1.8.

Remarks. (a) $\mathrm{Ch}_+\mathsf{Ab}$ is a full subcategory of the category of diagrams $\mathsf{Ab}^{\omega^{\mathrm{op}}}$, where ω is the first infinite ordinal (i.e. the ordered set of natural numbers), and an object $A = ((A_n)_{n\geqslant 0}, (u_n)_{n\geqslant 1})$ is a sequence of consecutive homomorphisms of Ab (under no condition). Cochain complexes, in 3.5.1, will form a full subcategory of Ab^{ω}. For unbounded chain complexes, see 2.8.1.

(b) The family of functors $(U_n)_{n\geqslant 0}$ is *jointly faithful*: if f, g are parallel chain morphism and $U_n(f) = U_n(g)$, for every $n \geqslant 0$, then $f = g$. This family can also be viewed as a faithful functor $U\colon \mathrm{Ch}_+\mathsf{Ab} \to \mathsf{Ab}^{\mathbb{N}}$ with values in a cartesian power of Ab, or a full embedding $\mathrm{Ch}_+\mathsf{Ab} \to \mathsf{Ab}^{\omega^{\mathrm{op}}}$. (Here we are considering $\mathbb{N}$ as a mere set, and a discrete category, while ω and ω^{op} are ordered sets.)

(c) The differential of chain complexes gives a natural transformation $\partial_n\colon U_n \to U_{n-1}$.

2.1.6 Chain homology

For a chain complex A, an element $a \in A_n$ is called a *chain of degree n* (or dimension n, when appropriate). We are interested in two subgroups of A_n

$$Z_n(A) = \mathrm{Ker}\,(\partial_n\colon A_n \to A_{n-1}) \subset A_n,$$
$$B_n(A) = \mathrm{Im}\,(\partial_{n+1}\colon A_{n+1} \to A_n) \subset A_n, \tag{2.20}$$

whose elements are called *n-cycles* and *n-boundaries*, respectively. Again, these names will acquire a concrete sense in Section 2.2.

This includes $Z_0(A) = \mathrm{Ker}\,(\partial_0\colon A_0 \to 0) = A_0$. The condition $\partial_n \partial_{n+1} = 0$ is equivalent to saying that $B_n(A) \subset Z_n(A)$ (for $n \geqslant 0$): *every boundary is a cycle.*

We have thus a quotient, the *n-th homology group* of the complex A

$$H_n(A) = Z_n(A)/B_n(A) = \mathrm{Ker}\,(\partial_n)/\mathrm{Im}\,(\partial_{n+1}) \quad (n \geqslant 0). \tag{2.21}$$

Every cycle $z \in Z_n(A)$ has a *homology class* $[z] \in H_n(A)$, written as $[z]_H$ when useful. Two cycles $z, z' \in A_n$ are *homologous* if $z - z'$ is a boundary, i.e. $[z] = [z']$.

Saying that $H_n(A) = 0$ amounts to the condition $\mathrm{Im}\,\partial_{n+1} = \mathrm{Ker}\,\partial_n$, also expressed saying that the complex A, as a sequence in Ab (see (2.15)), is exact at A_n, or — more precisely — in degree n. Chain homology measures *the failure of exactness* of a chain complex.

It is easy to see that we have functors

$$Z_n\colon \mathrm{Ch}_+\mathsf{Ab} \to \mathsf{Ab}, \qquad B_n\colon \mathrm{Ch}_+\mathsf{Ab} \to \mathsf{Ab} \qquad (n \geqslant 0), \tag{2.22}$$

where $Z_n(f)$ and $B_n(f)$ are restrictions of the component $f_n\colon A_n \to B_n$ (see Exercise (a)).

The morphism $f = (f_n)\colon A \to B$ gives thus a commutative diagram (for each $n \geqslant 0$)

$$
\begin{array}{ccccc}
B_n(A) & \rightarrowtail & Z_n(A) & \twoheadrightarrow & H_n(A) \\
{\scriptstyle B_n(f)}\big\downarrow & & {\scriptstyle Z_n(f)}\big\downarrow & & \big\downarrow{\scriptstyle H_n(f)} \\
B_n(B) & \rightarrowtail & Z_n(B) & \twoheadrightarrow & H_n(B)
\end{array}
\tag{2.23}
$$

where the rows are formed of inclusions and projections, while the left and right columns are induced homomorphisms (Exercise (b)).

Finally we have defined the *chain homology functor* of degree n

$$H_n\colon \mathrm{Ch}_+\mathsf{Ab} \to \mathsf{Ab}, \qquad H_n(A) = (\mathrm{Ker}\,\partial_n)/(\mathrm{Im}\,\partial_{n+1}). \tag{2.24}$$

The homomorphism $H_n(f)\colon H_n(A) \to H_n(B)$ is traditionally written as f_{*n}.

The whole construction can be extended from Ab to every abelian category A, giving an abelian category $\mathrm{Ch}_+(\mathsf{A})$, with homology functors $H_n \colon \mathrm{Ch}_+(\mathsf{A}) \to \mathsf{A}$.

Exercises and complements. (a) The homomorphism $f_n \colon A_n \to B_n$ does have restrictions $Z_n(f)$ and $B_n(f)$, as defined in 2.1.2.

(b) The homomorphism $H_n(f)$ is well defined, in diagram (2.23).

(c) A (positive) *graded abelian group* $A = (A_n)_{n \geqslant 0}$ is a sequence of abelian groups, and can be identified with a chain complex with trivial differentials. These objects form a full subcategory $\mathrm{Gr}_+\mathsf{Ab} \subset \mathrm{Ch}_+\mathsf{Ab}$.

Chain homology can be viewed as a single functor $H_+ \colon \mathrm{Ch}_+\mathsf{Ab} \to \mathrm{Gr}_+\mathsf{Ab}$.

2.1.7 Chain homotopy

The category $\mathrm{Ch}_+\mathsf{Ab}$ has a notion of homotopy, well related to topological homotopy (as we shall see). For two morphisms $f, g \colon A \to B$ in $\mathrm{Ch}_+\mathsf{Ab}$, a *homotopy* $\varphi \colon f \simeq g$ is a sequence of homomorphisms $\varphi_n \colon A_n \to B_{n+1}$ $(n \geqslant 0)$ such that

$$\partial_{n+1}\varphi_n + \varphi_{n-1}\partial_n = g_n - f_n, \tag{2.25}$$

where, as usual, we are letting $\varphi_{-1} = 0$.

Then the morphisms $f, g \colon A \to B$ are said to be *homotopic* in $\mathrm{Ch}_+\mathsf{Ab}$, written as $f \simeq g$.

More precisely, we should define the homotopy as the triple (f, φ, g), because the family (φ_n) does not determine its faces f, g; this will become clearer in 2.1.9.

Exercises and complements. (a) The homotopy relation $f \simeq g$ is a congruence of categories, in $\mathrm{Ch}_+\mathsf{Ab}$.

(b) (*Homotopy invariance of chain homology*) If two morphisms $f, g \colon A \to B$ in $\mathrm{Ch}_+\mathsf{Ab}$ are homotopic, then $H_n(f) = H_n(g)$, for all $n \geqslant 0$.

2.1.8 Direct sum of chain complexes

A family of chain complexes

$$A^i = ((A^i_n)_{n \geqslant 0}, (\partial^i_n)_{n \geqslant 1}) \qquad (i \in I), \tag{2.26}$$

has an obvious cartesian product and direct sum, defined componentwise in Ab

$$B = \prod_{i \in I} A^i, \qquad B_n = \prod_i A^i_n, \qquad \partial_n = \prod_i \partial^i_n, \tag{2.27}$$

$$A = \bigoplus_{i \in I} A^i, \qquad A_n = \bigoplus_i A^i_n, \qquad \partial_n = \bigoplus_i \partial^i_n. \tag{2.28}$$

These are obviously the categorical product and sum in $\mathrm{Ch_+Ab}$; they coincide on finite families. Taking into account the addition $f+g$ of parallel morphisms (in (2.19)), $\mathrm{Ch_+Ab}$ is an additive category (as defined in 1.6.5).

Here we are mainly interested in arbitrary direct sums.

An exercise. (a) Chain homology preserves direct sums: in the previous situation, there is a canonical isomorphism, for each $n \geqslant 0$

$$H_n(A) \cong \oplus_{i \in I} H_n(A^i), \qquad [(z^i)_{i \in I}] \mapsto ([z^i])_{i \in I}. \tag{2.29}$$

The reader is invited to carefully write down the proof, which is perhaps too easy, and may lead to overlooking a point. This is why we write a solution here.

Solution. Let $n \geqslant 0$. Plainly the functors B_n and Z_n preserve direct sums

$$Z_n(A) = \{(z^i) \in A \mid \partial_n^i(z^i) = 0, \ \forall\, i \in I\} = \oplus_{i \in I} Z_n(A^i),$$
$$B_n(A) = \{(b^i) \in A \mid \exists\,(c^i) \in A \colon \partial_n^i(c^i) = b^i, \ \forall\, i \in I\} \tag{2.30}$$
$$= \oplus_{i \in I} B_n(A^i).$$

The obvious epimorphism

$$p \colon Z_n(A) \to \oplus_{i \in I} H_n(A^i), \qquad p((z^i)_{i \in I}) = ([z^i])_{i \in I}, \tag{2.31}$$

gives a sequence

$$0 \longrightarrow B_n(A) \xrightarrow{\ u\ } Z_n(A) \xrightarrow{\ p\ } \oplus_{i \in I} H_n(A^i) \longrightarrow 0 \tag{2.32}$$

where u is the inclusion. The sequence is short exact in each component of index i, and short exact as a whole. Therefore u is the kernel of p, which induces the isomorphism (2.29).

Reconsidering the proof, after (2.30) one is tempted to jump to the conclusion, writing

$$Z_n(A)/B_n(A) \cong \oplus_{i \in I} (Z_n(A^i)/B_n(A^i)),$$

which is correct *if* we note that the formulas (2.29) describe $Z_n(A)$ and $B_n(A)$ as *subgroups* of A_n, not as 'independent' groups.

Indeed, $2\mathbb{Z}$ and $3\mathbb{Z}$ are isomorphic groups, but $\mathbb{Z}/2\mathbb{Z}$ and $\mathbb{Z}/3\mathbb{Z}$ are not!

Alternatively, and more formally, one can prove that the canonical injections $A^i \to A$ induce a family of homomorphisms $(H_n(A^i) \to H_n(A))_{i \in I}$ that satisfies the universal property of the categorical sum, in Ab.

2.1.9 *The cylinder functor*

Homotopies of chain complexes can be defined by a cylinder endofunctor

$$I \colon \mathrm{Ch}_+\mathsf{Ab} \to \mathrm{Ch}_+\mathsf{Ab}, \qquad (IA)_n = A_n \oplus A_{n-1} \oplus A_n,$$

$$\partial_n(a, x, b) = (\partial_n a - x, \ -\partial_{n-1}x, \ \partial_n b + x), \qquad (2.33)$$

$$(If)_n = f_n \oplus f_{n-1} \oplus f_n \qquad (\text{for } f \colon A \to B \text{ in } \mathrm{Ch}_+\mathsf{Ab}),$$

as verified in the exercises below.

This functor has a structure analogous to that of the cylinder functor of Top, in 1.4.3 with the following *faces, degeneracy* and *reversion*:

$$\partial^\alpha \colon A \to IA, \qquad \partial^-(a) = (a, 0, 0), \quad \partial^+(a) = (0, 0, a),$$

$$e \colon IA \to A, \qquad e(a, x, b) = a + b, \qquad\qquad (2.34)$$

$$r \colon IA \to IA, \qquad r(a, x, b) = (b, -x, a).$$

This structure is studied in [G1], Section 4.4, in a more general context.

Exercises and complements. (a) The differential defined in (2.33) satisfies the condition $\partial_n \partial_{n+1} = 0$.

(b) For a chain morphism $f \colon A \to B$, the components $(If)_n$ commute with the differential.

(c) The definitions (2.34) give indeed chain morphisms.

(d) A chain morphism $\Phi \colon IA \to B$ amounts to a chain homotopy from $f = \Phi\partial^-$ to $g = \Phi\partial^+$.

2.2 The singular homology groups

We now define the singular homology groups of a space X, using the cubical models $\mathbb{I}^n$; this requires three steps:

(i) a functor $\mathrm{Cub}\colon \mathsf{Top} \to \mathsf{CubSet}$, which transforms spaces into cubical sets,

(ii) a functor $\mathrm{Ch}_+\colon \mathsf{CubSet} \to \mathrm{Ch}_+\mathsf{Ab}$, which transforms the latter into chain complexes,

(iii) the chain homology functors $H_n\colon \mathrm{Ch}_+\mathsf{Ab} \to \mathsf{Ab}$, already defined in 2.1.6.

After introducing the first two steps, we define singular homology (in 2.2.4) and begin studying its basic properties.

The Betti numbers and torsion coefficients of a space (in 2.2.7) are numerical invariants, derived from the homology groups (under convenient hypotheses). We already remarked that the historical development followed the reversed path: these invariants came before, and were later reorganised in algebraic structures. Emmy Noether played an important role, in the first half of the XXth century, promoting the importance of these structures.

In Section 2.8 we shall see that singular homology can also be constructed using the tetrahedral models Δ^n, instead of the cubical ones.

2.2.1 The geometry of the cubes

We come back to the basic structure of the standard interval $\mathbb{I} = [0, 1]$, formed of two faces and a degeneracy, and already considered in 1.4.2 (with a different notation):

$$\delta^\alpha \colon \{*\} \rightleftarrows \mathbb{I} \colon \varepsilon, \qquad \delta^\alpha(*) = \alpha, \qquad \varepsilon.\delta^\alpha = \mathrm{id} \qquad (\alpha = 0, 1). \tag{2.35}$$

Higher faces and degeneracies of the standard cubes $\mathbb{I}^n$ are built with the previous ones, by means of the cartesian product, for $1 \leqslant i \leqslant n$ and $\alpha = 0, 1$

$$\begin{aligned}
\delta^\alpha_{ni} &= \mathbb{I}^{i-1} \times \delta^\alpha \times \mathbb{I}^{n-i} \colon \mathbb{I}^{n-1} \to \mathbb{I}^n, \\
\delta^\alpha_{ni}(t_1, ..., t_{n-1}) &= (t_1, ..., t_{i-1}, \alpha, t_i, ..., t_{n-1}), \\
\varepsilon_{ni} &= \mathbb{I}^{i-1} \times \varepsilon \times \mathbb{I}^{n-i} \colon \mathbb{I}^n \to \mathbb{I}^{n-1}, \\
\varepsilon_{ni}(t_1, ..., t_n) &= (t_1, ..., t_{i-1}, t_{i+1}, ..., t_n).
\end{aligned} \tag{2.36}$$

(We shall generally omit the degree n.) For instance, the standard square $\mathbb{I}^2$ has four faces $\delta^\alpha_i \colon \mathbb{I} \to \mathbb{I}^2$, which are parametrisations of its edges on the standard interval

$$\tag{2.37}$$

Faces and degeneracies satisfy the *cocubical relations* (proved in an exercise below):

$$\begin{aligned}
\delta^\beta_j \delta^\alpha_i &= \delta^\alpha_{i+1} \delta^\beta_j, \qquad \text{for } j \leqslant i, \\
\varepsilon_i \varepsilon_j &= \varepsilon_j \varepsilon_{i+1}, \qquad \text{for } j \leqslant i, \\
\varepsilon_j \delta^\alpha_i &= \begin{cases} \delta^\alpha_{i-1} \varepsilon_j, & \text{for } j < i, \\ \mathrm{id}, & \text{for } j = i, \\ \delta^\alpha_i \varepsilon_{j-1}, & \text{for } j > i. \end{cases}
\end{aligned} \tag{2.38}$$

The system $((\mathbb{I}^n), (\delta_i^\alpha), (\varepsilon_i))$ is a *cocubical object* in Top.

Exercises and comments. (a) Prove the cocubical relations.

(b) Geometrically, the cocubical relation of the faces express the fact that each vertex of a square belongs to two edges, each edge of a 3-cube belongs to two faces, and so on. The other relations have also a clear geometrical interpretation.

2.2.2 *The singular cubical set of a space*

For a space X, we apply the contravariant functor

$$\mathsf{Top}(-, X)\colon \mathsf{Top}^{\mathrm{op}} \to \mathsf{Set}$$

to the system $((\mathbb{I}^n), (\delta_i^\alpha), (\varepsilon_i))$, obtaining the *singular cubical set* $\mathrm{Cub}(X)$ of the space X.

This is a system of sets and mappings, still called *faces* and *degeneracies*

$$\mathrm{Cub}(X) = ((\mathrm{Cub}_n X), (\partial_i^\alpha), (e_i)), \quad \mathrm{Cub}_n X = \mathsf{Top}(\mathbb{I}^n, X),$$

$$\partial_i^\alpha\colon \mathrm{Cub}_n X \to \mathrm{Cub}_{n-1} X, \qquad \partial_i^\alpha(a\colon \mathbb{I}^n \to X) = a\delta_i^\alpha, \qquad (2.39)$$

$$e_i\colon \mathrm{Cub}_{n-1} X \to \mathrm{Cub}_n X, \qquad e_i(a\colon \mathbb{I}^{n-1} \to X) = a\varepsilon_i.$$

The arrows satisfy now the *cubical relations*, dual to the cocubical ones (as we are applying a contravariant functor)

$$\partial_i^\alpha \partial_j^\beta = \partial_j^\beta \partial_{i+1}^\alpha, \qquad \text{for } j \leqslant i,$$

$$e_j e_i = e_{i+1} e_j, \qquad \text{for } j \leqslant i,$$

$$\partial_i^\alpha e_j = \begin{cases} e_j \partial_{i-1}^\alpha, & \text{for } j < i, \\ \mathrm{id}, & \text{for } j = i, \\ e_{j-1} \partial_i^\alpha, & \text{for } j > i. \end{cases} \qquad (2.40)$$

For instance: $\partial_i^\alpha \partial_j^\beta(a) = a\delta_j^\beta \delta_i^\alpha = a\delta_{i+1}^\alpha \delta_j^\beta = \partial_j^\beta \partial_{i+1}^\alpha(a)$, for $j \leqslant i$.

In general a *cubical set* $K = ((K_n), (\partial_i^\alpha), (e_i))$ is a sequence of sets $(K_n)_{n \geqslant 0}$ equipped with faces and degeneracies

$$\partial_i^\alpha\colon K_n \to K_{n-1}, \quad e_i\colon K_{n-1} \to K_n \qquad (i = 1, ..., n,\ \alpha = 0, 1), \qquad (2.41)$$

$$K_0 \rightleftarrows K_1 \Rrightarrow K_2 \quad \ldots$$

that satisfy the cubical relations (2.40).

When useful, we use a more explicit notation: ∂_{ni}^α and e_{ni}, without omitting the degree n.

The elements of K_n are called *n-cubes*, and *vertices* or *edges* when n is 0 or 1, respectively. Every n-cube a has 2^n vertices: for instance a 3-cube $a\colon \mathbb{I}^3 \to X$ has eight of them, namely $\partial_1^\alpha \partial_2^\beta \partial_3^\gamma (a) = a(\alpha, \beta, \gamma)$, for $\alpha, \beta, \gamma = 0, 1$.

A *morphism* of cubical sets $f = (f_n)\colon K \to L$ is a sequence of mappings $f_n\colon K_n \to L_n$ that commute with faces and degeneracies, giving commutative squares (for $n \geqslant 1$)

$$\begin{array}{ccc}
K_n \xrightarrow{\ f_n\ } L_n & \qquad & K_{n-1} \xrightarrow{\ f_{n-1}\ } L_{n-1} \\
\partial_i^\alpha \downarrow \qquad \downarrow \partial_i^\alpha & & e_i \downarrow \qquad \downarrow e_i \\
K_{n-1} \xrightarrow[\ f_{n-1}\ ]{} L_{n-1} & & K_n \xrightarrow[\ f_n\ ]{} L_n
\end{array} \qquad (2.42)$$

Cubical sets and their morphisms, with the obvious componentwise composition, form the category CubSet.

Finally, we have the *singular cubical functor*, which acts on a space X as we have seen, and on a map $f\colon X \to Y$ by covariant composition

$$\mathrm{Cub}\colon \mathsf{Top} \to \mathsf{CubSet}, \qquad X \mapsto \mathrm{Cub}(X),$$

$$\mathrm{Cub}(f)\colon \mathrm{Cub}(X) \to \mathrm{Cub}(Y), \qquad (2.43)$$

$$f_{\sharp n}(a\colon \mathbb{I}^n \to X) = fa\colon \mathbb{I}^n \to Y.$$

This achieves the first step to define singular homology.

Complements. Cubical sets can be defined as presheaves $K\colon \underline{\mathrm{I}}^{\mathrm{op}} \to \mathsf{Set}$ on the *basic cubical site* $\underline{\mathrm{I}}$. The latter is the small subcategory of Set consisting of the *elementary cubes* $2^n = \{0, 1\}^n$ (for $n \geqslant 0$), together with the mappings $2^m \to 2^n$ which delete some coordinates and insert some 0's and 1's (without modifying the order of the remaining coordinates). This can be seen in [G5], Section 5.4, or in the original paper [GM], in a more extended form.

2.2.3 The chain complex of a cubical set

For the second step, we construct the chain complex $\mathrm{Ch}_+(K)$ associated to a cubical set K.

First, an n-cube $a \in K_n$ is said to be *degenerate* if $a = e_i(b)$, for some cube $b \in K_{n-1}$ and some index i

$$\mathrm{Deg}_n K = \bigcup_i \mathrm{Im}\, (e_i\colon K_{n-1} \to K_n), \qquad \mathrm{Deg}_0 K = \emptyset \qquad (n > 0). \qquad (2.44)$$

Concretely, this means that a does not depend on the i-th coordinate. Then, because of the cubical relations, $\partial_i^\alpha e_j(b)$ is degenerate for $i \neq j$, while $\partial_i^\alpha e_i(b) = b$. We only register that:

$$a \in \mathrm{Deg}_n K \;\Rightarrow\; (\partial_i^\alpha a \in \mathrm{Deg}_{n-1} K \ \text{ or } \ \partial_i^- a = \partial_i^+ a). \qquad (2.45)$$

Now the (*normalised*) *chain complex* $\mathrm{Ch}_+(K) = ((C_n(K)), (\partial_n))$ is defined as follows

$$\ldots C_{n+1}(K) \xrightarrow{\partial_{n+1}} C_n(K) \xrightarrow{\partial_n} C_{n-1}(K) \ldots C_1(K) \xrightarrow{\partial_1} C_0(K) \qquad (2.46)$$

$$C_n(K) = (\mathbb{Z}K_n)/(\mathbb{Z}\mathrm{Deg}_n K),$$

$$\partial_n \colon C_n(K) \to C_{n-1}(K),$$

$$\partial_n(\hat{a}) = \Sigma_{i,\alpha} \, (-1)^{i+\alpha} \, (\partial_i^\alpha a)\hat{\ } \qquad (a \in K_n).$$

Thus $C_n(K)$ is the quotient of the free abelian group $\mathbb{Z}K_n$ generated by the set K_n (see 1.2.5(c)), modulo the subgroup generated by the degenerate n-cubes. For an n-cube $a \in K_n$ we write as $\hat{a}$ its *normalised class* in $C_n(K)$, up to degenerate cubes, called a *normalised cube*.

As specified above, the boundary ∂_n takes the class $\hat{a}$ to a linear combination of the classes of its faces $\partial_i^\alpha a$, which are normalised cubes of dimension $n - 1$. The exercises below prove that this is correct, and that $\partial_n \partial_{n+1} = 0$.

In computations, we shall usually write the class $\hat{a}$ as a, in order to simplify notation; of course we have to check that a definition involving this class is invariant up to adding degenerate cubes.

The second step is now complete: we have a functor

$$\mathrm{Ch}_+ \colon \mathsf{CubSet} \to \mathrm{Ch}_+\mathsf{Ab},$$

$$K \mapsto \mathrm{Ch}_+(K) = ((C_n(K))_{n\geqslant 0}, (\partial_n)_{n\geqslant 1}), \qquad (2.47)$$

$$f_{\sharp n} = C_n(f) \colon C_n(K) \to C_n(L), \quad \Sigma \lambda_i a_i \mapsto \Sigma \lambda_i f_n(a_i).$$

On a morphism $f \colon K \to L$ of cubical sets, the n-component of the chain morphism $\mathrm{Ch}_+(f)$ is induced by the linear extension of f_n; this is legitimate, because a degenerate n-cube $a = e_i b$ is sent to the degenerate cube $f_n(a) = e_i f_n(b)$. Plainly, this procedure preserves composition and identities.

Composing with the functor $\mathsf{Cub}\colon \mathsf{Top} \to \mathsf{CubSet}$ of (2.43), we get the functor *of the singular (cubical) chain complex*:

$$C_+ \colon \mathsf{Top} \to \mathrm{Ch}_+\mathsf{Ab}, \qquad C_+(X) = \mathrm{Ch}_+(\mathsf{Cub}(X)),$$

$$f_{\sharp n}(a) = fa \qquad (\text{for } f \colon X \to Y \text{ and } a \colon \mathbb{I}^n \to X \text{ in } \mathsf{Top}). \qquad (2.48)$$

Here the name of 'boundary' has a geometrical sense: for a map $a \colon \mathbb{I}^n \to X$, the boundary $\partial_n a$ is a (formal) linear combination of the faces of a, as

in the following figure, for $n = 2$

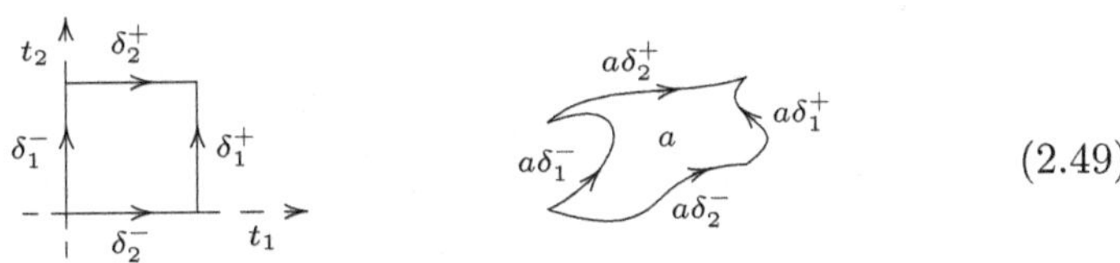

$$\partial_2 a = a\delta_2^- + a\delta_1^+ - a\delta_2^+ - a\delta_1^-.$$

Taking account of signs, we can view the chain $\partial_2 a$ as representing a path around a.

Exercises and complements. (a) The boundary homomorphism ∂_n is well defined in (2.46). *Hints:* use the Induction Lemma 2.1.4.

(b) Prove that $\partial_n \partial_{n+1} = 0$, in (2.46). *Hints:* use the cubical relations of the faces.

(c) $C_n(K)$ is isomorphic to the subgroup of $\mathbb{Z}K_n$ generated by the non-degenerate cubes $a \in K_n \setminus \mathrm{Deg}_n K$, and *is a free abelian group*.

This fact will be important in the sequel, but is not relevant here. In fact, the faces $\partial_i^\alpha a$ of a non-degenerated cube a can be degenerate (for instance, a non-degenerate 2-cube can have constant edges), and the boundary $\partial_n \colon \mathbb{Z}K_n \to \mathbb{Z}K_{n-1}$ *cannot be restricted* to the subgroups spanned by the non-degenerated cubes; we really have to take an induced morphism modulo degenerated cubes.

2.2.4 Singular homology

We are now ready to define the singular homology functor $H_n \colon \mathsf{Top} \to \mathsf{Ab}$ as the composite of three functors, defined in (2.43), in (2.47) and in (2.24)

$$H_n = (\mathsf{Top} \xrightarrow{\mathrm{Cub}} \mathsf{CubSet} \xrightarrow{\mathrm{Ch}_+} \mathsf{Ch}_+\mathsf{Ab} \xrightarrow{H_n} \mathsf{Ab}),$$

$$H_n(X) = H_n(\mathrm{Ch}_+(\mathrm{Cub}(X))) = H_n(C_+(X)), \qquad (2.50)$$

$$f_{*n} = H_n(f) \colon H_n(X) \to H_n(Y), \qquad f_{*n}([z]) = [f_{\sharp n}(z)],$$

for a map $f \colon X \to Y$ and a cycle $z \in C_n(X)$.

The n-th *singular homology group* $H_n(X)$ of the space X is thus a homology group of the (normalised) chain complex associated to the cubical set $\mathrm{Cub}(X)$.

We write:

$$Z_n(X) = Z_n(C_+(X)) = \mathrm{Ker}\,(\partial_n \colon C_n(X) \to C_{n-1}(X))$$
$$\text{(the group of } n\text{-}cycles),$$

$$B_n(X) = B_n(C_+(X)) = \mathrm{Im}\,(\partial_{n+1} \colon C_{n+1}(X) \to C_n(X)) \qquad (2.51)$$
$$\text{(the group of } n\text{-}boundaries),$$

$$H_n(X) = Z_n(X)/B_n(X).$$

A *singular n-chain* $c \in C_n(X)$ is a linear combination of n-cubes $\mathbb{I}^n \to X$ (up to degenerate cubes); it is an *n-cycle* if $\partial_n(c) = 0$ in $C_{n-1}(X)$, and an *n-boundary* if $c \in \mathrm{Im}\,\partial_{n+1}$. Every cycle z determines a homology class $[z] \in H_n(X)$, also written as $[z]_H$, which vanishes if and only if z is a boundary. Two n-cycles z, z' are homologous if $z - z'$ is a boundary.

As already stated in (2.50), a map $f \colon X \to Y$ in Top has an associated homomorphism $H_n(f) \colon H_n(X) \to H_n(Y)$, usually written as f_{*n}.

Remarks and complements. (a) A 0-chain is a formal linear combination $\sum \lambda_i x_i$ of points $x_i \colon \{*\} \to X$; it is always a 0-cycle.

A 1-chain is a formal linear combination of paths $\mathbb{I} \to X$. The boundary $\partial_1(a) = a(1) - a(0)$ of a path is the formal difference of its endpoints; the path a is a 1-cycle if and only if it is a loop.

(b) A chain $c = \sum \lambda_i a_i \in C_n(X)$ of the space X has a *support*, the (compact) subspace

$$\mathrm{supp}(c) = \bigcup \mathrm{Im}\,a_i \subset X, \qquad (2.52)$$

where we are supposing that all a_i are distinct non-degenerate cubes, and all λ_i are non-zero. Plainly

$$\mathrm{supp}(\partial_n c) \subset \mathrm{supp}(c), \qquad (2.53)$$

because the image of each face $\partial_i^\alpha a = a\delta_i^\alpha \colon \mathbb{I}^{n-1} \to X$ of a cube a is contained in $\mathrm{Im}\,a$.

*(c) The set $\mathcal{K}$ of compact subspaces of X, ordered by inclusion, is a filtered set: any pair $K, K' \in \mathcal{K}$ is contained in a larger element, e.g. $K \cup K'$. The previous point shows that $H_n(X)$ is the direct limit of the system of abelian groups $H_n(K)$ ($K \in \mathcal{K}$), with respect to the system of homomorphisms $H_n(K) \to H_n(K')$ induced by the inclusions.

Direct limits of abelian groups are a classical tool. Categorically, $H_n(X)$ is the colimit of the functor $H_n \colon \mathcal{K} \to \mathsf{Ab}$, where the ordered set $\mathcal{K}$ is viewed as a filtered category. An elementary exposition of filtered colimits can be found in [G5], Section 2.6.

2.2.5 *Exercises and complements* (Basic results)

A few very elementary computations of homology groups can be done directly. We shall use the partition of a space X in path components

$$X = \bigcup_{i \in I} X_i, \qquad (2.54)$$

and we refer to I as the *set of path components* of X (written as $\Pi_0(X)$ in (1.69)).

(a) The homology groups of the empty space and the singleton are:

$$H_k(\emptyset) = 0 \qquad (k \geqslant 0), \qquad (2.55)$$

$$H_0(\{*\}) \cong \mathbb{Z}, \qquad H_k(\{*\}) = 0 \quad (k > 0). \qquad (2.56)$$

(b) (*Path-connected spaces*) For a path-connected space $X \neq \emptyset$ we have

$$H_0(X) \cong \mathbb{Z}. \qquad (2.57)$$

More precisely, $H_0(X)$ is the free abelian group generated by the homology class $[x]$ of any point $x \in X$ (all these classes are the same).

(c) Using the partition (2.54) in path components, there are canonical isomorphisms

$$\begin{aligned}
\bigoplus_{i \in I} C_+(X_i) \to C_+(X), \qquad & (c_i)_{i \in I} \mapsto \Sigma_{i \in I}\, c_i, \\
\bigoplus_{i \in I} H_n(X_i) \to H_n(X), \qquad & ([z_i])_{i \in I} \mapsto \Sigma_{i \in I}\, [z_i],
\end{aligned} \qquad (2.58)$$

where a cube $a \colon \mathbb{I}^n \to X_i$ is identified with its 'extension' to X, and a chain in X_i with the corresponding chain in X. (This issue will be better examined in 2.3.4.)

More formally, the inclusions $u_i \colon X_i \to X$ give homomorphisms

$$u_{i\sharp} \colon C_+(X_i) \to C_+(X), \qquad u_{i*} \colon H_n(X_i) \to H_n(X),$$

which are the injections of a categorical sum, for every $n \geqslant 0$.

(d) Loosely speaking, the functor H_0 'counts' the path components of the space: the group $H_0(X)$ is isomorphic to the free abelian group $\mathbb{Z}I = \bigoplus_{i \in I} \mathbb{Z}$ generated by the set I of path components of $X = \bigcup_{i \in I} X_i$. More precisely, $H_0(X)$ has a canonical basis $[x_i]_H$ $(i \in I)$, where x_i is any point of the path component X_i.

There is thus a natural isomorphism

$$\begin{aligned}
\varphi \colon H_0 \to F\Pi_0 \colon \mathsf{Top} \to \mathsf{Ab}, \\
\varphi X \colon H_0(X) \to F\Pi_0(X), \qquad [x]_H \mapsto [x]_P,
\end{aligned} \qquad (2.59)$$

with values in the composite of $\Pi_0 \colon \mathsf{Top} \to \mathsf{Set}$ (defined in (1.14)) with the free-abelian-group functor $F \colon \mathsf{Set} \to \mathsf{Ab}$, $F(I) = \mathbb{Z}I$ (in (1.25)).

(e) A space X whose path components (if any) are singletons has $H_n(X) = 0$ for all $n > 0$, while $H_0(X)$ is determined in (d).

This includes all discrete spaces and the rational line $\mathbb{Q}$. Note that singular homology does not distinguish the euclidean spaces $\mathbb{Q}$ and $\mathbb{Z}$.

2.2.6 Homotopy Invariance Theorem

(a) The functors $H_n \colon$ Top $\to$ Ab are homotopy invariant: if $f \simeq g \colon X \to Y$ in Top, then $H_n(f) = H_n(g) \colon H_n(X) \to H_n(Y)$ (for all $n \geqslant 0$).

(b) If the spaces X, Y are homotopy equivalent, then $H_n(X) \cong H_n(Y)$, for all $n \geqslant 0$.

(c) Contractible spaces have the same homology groups as the singleton, see (2.56). In particular, this applies to all euclidean spaces $\mathbb{R}^m$.

Proof It is sufficient to prove (a). We want to prove that a homotopy $\varphi \colon f \to g$ between two maps $f, g \colon X \to Y$ has an associated chain homotopy Φ between the corresponding morphisms $f_\sharp$ and $g_\sharp$ of chain complexes

$$\Phi \colon f_\sharp \simeq g_\sharp \colon \mathrm{Ch}_+(\mathrm{Cub}(X)) \longrightarrow \mathrm{Ch}_+(\mathrm{Cub}(Y)),$$
$$(\partial_{n+1}\Phi_n + \Phi_{n-1}\partial_n)(a) = ga - fa \qquad (\text{for } a \colon \mathbb{I}^n \to X). \tag{2.60}$$

This will imply that $H_n(f) = H_n(g)$ (for $n \geqslant 0$), by the homotopy invariance of chain homology (in 2.1.7(b)).

We write the homotopy $\varphi \colon f \to g \colon X \to Y$ as a continuous mapping $\varphi \colon \mathbb{I} \times X \to Y$. For every $n \geqslant 0$ we define the homomorphism

$$\Phi_n \colon C_n(X) \to C_{n+1}(Y), \qquad \Phi_n(a) = \varphi(\mathbb{I} \times a), \tag{2.61}$$

sending an n-cube $a \colon \mathbb{I}^n \to X$ to the composite $\varphi(\mathbb{I} \times a) \colon \mathbb{I}^{n+1} \to Y$, an $(n+1)$-cube of Y. This is legitimate, because a degenerate cube $a = b\varepsilon_i \colon \mathbb{I}^n \to \mathbb{I}^{n-1} \to X$ is sent to the cube

$$\varphi(\mathbb{I} \times b\varepsilon_i) = \varphi(\mathbb{I} \times b)\varepsilon_{i+1} \colon \mathbb{I}^{n+1} \to Y,$$

which is also degenerate. Finally (with $\alpha = 0, 1$ in each indexed sum)

$$\partial_{n+1}\Phi_n(a) = \partial_{n+1}(\varphi(\mathbb{I} \times a)) = \sum_{1 \leqslant i \leqslant n+1} (-1)^{i+\alpha}\, \varphi(\mathbb{I} \times a)\delta_i^\alpha$$
$$= \sum_{i=1} (-1)^{1+\alpha}\, \varphi(\delta_i^\alpha \times a) + \sum_{2 \leqslant i \leqslant n+1} (-1)^{i+\alpha}\, \varphi(\mathbb{I} \times a\delta_{i-1}^\alpha)$$
$$= ga - fa - \sum_{1 \leqslant i \leqslant n}(-1)^{i+\alpha}\, \varphi(\mathbb{I} \times a\delta_i^\alpha)$$
$$= ga - fa - \Phi_{n-1}\partial_n(a).$$

Remarks. (i) If we use the cylinder $X \times \mathbb{I}$ (instead of $\mathbb{I} \times X$) the computation is slightly more complex: we should single out the last faces instead of the first ones, and apply a sign change depending on the degree n.

(ii) Let us note that the construction of Φ is easy because the cylinder functor $\mathbb{I} \times -: \mathsf{Top} \to \mathsf{Top}$ takes cubes to cubes (from dimension n to dimension $n + 1$).

The corresponding simplicial construction will be sketched in 2.8.8(a).

$\square$

2.2.7 Numerical invariants

Let X be a space whose singular homology groups are finitely generated, as is often the case when homology theory is used.

Applying the Structure Theorem recalled in 1.2.6(c), the n-th homology group $H_n(X)$ gives a numerical invariant (up to homotopy equivalence, by Theorem 2.2.6)

$$\beta_n(X) = \mathrm{rk}(H_n(X)) \in \mathbb{N} \qquad (n \geqslant 0) \tag{2.62}$$

called the *n-th Betti number* of X, or the *Betti number of dimension n*. In particular, $\beta_0(X)$ is the number of path components of X.

If we also assume that these numbers are eventually zero (as will generally be the case, here), we can define another important numerical invariant, the *Euler–Poincaré characteristic* of X

$$\chi(X) = \sum_{n \geqslant 0} (-1)^n \beta_n(X) \in \mathbb{Z}. \tag{2.63}$$

One can organise the Betti numbers in a polynomial: see 5.6.7(c).

In fact, the Structure Theorem of finitely generated abelian groups gives, in each dimension n, a finite sequence of numerical invariants, that characterise the group $H_n(X)$ up to isomorphism: the Betti number $\beta_n(X)$, together with the *torsion coefficients of X of degree n*

$$\tau_{n1}(X) = \tau_1(H_n(X)), \quad \tau_{n1}(X) = \tau_2(H_n(X)), \quad \ldots \tag{2.64}$$

2.2.8 Paths and homology in degree 1

Let X be a space. We are not yet able to compute $H_1(X)$, except very simple cases, but we can note that this group is formed of equivalence classes $[\sum \lambda_i a_i]$ of 1-cycles of X

$$0 = \partial_1 (\textstyle\sum \lambda_i a_i) = \sum \lambda_i a_i(1) - \sum \lambda_i a_i(0). \tag{2.65}$$

Here each $a_i \colon \mathbb{I} \to X$ is a path; by abuse of notation, it represents the associated normalised 1-cube $\hat{a}_i \in C_1(X) = \mathbb{Z}\mathrm{Cub}_1(X)/\mathbb{Z}\mathrm{Deg}_1(X)$.

Exercises and complements. All the following exercises are about paths in X, written as $a \colon x \simeq y$ to denote the endpoints.

(a) We already remarked that a single path $a \colon x \simeq y$ is a cycle if and only if

$$\partial(a) = \partial_1^+ a - \partial_1^- a = y - x = 0,$$

if and only if a is a loop. In this case there is a homology class $[a]$, which can be of interest or trivial — the zero element of $H_1(X)$.

(b) If the paths $a, b \colon x \simeq y$ are homotopic with fixed endpoints (as defined in 1.4.7), the chain $a - b \in C_1(X)$ is a boundary: it belongs to $B_1(X)$. Therefore, in every cycle $z = \sum \lambda_i a_i$ where a is present, we can replace a with b, obtaining a homologous cycle z'.

(c) If two loops a, b are homotopic as loops (as defined in 1.4.8), then $[a] = [b]$ in $H_1(X)$.

(d) If the loop a is loop homotopic to a constant loop, then $[a] = 0$ in $H_1(X)$. On the other hand, if $[a] \neq 0$, the loop a cannot be contracted to a point. A fortiori, it is not 2-homotopic to a constant loop and X is not simply connected.

(e) If $c = a * b$ is the concatenation of two consecutive paths in X (see (1.54)), then $a + b - c$ is a boundary: in every cycle in which $a + b$ occurs, we can replace it with c, without changing the homology class; or vice versa.

(f) The reversed path $a^\sharp$ of a (see 1.4.2(iii)) gives a chain $a + a^\sharp$ which is a boundary: $[a + a^\sharp] = 0$. In every cycle where $a^\sharp$ occurs, we can replace it with $-a$, obtaining a cycle with the same homology class; or vice versa.

2.3 Mayer–Vietoris and subdivision

The exact sequence of Mayer–Vietoris is the main tool to compute the homology of a space, knowing the homology of suitable subspaces.

We begin by introducing some further elements of basic Homological Algebra, about exact sequences of chain complexes, culminating in Theorem 2.3.3 on the exact sequence of chain homology.

The Subdivision Theorem, in 2.3.5, says that homology can be computed by 'small chains' — an essential issue of singular homology. It opens the way to the Mayer–Vietoris sequence, in 2.3.7, and to basic properties of relative homology, in Chapter 3.

The trivial group and the trivial chain complex are always written as 0.

2.3.1 Exact sequences of chain complexes

All issues about exactness can be lifted from Ab to $\mathsf{Ch}_+\mathsf{Ab}$, using the family of forgetful functors $U_n\colon \mathsf{Ch}_+\mathsf{Ab} \to \mathsf{Ab}$, $A \mapsto A_n$ of (2.17).

For a chain morphism $f\colon A \to B$, the kernel and cokernel are defined componentwise, on a chain morphism $f\colon A \to B$

$$\mathrm{Ker}\, f = ((\mathrm{Ker}\, f_n), (\partial_n)), \qquad \mathrm{Cok}\, f = ((\mathrm{Cok}\, f_n), (\partial_n)), \qquad (2.66)$$

where the differentials are induced by the differentials of A and B, applying 2.1.2(c)

$$\begin{array}{ccccccc}
\mathrm{Ker}\, f_n & \rightarrowtail & A_n & \xrightarrow{f_n} & B_n & \twoheadrightarrow & \mathrm{Cok}\, f_n \\
{\scriptstyle\partial}\downarrow & & {\scriptstyle\partial}\downarrow & & {\scriptstyle\partial}\downarrow & & {\scriptstyle\partial}\downarrow \\
\mathrm{Ker}\, f_{n-1} & \rightarrowtail & A_{n-1} & \xrightarrow[f_{n-1}]{} & B_{n-1} & \twoheadrightarrow & \mathrm{Cok}\, f_{n-1}
\end{array} \qquad (2.67)$$

The chain complexes $\mathrm{Coim}\, f$ and $\mathrm{Im}\, f$ are similarly defined, applying 2.1.2(d)

$$\mathrm{Coim}\, f = ((\mathrm{Coim}\, f_n), (\partial_n)), \qquad \mathrm{Im}\, f = ((\mathrm{Im}\, f_n), (\partial_n)). \qquad (2.68)$$

The chain morphism $f\colon A \to B$ is said to be:

- a *monomorphism* (or *mono*) if each component f_n is mono, or equivalently $\mathrm{Ker}\, f = 0$,

- an *epimorphism* (or *epi*) if each component f_n is epi, or equivalently $\mathrm{Cok}\, f = 0$ (i.e. $\mathrm{Im}\, f = B$).

The arrows $\rightarrowtail$ and $\twoheadrightarrow$ will also denote monomorphisms and epimorphisms of chain complexes, respectively.

The *exactness* of a sequence $A \to B \to C$ in $\mathsf{Ch}_+\mathsf{Ab}$ is defined as in Ab, and amounts to the exactness of all its components $A_n \to B_n \to C_n$.

A short exact sequence $A \rightarrowtail B \twoheadrightarrow C$ of chain complexes is again detected by its components. Up to isomorphism, it can always be realised in standard form, taking:

- a *chain subcomplex* $A \subset B$, which means a family of subgroups $A_n \subset B_n$ such that $\partial_n(A_n) \subset A_{n-1}$ (so that we can define the differentials of A as restrictions of the differentials of B),

- the *quotient chain complex* $C = B/A$, with components $C_n = B_n/A_n$ and induced differentials

$$\begin{array}{ccccc}
A_n & \rightarrowtail & B_n & \twoheadrightarrow & C_n \\
{\scriptstyle\partial}\downarrow & & {\scriptstyle\partial}\downarrow & & {\scriptstyle\partial}\downarrow \\
A_{n-1} & \rightarrowtail & B_{n-1} & \twoheadrightarrow & C_{n-1}
\end{array} \qquad (2.69)$$

Exercises and complements. Kernels and cokernels of chain complexes agree with the abstract definition of 2.1.2, as shown by the following easy points, left to the reader.

(a) The embedding $\ker f \colon \mathrm{Ker}\, f \rightarrowtail A$ satisfies the universal property 2.1.2(i), in $\mathrm{Ch}_+\mathsf{Ab}$.

(b) The projection $\mathrm{cok}\, f \colon B \twoheadrightarrow \mathrm{Cok}\, f$ satisfies the universal property 2.1.2(i*), in $\mathrm{Ch}_+\mathsf{Ab}$.

(c) The Induction Lemma 2.1.4, written in Ab, is readily extended to $\mathrm{Ch}_+\mathsf{Ab}$.

2.3.2 A category of short exact sequences

We form the category $\mathrm{ShCh}_+\mathsf{Ab}$ *of short exact sequences of chain complexes*, as a full subcategory of the category $(\mathrm{Ch}_+\mathsf{Ab})^3$ of pairs of consecutive chain morphisms.

A general morphism $(u, v, w) \colon (f, g) \to (f', g')$ is a commutative diagram of chain complexes, whose rows are short exact

$$
\begin{array}{ccccc}
A & \xrightarrow{\;f\;} & B & \xrightarrow{\;g\;} & C \\
\downarrow{\scriptstyle u} & & \downarrow{\scriptstyle v} & & \downarrow{\scriptstyle w} \\
A' & \xrightarrow{\;f'\;} & B' & \xrightarrow{\;g'\;} & C'
\end{array}
\qquad (2.70)
$$

We are also interested in the functors

$$
\begin{aligned}
P' \colon \mathrm{ShCh}_+\mathsf{Ab} &\to \mathrm{Ch}_+\mathsf{Ab}, & P''' \colon \mathrm{ShCh}_+\mathsf{Ab} &\to \mathrm{Ch}_+\mathsf{Ab}, \\
P'(A \rightarrowtail B \twoheadrightarrow C) &= A, & P'''(A \rightarrowtail B \twoheadrightarrow C) &= C, \\
P'(u, v, w) &= u, & P'''(u, v, w) &= w
\end{aligned}
\qquad (2.71)
$$

which project a short exact sequence to its first or third issue.

2.3.3 Theorem (The exact sequence of chain homology)

A short exact sequence of chain complexes, in $\mathrm{Ch}_+\mathsf{Ab}$

$$
A \xrightarrow{\;f\;} B \xrightarrow{\;g\;} C
\qquad (2.72)
$$

has an associated long exact homology sequence *in* Ab

$$\cdots \longrightarrow H_{n+1}(C)$$
$$\overset{D_{n+1}}{\nearrow}$$
$$H_n(A) \overset{f_{*n}}{\longrightarrow} H_n(B) \overset{g_{*n}}{\longrightarrow} H_n(C) \qquad (2.73)$$
$$\overset{D_n}{\nearrow}$$
$$H_{n-1}(A) \longrightarrow \cdots \longrightarrow H_0(C) \longrightarrow 0$$

natural for morphisms of short exact sequences of chain complexes.

The homomorphisms $D_n \colon H_n(C) \to H_{n-1}(A)$ are called connecting morphisms. *For $[c] \in H_n(C)$, the value $D_n[c] = [a] \in H_{n-1}(A)$ is obtained with
the following diagram, however we choose the representative cycle c and the
elements $b \in B_n$, $a \in A_{n-1}$ (under the following constraints)*

$$
\begin{array}{ccc}
b & \overset{g_n}{\longmapsto} & c \\
\partial \uparrow & & \uparrow \partial \\
a \overset{f_{n-1}}{\longmapsto} \partial b & \overset{g_{n-1}}{\longmapsto} & 0 \\
\partial \uparrow & & \uparrow \partial \\
\partial a & \overset{f_{n-2}}{\longmapsto} & 0
\end{array}
\qquad
\begin{array}{l}
g_n(b) = c, \\[2ex]
f_{n-1}(a) = \partial_n(b).
\end{array}
\qquad (2.74)
$$

Note. *The naturality of the sequence amounts to saying that the connecting morphisms $D_n \colon H_n(C) \to H_{n-1}(A)$ are the components of a natural transformation:*

$$D_n \colon H_n P''' \to H_{n-1} P' \colon \mathrm{ShCh}_+ \mathsf{Ab} \to \mathsf{Ab}, \qquad (2.75)$$

where P''' and P' are the projections defined in (2.71).

Proof The argument uses a procedure called *diagram chasing*, exemplified
in diagram (2.74) and in many similar ones, below.

We note that, after the last term $H_0(C)$ of sequence (2.73), there is
nothing special to prove: by the usual extension of chain complexes on
negative degrees, all homology groups $H_k(-)$ are zero for $k < 0$.

(a) We begin by proving that the connecting homomorphism D_n is well
defined, constructing diagram (2.74) from the upper right corner to the
lower left one.

We start from the cycle $c \in Z_n(C)$, with $\partial_n(c) = 0$ in C_{n-1}. The
homomorphism g_n is surjective, and there is some $b \in B_n$ such that $g_n(b) =
c$. Now $g_{n-1}\partial_n(b) = \partial_n g_n(b) = \partial_n(c) = 0$, which means that $\partial_n(b) \in
\mathrm{Ker}\, g_{n-1} = \mathrm{Im}\, f_{n-1}$: there is some $a \in A_{n-1}$ such that $f_{n-1}(a) = \partial_n(b)$
(and in fact only one, because f_{n-1} is mono). This element a is a cycle,
because $f_{n-2}\partial_{n-1}(a) = \partial_{n-1}f_{n-1}(a) = \partial_{n-1}\partial_n(b) = 0$, and f_{n-2} is mono.
We have obtained a homology class $[a] \in H_{n-1}(A)$.

We prove now that this class does not depend on the choice of the representative c (in its homology class), and of the element b.

Suppose we have a homologous cycle $c' \in Z_n(C)$, with $c - c' = \partial_{n+1}(\hat{c})$. We choose a new $b' \in B_n$ and form a second commutative diagram

$$
\begin{array}{ccccc}
b' & \longmapsto & c' & \quad & (n) \\
\downarrow & & \downarrow & & \\
a' & \longmapsto & \partial b' \longmapsto 0 & \quad & (n-1)
\end{array}
\tag{2.76}
$$

omitting the names of arrows, but writing the degree in the right-hand column.

We have to prove that $a - a'$ is a boundary. In the right part below, $g_n(b - b') = c - c'$. Moreover there is some $\hat{b} \in B_{n+1}$ such that $g_{n+1}(\hat{b}) = \hat{c}$. Its boundary need not be $b - b'$, but $g_n\partial_{n+1}(\hat{b}) = \partial_{n+1}(\hat{c}) = g_n(b - b')$. Thus $g_n(b - b' - \partial_{n+1}\hat{b}) = 0$

$$
\begin{array}{ccccc}
 & & \hat{b} \longmapsto \hat{c} & \quad & (n+1) \\
 & & \downarrow & & \\
\hat{a} \longmapsto b - b' - \partial\hat{b} & \quad b - b' \longmapsto c - c' & \quad & (n) \\
\downarrow \qquad \downarrow & & & \\
\partial\hat{a} \longmapsto \partial(b - b') & & & (n-1)
\end{array}
\tag{2.77}
$$

Moving to the left part of the previous diagram, exactness in degree n gives some $\hat{a} \in A_n$ such that $f_n(\hat{a}) = b - b' - \partial_{n+1}(\hat{b})$. Since f commutes with differentials, $f_{n-1}\partial_n(\hat{a}) = \partial_n(b-b') = f_{n-1}(a-a')$, and $\partial_n(\hat{a}) = a-a'$.

Finally, the mapping D_n is a homomorphism of abelian groups: for two cycles $c, c' \in Z_n(C)$, we construct $D_n[c] = [a]$ and $D_n[c'] = [a']$ as in (2.74) and (2.76). All the arrows of these diagrams are homomorphisms, and the following diagram says that $D_n[c + c'] = [a + a'] = [a] + [a']$

$$
\begin{array}{ccccc}
b + b' & \longmapsto & c + c' & \quad & (n) \\
\downarrow & & \downarrow & & \\
a + a' & \longmapsto \partial(b + b') \longmapsto 0 & \quad & (n-1)
\end{array}
\tag{2.78}
$$

(b) The exactness of the homology sequence is to be verified at $H_n(A)$, $H_n(B)$ and $H_n(C)$, for each $n \geqslant 0$. We prove the first claim; the rest will be the subject of two exercises, in 2.3.8.

To show that $\operatorname{Im} D_{n+1} \subset \operatorname{Ker} f_{*n}$ it is sufficient to note that the composite $f_{*n}D_{n+1}\colon H_{n+1}(C) \to H_n(B)$ is zero. In fact diagram (2.74), lifted one degree up, shows that

$$
f_{*n}D_{n+1}[c] = f_{*n}[a] = [\partial_{n+1}(b)] = 0.
$$

Suppose now that $a \in Z_n(A)$ and $f_{*n}[a] = [f_n(a)] = 0$. Thus $f_n(a) = \partial_{n+1}(\hat{b})$, for some $\hat{b} \in B_{n+1}$, and we form the diagram

$$
\begin{array}{ccccc}
\hat{b} & \mapsto & \hat{c} & & (n+1) \\
\downarrow & & \downarrow & & \\
a & \mapsto & b \mapsto 0 & & (n)
\end{array}
\tag{2.79}
$$

where $\hat{c}$ is a cycle (because $g_n f_n(a) = 0$), and $D_{n+1}[\hat{c}] = [a]$.

(c) We end by checking the naturality of the homology sequence, for a morphism in $\mathrm{ShCh}_+\mathrm{Ab}$

$$
(u, v, w)\colon (A \rightarrowtail B \twoheadrightarrow C) \to (A' \rightarrowtail B' \twoheadrightarrow C'),
$$

as represented in diagram (2.70). We are to prove the commutativity of three squares (for each $n \geqslant 0$)

$$
\begin{array}{ccccccc}
H_n(A) & \xrightarrow{f_{*n}} & H_n(B) & \xrightarrow{g_{*n}} & H_n(C) & \xrightarrow{D_n} & H_{n-1}(A) \\
\downarrow{\scriptstyle u_{*n}} & & \downarrow{\scriptstyle v_{*n}} & & \downarrow{\scriptstyle w_{*n}} & & \downarrow{\scriptstyle u_{*n-1}} \\
H_n(A') & \xrightarrow[f'_{*n}]{} & H_n(B') & \xrightarrow[g'_{*n}]{} & H_n(C') & \xrightarrow[D'_n]{} & H_{n-1}(A')
\end{array}
\tag{2.80}
$$

This is automatically true for the left and central square, applying the functor H_n to the commutative diagram (2.70). For the right square, we take a class $[c] \in H_n(C)$ and construct $D_n[c] = [a]$ as in diagram (2.74)

$$
\begin{array}{ccc}
b & \mapsto & c \\
\downarrow & & \downarrow \\
a & \mapsto & \partial b \mapsto 0
\end{array}
\qquad\qquad
\begin{array}{cccc}
v_n b & \mapsto & w_n c & (n) \\
\downarrow & & \downarrow & \\
u_{n-1} a & \mapsto & \partial v_n b \mapsto 0 & (n-1)
\end{array}
$$

Applying the three homomorphisms u_{n-1}, v_n and w_n, the right diagram above shows that: $D_n w_{*n}[c] = D_n[w_n c] = [u_{n-1} a] = u_{*n-1} D_n[c]$. $\qquad\square$

2.3.4 Small chains

Let X be a topological space. It will be useful to fix notation for chains of X 'living' in a subspace: in fact, as established in the next theorem, 'small chains are sufficient to compute homology'.

(i) Let U be a subspace of X (not necessarily open), with inclusion $u\colon U \to X$.

It can be convenient to view a cube $a\colon \mathbb{I}^n \to U$ of U as a cube of X with image contained in U (formally, the map ua), and a chain $c = \sum \lambda_i a_i \in C_n(U)$ of U as a U-*small* chain of X, with support contained in U (see (2.52)).

This is correct, because a cube of U is degenerate if and only if it is degenerate as a cube of X.

In this way, $C_+(U)$ is viewed as a subcomplex of $C_+(X)$, with inclusion $u_\sharp\colon C_+(U) \to C_+(X)$.

On the other hand, the homomorphism $u_{*n}\colon H_n(U) \to H_n(X)$ induced in homology need not be injective (as already remarked in 1.1.6(b)): our abuse of notation on chains requires a more specific notation for homology. Thus, for a cycle $z \in C_n(U)$, we shall distinguish its homology class $[z]_U \in H_n(U)$ from the homology class $[z]_X = u_{*n}[z]_U \in H_n(X)$.

Actually, for an n-chain c of U:

- c is a cycle in U if and only if it is in X,

- if c is a boundary $\partial\hat{c}$ in U, then it is also a boundary in X, but the converse fails: the support of any such $\hat{c} \in C_{n+1}(X)$ need not be contained in U.

(ii) Extending this issue, the next theorem is concerned with a *generalised open cover* $\mathcal{U} = (U_i)$ of the space X, which means that $X = \bigcup_i \operatorname{int}(U_i)$.

Again, we are not assuming that the subspaces U_i are open in X; this will generally be the case, in the applications of this chapter — but will not be in the applications of Chapter 3.

We denote as $C_+(X;\mathcal{U})$ the subcomplex of $C_+(X)$ of $\mathcal{U}$-*small chains* (or *small chains*, for short): $C_n(X;\mathcal{U})$ is generated by the cubes $a\colon \mathbb{I}^n \to X$ whose image is contained in some U_i (up to degenerate small cubes).

Exercises. (a) $C_+(X;\mathcal{U})$ is indeed a chain subcomplex of $C_+(X)$.

(b) Let $\mathcal{U} = (U_i)$ be a generalised open cover of the space X. Then X has the finest topology making all the inclusions $U_i \to X$ continuous: a subset $W \subset X$ is open if and only if every trace $W \cap U_i$ is open in U_i.

2.3.5 Subdivision Theorem

For a space X and a generalised open cover $\mathcal{U} = (U_i)$ of X, the inclusion $j\colon C_+(X;\mathcal{U}) \to C_+(X)$ induces isomorphism in homology, in every degree:

$$j_{*n}\colon H_n(X;\mathcal{U}) \to H_n(X).$$

Proof The general idea is to subdivide cubes, replacing them by $\mathcal{U}$-small chains. The proof, which is long and interesting, is deferred to 2.3.9.

This result will be applied, in Theorems 2.3.7 and 3.2.1, to a generalised open cover (U, V) formed of two subspaces, but this particular case would reduce the interest of the statement, without simplifying the proof. $\qquad\square$

2.3.6 Pasting two spaces

The next theorem links the homology of a space X with that of three subspaces U, V, A, where

$$X = \operatorname{int} U \cup \operatorname{int} V, \qquad A = U \cap V. \tag{2.81}$$

Again, the subspaces U, V are not assumed to be open in X. The topology of X is determined by that of U and V, as proved in Exercise 2.3.4(b). We shall say that X is the *pasting* of U and V over A, written as $X = U +_A V$; this terminology is explained more clearly in the exercises below.

The inclusions between these spaces form a commutative square $ui = vj$

$$
\begin{array}{ccc}
A & \xrightarrow{\ i\ } & U \\
\downarrow{\scriptstyle j} & & \downarrow{\scriptstyle u} \\
V & \xrightarrow{\ v\ } & X
\end{array}
\qquad
\begin{array}{ccc}
C_+(A) & \xrightarrow{\ i_{\sharp n}\ } & C_+(U) \\
\downarrow{\scriptstyle j_{\sharp n}} & & \downarrow{\scriptstyle u_{\sharp n}} \\
C_+(V) & \xrightarrow[\ v_{\sharp n}\ ]{} & C_+(X)
\end{array}
\tag{2.82}
$$

and we agreed in 2.3.4 to consider the associated monomorphisms of $\mathsf{Ch_+Ab}$, in the right square above, as inclusions.

We have already noted that a cycle $z \in C_n(A)$ has also a homology class in the other spaces

$$[z]_U = i_{*n}[z]_A, \qquad [z]_V = j_{*n}[z]_A, \qquad [z]_X = u_{*n}[z]_U = v_{*n}[z]_V,$$

while a cycle z of U or V also has a homology class $[z]_X \in H_n(X)$.

Exercises and complements. (a) The left square (2.82) is a 'pushout' in Top. This means that the square commutes, and satisfies the following universal property: for every commutative square $fi = gj$ there is a unique map h such that $f = hu$ and $g = hv$

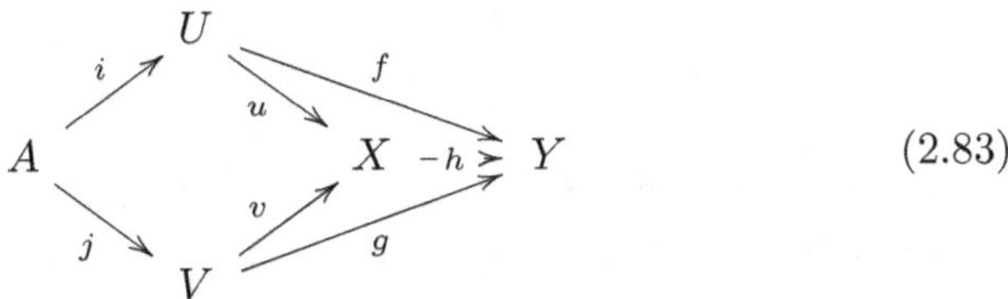

$$\tag{2.83}$$

(b) The universal property determines the space X, up to homeomorphism.

*(c) Two arbitrary maps $i\colon A \to U$, $j\colon A \to V$ have a pushout $(X; u, v)$, which is constructed as a topological quotient $(U+V)/R$, by the equivalence relation generated by identifying each $i(a)$ with $j(a)$, for $a \in A$. If i and j are injective maps (resp. topological embeddings), so are u and v.

Categorically, a pushout is a particular colimit: see 7.1.2(b).

2.3.7 Theorem (The Mayer–Vietoris sequence)

Let X be a topological space, U and V subspaces of X such that $X = \operatorname{int} U \cup \operatorname{int} V$, and $A = U \cap V$.

There is an exact sequence of singular homology groups

$$\ldots \to H_n(A) \xrightarrow{\ h_n\ } H_n(U) \oplus H_n(V) \xrightarrow{\ k_n\ } H_n(X) \xrightarrow{\ D_n\ } H_{n-1}(A) \to \ldots \qquad (2.84)$$

called the exact sequence of Mayer–Vietoris, *or* MV-sequence, *of the space X with respect to the generalised open cover (U,V).*

With the notation of 2.3.6, these homomorphisms are computed as follows:

$$h_n[z]_A = ([z]_U, [z]_V),$$

$$k_n([z]_U, [z']_V) = [z]_X - [z']_X = [z - z']_X, \qquad (2.85)$$

$$D_n[z]_X = [\partial_n c]_A = [\partial_n c']_A,$$

where $z \in Z_n(X)$ is expressed as $z = c - c'$, with $c \in C_n(U)$ and $c' \in C_n(V)$.

The sequence is natural for continuous mappings $f \colon X \to X'$, where

$$X' = \operatorname{int} U' \cup \operatorname{int} V', \qquad f(U) \subset U', \qquad f(V) \subset V'.$$

Notes. The definition of the connecting morphism D_n will be shown to make sense in the proof.

Typically, this sequence will be used to compute the homology of X, knowing the homology of the other spaces. The reader can keep in mind a simple case, where X is the circle covered by two open (non-total) arcs U, V which meet at A, the union of two disjoint arcs. This will be our first application, in 2.4.1.

Proof (a) We use the notation of 2.3.6 for the inclusion maps i, j, u, v

$$
\begin{array}{ccc}
A \xrightarrow{\ i\ } U & \qquad & C_+(A) \xrightarrow{\ i_{\sharp n}\ } C_+(U) \\
\ \ \downarrow{\scriptstyle j} \qquad \downarrow{\scriptstyle u} & & \ \ \downarrow{\scriptstyle j_{\sharp n}} \qquad\qquad \downarrow{\scriptstyle u_{\sharp n}} \\
V \xrightarrow[\ v\ ]{} X & & C_+(V) \xrightarrow[\ v_{\sharp n}\ ]{} C_+(X)
\end{array}
\qquad (2.86)
$$

and view the associated monomorphisms of chain complexes as inclusions: a chain of A is identified with its image in $C_+(U)$, and so on (as in 2.3.4(i)).

(b) We form a short exact sequence of chain complexes

$$C_+(A) \xrightarrow{\ h'\ } C_+(U) \oplus C_+(V) \xrightarrow{\ k'\ } C_+(X; \mathcal{U}) \qquad (2.87)$$

$$h'_n(c) = (i_{\sharp n}(c), j_{\sharp n}(c)) = (c, c), \quad k'_n(c, c') = u_{\sharp n}(c) - v_{\sharp n}(c) = c - c',$$

where $\mathcal{U}$ is the generalised open cover (U, V) of X, and the direct sum of chain complexes is defined componentwise (see 2.1.8).

In fact:

- every component h'_n is obviously mono,

- every component k'_n is epi: a small generator $a \colon \mathbb{I}^n \to X$ of $C_n(X;\mathcal{U})$ has image contained in U or V, and can be obtained as $k'_n(a,0)$ or $k'_n(0,-a)$ (in both ways if $\operatorname{Im} a \subset A$),

- plainly, $k'_n h'_n = 0$ (for all n).

The last point, $\operatorname{Ker} k' \subset \operatorname{Im} h'$, is slightly more complex. Take a pair $(c,c') \in \operatorname{Ker} k'_n$: then c and c' are the same chain of X, and their support is contained in $U \cap V = A$; in other words, $c = c' \in C_n(A)$ and $(c,c') = h'_n(c)$.

(c) Applying the Subdivision Theorem 2.3.5, the inclusion $C_+(X;\mathcal{U}) \subset C_+(X)$ induces isomorphism in homology, in every degree. Taking a class $[z] \in H_n(X)$, we can always replace the cycle z with a homologous cycle $c - c' \in C_n(X;\mathcal{U})$, where $c \in C_n(U)$ and $c' \in C_n(V)$; neither of them has to be a cycle.

From now on, we shall identify the homology groups of $C_+(X;\mathcal{U})$ and $C_+(X)$.

(d) We also identify the group $H_n(C_+(U) \oplus C_+(V))$ with $H_n(U) \oplus H_n(V)$, using the canonical isomorphism (2.29).

Applying Theorem 2.3.3 to the sequence (2.87), we get the exact sequence of chain homology

$$\dots \to H_n(A) \xrightarrow{h_n} H_n(U) \oplus H_n(V) \xrightarrow{k_n} H_n(X) \xrightarrow{D_n} H_{n-1}(A) \to \dots \quad (2.88)$$

where the homomorphisms h_n and k_n, induced in chain homology by h' and k', act as claimed in the statement.

(e) We still have to compute $D_n[z]_X$ for a cycle $z \in Z_n(X)$; as remarked above, we can assume that the representative z can be rewritten as $c - c'$, with $c \in C_n(U)$ and $c' \in C_n(V)$. Following the definition in diagram (2.74)

$$
\begin{array}{ccccc}
(c,c') & \longmapsto & z & & (n) \\
\downarrow & & \downarrow & & \\
\partial c \longmapsto (\partial c, \partial c') & \longmapsto & 0 & & (n-1)
\end{array}
\qquad (2.89)
$$

we lift z to the pair $(c,c') \in C_n(U) \oplus C_n(V)$. The relation

$$\partial_n c - \partial_n c' = k'_{n-1}\partial_n(c,c') = 0 \in C_{n-1}(X)$$

means that $\partial_n c = \partial_n c' \in C_{n-1}(A)$, and

$$h'_{n-1}(\partial_n c) = (\partial_n c, \partial_n c) = (\partial_n c, \partial_n c'),$$

proving that $D_n[z]_X = [\partial_n c] = [\partial_n c']$.

(f) Finally, given a generalised open cover $\mathcal{U}' = (U', V')$ of X', and a map respecting the given covers:

$$f\colon X \to X', \qquad f(U) \subset U', \qquad f(V) \subset V',$$

we also have $f(A) \subset A' = U' \cap V'$. The map f (with its restrictions, still written as f) gives a morphism of short exact sequences

$$
\begin{array}{ccccc}
C_+(A) & \rightarrowtail & C_+(U) \oplus C_+(V) & \twoheadrightarrow & C_+(X;\mathcal{U}) \\
\Big\downarrow{\scriptstyle f_\sharp} & & \Big\downarrow{\scriptstyle f_\sharp} & & \Big\downarrow{\scriptstyle f_\sharp} \\
C_+(A') & \rightarrowtail & C_+(U') \oplus C_+(V') & \twoheadrightarrow & C_+(X';\mathcal{U}')
\end{array}
\tag{2.90}
$$

and this gives a natural transformation of the corresponding homology sequences. $\qquad\square$

2.3.8 *Exercises* (Completing the sequence of chain homology)

For a short exact sequence of chain complexes $A \rightarrowtail B \twoheadrightarrow C$, we have formed the homology sequence (2.73). There are still two points to prove.

(a) The homology sequence is exact at $H_n(B)$, for each $n \geqslant 0$.

(b) The homology sequence is exact at $H_n(C)$, for each $n \geqslant 0$.

2.3.9 *Proof of the Subdivision Theorem*

The general idea is to subdivide cubes, as many times as needed, to replace an n-chain of the space X by a $\mathcal{U}$-small chain; this is based on the Lebesgue Number Lemma 1.4.9, applied to the compact metric space $\mathbb{I}^n$.

We give a nearly complete account of the proof, which has many interesting aspects.

(a) First we build the *subdivision operator*, a natural morphism of chain complexes

$$
\begin{aligned}
\mathrm{Sd}\colon C_+(X) \to C_+(X), \quad \mathrm{Sd}_n(a) &= \Sigma_v\, au_v \quad (v \in \{0,1\}^n), \\
u_v\colon \mathbb{I}^n \to \mathbb{I}^n, \qquad u_v(t) &= (t+v)/2 \qquad (t \in \mathbb{I}^n),
\end{aligned}
\tag{2.91}
$$

which subdivides any n-cube into a chain of 2^n cubes, indexed by the vertices $v \in \{0,1\}^n$ of $\mathbb{I}^n$, as in the following examples in dimension 2

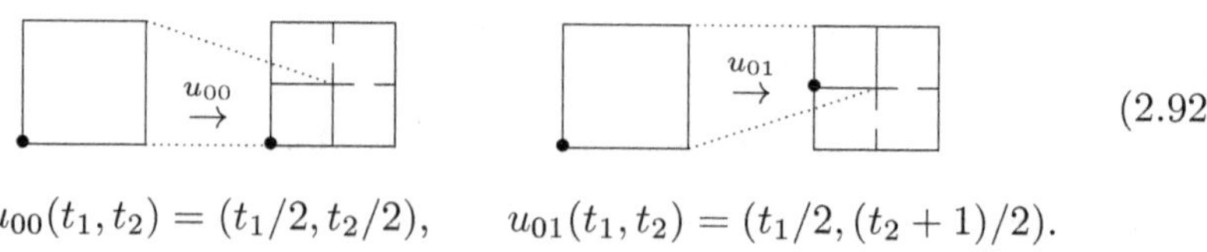

$$u_{00}(t_1, t_2) = (t_1/2, t_2/2), \qquad u_{01}(t_1, t_2) = (t_1/2, (t_2+1)/2).$$

The fact that Sd commutes with the differential is geometrically evident, looking at the previous picture: in $\partial_n \operatorname{Sd}_n(a)$, the 'internal faces' cancel, and the 'external faces' give $\operatorname{Sd}_{n-1}\partial_n(a)$. (One can find the details in [Mas3], Section II.7.)

Repeating subdivision, $\operatorname{Sd}^k(a)$ is a chain of cubes of X, each parametrised on a 'subcube' of $\mathbb{I}^n$, of edge 2^{-k}.

(b) This morphism Sd is homotopic to $\operatorname{id} C_+(X)$, by a chain homotopy $\varphi = (\varphi_n)$

$$\varphi_n \colon C_n(X) \to C_{n+1}(X), \qquad \operatorname{Sd}_n - \operatorname{id}_n = \partial_{n+1}\varphi_n + \varphi_{n-1}\partial_n, \qquad (2.93)$$

that preserves small chains:

$$\varphi_n(C_n(X;\mathcal{U})) \subset C_{n+1}(X;\mathcal{U}). \qquad (2.94)$$

We shall say something of this construction in the last point (e).

(c) We now prove that the homomorphism $j_{*n} \colon H_n(X;\mathcal{U}) \to H_n(X)$ is surjective.

For every cube $a \colon \mathbb{I}^n \to X$, we consider the following open cover of $\mathbb{I}^n$

$$V_i = a^{-1}(\operatorname{int} U_i) \qquad (i \in I). \qquad (2.95)$$

Applying the Lebesgue Number Lemma on open covers of compact metric spaces, there is some $k \in \mathbb{N}$ such that any 'subcube' K of $\mathbb{I}^n$ with edge 2^{-k} is contained in some V_{i_K}. It follows that

$$a(K) \subset a(V_{i_K}) \subset \operatorname{int} U_{i_K} \subset U_{i_K}, \quad \operatorname{Sd}^k(a) \in C_n(X;\mathcal{U}). \qquad (2.96)$$

We take a cycle $z \in C_n(X)$. For every cube $a \colon \mathbb{I}^n \to X$ which appears in z, we can proceed as above. There is thus some $k \in \mathbb{N}$ such that $z' = \operatorname{Sd}^k(z) \in C_n(X;\mathcal{U})$. The composed chain homotopy $\psi \colon \operatorname{Sd}^k \simeq \operatorname{id} : C_+(X) \to C_+(X)$ gives in $C_n(X)$

$$z - z' = \partial_{n+1}\psi_n(z) + \psi_{n-1}\partial_n(z) = \partial_{n+1}\psi_n(z), \qquad (2.97)$$

so that $[z] = j_{*n}[z']$.

(d) We prove that the induced homomorphism $j_{*n} \colon H_n(X;\mathcal{U}) \to H_n(X)$ is injective.

We take a cycle $z \in C_n(X;\mathcal{U})$ which vanishes in $H_n(X)$:

$$z = \partial_{n+1}c, \quad \text{for some chain } c \in C_{n+1}(X), \qquad (2.98)$$

and we prove that z is also a boundary of small chains.

As above, there is some $k \in \mathbb{N}$ such that $c' = \operatorname{Sd}^k(c) \in C_{n+1}(X;\mathcal{U})$.

The composed chain homotopy $\psi \colon \operatorname{Sd}^k \simeq \operatorname{id} : C_+(X) \to C_+(X)$ gives

$$c - c' = \partial_{n+2}\psi_{n+1}(c) + \psi_n\partial_{n+1}(c) = \partial_{n+2}\psi_{n+1}(c) + \psi_n(z). \qquad (2.99)$$

We have proved that $z = \partial_{n+1} c = \partial_{n+1} c' + \partial_{n+1} \psi_n(z)$. This is a boundary in $C_n(X;\mathcal{U})$, because φ takes $C_n(X;\mathcal{U})$ into $C_{n+1}(X;\mathcal{U})$, by (2.94), and also the composed homotopy ψ does.

(e) Finally, we describe the chain homotopy $\varphi \colon \mathrm{id} \simeq \mathrm{Sd} \colon C_+(X) \to C_+(X)$, as constructed in [Mas3], Section II.7.

The homomorphism φ_n is obtained by a family of maps $\eta_v \colon \mathbb{I}^{n+1} \to \mathbb{I}^n$ indexed by the vertices $v = (v_1, ..., v_n) \in \{0,1\}^n$ of $\mathbb{I}^n$

$$\varphi_n \colon C_n(X) \to C_{n+1}(X), \qquad \varphi_n(a) = (-1)^{n+1} \textstyle\sum_v a\eta_v. \qquad (2.100)$$

The mappings η_v are defined as follows.

- For $n = 0$, $\eta \colon \mathbb{I} \to \{*\}$ is the degeneracy, and $\varphi_0 = 0 \colon C_0(X) \to C_1(X)$.
- For $n = 1$: $\quad \eta_0(t_1, t_2) = t_1/(2 - t_2)$,

$$\eta_1(t_1, t_2) = \begin{cases} (t_1 + 1)/(2 - t_2), & \text{if } t_1 + t_2 \leqslant 1, \\ 1, & \text{if } t_1 + t_2 \geqslant 1. \end{cases}$$

- For $n \geqslant 2$:

$$\eta_v(t_1, t_2, ..., t_{n+1}) = (\eta_{v_1}(t_1, t_{n+1}), \eta_{v_2}(t_2, t_{n+1}), ..., \eta_{v_n}(t_n, t_{n+1})).$$

Condition (2.94) on the preservation of small chains is plainly satisfied, because the support of the chain $\varphi_n(a)$ in (2.100) is contained in $\mathrm{Im}\, a$. We still have to verify the identity (2.93), for $n \geqslant 0$.

For $n = 1$, we check that $\partial_2 \varphi_1(a) = \mathrm{Sd}_1(a) - a$, on every 1-cube $a \colon \mathbb{I} \to X$

$$\partial_2 \varphi_1(a) = \partial_2(a\eta_0 + a\eta_1)$$

$$= a\eta_0(-\delta_1^- + \delta_1^+ + \delta_2^- - \delta_2^+) + a\eta_1(-\delta_1^- + \delta_1^+ + \delta_2^- - \delta_2^+)$$

$$= a\eta_0\delta_2^- - a\eta_0\delta_2^+ + a\eta_1\delta_2^- = au_0 - a + au_1 = \mathrm{Sd}_1(a) - a,$$

taking out three degenerate cubes and cancelling two terms. This follows from the following computations of the faces $\eta_j \delta_i^\alpha \colon \mathbb{I} \to \mathbb{I}$

- $\eta_0 \delta_1^-(t) = \eta_0(0, t) = 0$ is constant, and $a\eta_0 \delta_1^-(t)$ is a degenerate 1-cube,
- $\eta_0 \delta_1^+(t) = 1/(2 - t) = \eta_1 \delta_1^-(t)$, so that $a\eta_0 \delta_1^+ - a\eta_1 \delta_1^- = 0$,
- $\eta_0 \delta_2^-(t) = \eta_0(t, 0) = t/2 = u_0(t)$,
- $\eta_0 \delta_2^+(t) = t$,
- $\eta_1 \delta_1^+(t) = \eta_1(1, t) = 1$ and $\eta_1 \delta_2^+(t) = 1$ are constant,
- $\eta_1 \delta_2^-(t) = \eta_1(t, 0) = (t + 1)/2 = u_1(t)$.

The general case is similarly proved. An interested reader can fill the missing points, with the help of [Mas3], page 35. $\qquad\square$

2.4 The homology of the spheres

We compute the homology groups of the spheres $\mathbb{S}^n$, anticipated in (1.9), and explore several consequences.

The reader should find it useful to work on the exercises and applications. Easy arguments about homotopy equivalence and deformation retracts will now be left as understood.

2.4.1 The homology of the circle

We already know, from the basic results of 2.2.5, that the 0-dimensional sphere $\mathbb{S}^0 = \{-1, 1\}$ has homology

$$H_0(\mathbb{S}^0) \cong \mathbb{Z} \oplus \mathbb{Z}, \qquad H_k(\mathbb{S}^0) = 0 \quad (k > 0). \qquad (2.101)$$

More precisely, $H_0(\mathbb{S}^0)$ is the free abelian group generated by the two 0-cubes, identified with their homology classes

$$p, q \colon \mathbb{I}^0 \to \mathbb{S}^0, \qquad p(*) = 1, \quad q(*) = -1. \qquad (2.102)$$

A general element of $H_0(\mathbb{S}^0)$ is a linear combination $\lambda p + \mu q$, with $\lambda, \mu \in \mathbb{Z}$. (Of course, it is not convenient to write the points p, q as ± 1.)

Now we want to compute the homology of the circle $\mathbb{S}^1$. We choose an open cover formed by two open non-total arcs U, V, e.g. the complement of the South pole $\underline{s} = (0, -1)$ and the complement of the North pole $\underline{n} = (0, 1)$

$$U = \mathbb{S}^1 \setminus \{\underline{s}\} \simeq \{*\},$$
$$V = \mathbb{S}^1 \setminus \{\underline{n}\} \simeq \{*\}. \qquad (2.103)$$

U and V are contractible spaces; their intersection $A = \mathbb{S}^1 \setminus \{\underline{s}, \underline{n}\}$ admits $\mathbb{S}^0 = \{p, q\}$ as a deformation retract.

Exercises and complements. The following exercises compute the groups $H_k(\mathbb{S}^1)$, using the MV-sequence of the space $X = \mathbb{S}^1$, with respect to the open cover (U, V). A solution can be found below.

(a) $H_0(\mathbb{S}^1) \cong \mathbb{Z}, \qquad H_k(\mathbb{S}^1) = 0, \quad$ for $k > 1$.

(b) $H_1(\mathbb{S}^1) \cong \mathbb{Z}$.

These results say that the circle is path connected, has a hole of dimension 1 and no holes of higher dimension.

(c) As an infinite cyclic group, $H_1(\mathbb{S}^1)$ has a generator $\zeta = [z]$ of infinite order. A cycle $z \in Z_1(\mathbb{S}^1)$ satisfying this condition can be obtained from the definition of the connecting morphism D_1, in (2.85).

(d) One can find a simpler expression of the same generator $\zeta = [a]$, using the standard counterclockwise loop of the circle: $a(t) = e^{2\pi i t}$, for $t \in \mathbb{I}$.

(e) We have now a clear geometrical meaning for the elements of $H_1(\mathbb{S}^1)$. A homology class is expressed as $k\zeta = [ka]$, for a unique $k \in \mathbb{Z}$. But it can also be expressed as $k\zeta = [a_k]$, using the loop

$$a_k \colon \mathbb{I} \to \mathbb{S}^1, \qquad a_k(t) = e^{2k\pi i t}, \tag{2.104}$$

that turns k times around the origin; it is understood that $-n$ times means n times clockwise, for $n > 0$. Adding homology classes amounts to adding the number of turns (in $\mathbb{Z}$).

(f) It is also evident that choosing a generator of $H_1(\mathbb{S}^1)$, either ζ or $-\zeta$, amounts to choosing an orientation of the circle. We shall see how the orientation of manifolds can be based on homology, in Section 3.3.

Solutions. (a) For $H_0(\mathbb{S}^1)$ it is sufficient to apply Exercise 2.2.5(b), on the 0-th homology group of a path-connected space. (The MV-sequence would also give this result for $H_0(\mathbb{S}^1)$, as the cokernel of the homomorphism h_0 computed below, in (2.107).)

For $H_k(\mathbb{S}^1)$, with $k > 1$, we use a portion of the sequence of Mayer–Vietoris

$$H_k(U) \oplus H_k(V) \to H_k(\mathbb{S}^1) \to H_{k-1}(A). \tag{2.105}$$

As the left and the right term are 0, the central term is also (by 2.1.3(b)).

(b) For $H_1(\mathbb{S}^1)$ we use another portion of the sequence

$$H_1(U) \oplus H_1(V) \xrightarrow{\ k_1\ } H_1(\mathbb{S}^1) \xrightarrow{\ D_1\ } H_0(A) \xrightarrow{\ h_0\ } H_0(U) \oplus H_0(V). \tag{2.106}$$

Here $H_1(U) \oplus H_1(V) = 0$, whence D_1 is mono, and induces an isomorphism

$$D_1' \colon H_1(\mathbb{S}^1) \to \operatorname{Im} D_1 = \operatorname{Ker} h_0.$$

The homomorphism h_0 is computed as follows, on each point $x \in A$:

$$h_0 \colon H_0(A) \to H_0(U) \oplus H_0(V), \qquad h_0[x]_A = ([x]_U, [x]_V).$$

Now $H_0(A) \cong H_0(\mathbb{S}^0)$ is the free abelian group on two generators, $[p]_A$ and $[q]_A$, while $H_0(U)$ is a free abelian group on a single generator, $[p]_U = [q]_U$; similarly for V. Thus:

$$h_0(\lambda[p]_A + \mu[q]_A) = ((\lambda + \mu)[p]_U, (\lambda + \mu)[p]_V). \tag{2.107}$$

The outcome is 0 if and only if $\lambda + \mu = 0$. Finally the group

$$H_1(\mathbb{S}^1) \cong \operatorname{Ker} h_0 = \{\lambda([p]_A - [q]_A) \in H_0(A) \mid \lambda \in \mathbb{Z}\},$$

is isomorphic to the cyclic subgroup of $H_0(A)$ generated by the chain $[p]_A - [q]_A$. This is non-zero (because the 0-cubes p, q belong to different components of A), and infinite (because $H_0(A)$ is torsion-free).

(c) By the previous point, $H_1(\mathbb{S}^1)$ is generated by any homology class $[z]$ such that $D_1[z]$ is the generator $[q]_A - [p]_A$ of $\mathrm{Ker}\, h_0$. To compute $D_1[z]$ we write the 1-cycle z as

$$z = c + c', \quad c \in C_1(U), \quad c' \in C_1(V), \quad \partial_1 c + \partial_1 c' = 0.$$

Here this form is more convenient than the form $c - c'$ used in (2.85); of course, it gives $D_1[c + c'] = [\partial_1 c]_A = [-\partial_1 c']_A$.

Taking two semicircles $c, c' : \mathbb{I} \to \mathbb{S}^1$, with image in U, V, respectively

$$(2.108)$$

$$c(t) = (\cos \pi t, \sin \pi t), \qquad\qquad \partial_1 c = q - p,$$
$$c'(t) = (\cos(\pi(1 + t)), \sin(\pi(1 + t))), \qquad \partial_1 c' = p - q,$$

we do get a cycle $c + c'$ with

$$D_1[c + c'] = [\partial_1 c]_A = [q]_A - [p]_A,$$

proving that $\zeta = [c + c']$ is a generator of the infinite cyclic group $H_1(\mathbb{S}^1)$.

(d) By Exercise 2.2.8(e) we can replace $c + c'$ with the concatenation $a = c * c'$, which is the standard counterclockwise loop of the circle $a(t) = e^{2\pi i t}$.

(e) We want to show that the loop $a_k(t) = e^{2k\pi i t}$ represents the general homology class $[a_k] = k[a]$, for $k \in \mathbb{Z}$.

The case $k = 0$ is obvious, because a_0 is a degenerate loop.

If $k > 0$, a_k is the regular concatenation $a * a * \ldots * a$ (see (1.74)) of k loops equal to a. This loop is 2-homotopic to the iterated concatenation $(\ldots((a * a) * a)\ldots) * a$ (by Exercise 1.4.7(d)), which is homologous to the chain $a + a + \ldots + a = ka$ (by Exercise 2.2.8(e)). Therefore

$$[a_k] = [a * a * \ldots * a] = [(\ldots((a * a) * a)\ldots) * a] = [ka] = k[a].$$

The case $h = -1$ gives the reversed loop $a_{-1} = a^\sharp$, whose homology class is indeed $-[a]$, by Exercise 2.2.8(f). The case $k < 0$ is a consequence of the previous ones.

2.4.2 Theorem (The homology of the spheres)

For $n \geqslant 1$, the singular homology of $\mathbb{S}^n$ is:

$$H_0(\mathbb{S}^n) \cong H_n(\mathbb{S}^n) \cong \mathbb{Z}, \qquad H_k(\mathbb{S}^n) = 0 \qquad (0 \neq k \neq n). \qquad (2.109)$$

Moreover, for $n \geqslant 2$ there is an isomorphism

$$D'_n \colon H_n(\mathbb{S}^n) \to H_{n-1}(\mathbb{S}^{n-1}), \qquad (2.110)$$

natural for maps $f \colon \mathbb{S}^n \to \mathbb{S}^n$ satisfying the conditions

$$f(\mathbb{S}^{n-1}) \subset \mathbb{S}^{n-1}, \qquad f^{-1}\{\underline{s}\} = \{\underline{s}\}, \quad f^{-1}\{\underline{n}\} = \{\underline{n}\}. \qquad (2.111)$$

Notes. $\mathbb{S}^{n-1}$ is embedded as the equator of $\mathbb{S}^n$, in the hyperplane $t_{n+1} = 0$, as in (1.48). The points $\underline{s} = (0, ..., 0, -1)$ and $\underline{n} = (0, ..., 0, 1)$ are the poles of $\mathbb{S}^n$. A generator of $H_n(\mathbb{S}^n)$ will be constructed in 2.4.9.

Proof We know that $H_0(\mathbb{S}^n) \cong \mathbb{Z}$, and prove the other results of (2.109) by induction. The initial step, for $n = 1$, has been proved in 2.4.1. We suppose that (2.109) is true for $n-1$, and prove that it also holds for $n \geqslant 2$.

We choose an open cover of $\mathbb{S}^n$ formed by two open subsets U, V, the complements of the two poles; the figure represents the sphere $\mathbb{S}^2$, with the equator $\mathbb{S}^1$ in the plane $t_3 = 0$

$$U = \mathbb{S}^n \setminus \{\underline{s}\} \simeq \{*\},$$

$$V = \mathbb{S}^n \setminus \{\underline{n}\} \simeq \{*\}, \qquad (2.112)$$

$$A = U \cap V = \mathbb{S}^n \setminus \{\underline{s}, \underline{n}\} \simeq \mathbb{S}^{n-1}.$$

U and V are contractible spaces (they are homeomorphic to $\mathbb{R}^n$, by stereographic projection); their intersection A contains $\mathbb{S}^{n-1}$ as a deformation retract, with retraction

$$p \colon A \to \mathbb{S}^{n-1}, \qquad p(t_1, ..., t_{n+1}) = N(t_1, ..., t_n),$$

where $N \colon \mathbb{R}^n \setminus \{0\} \to \mathbb{S}^{n-1}$ is the normalisation map: indeed $t_{n+1} \neq \pm 1$ in A, and $(t_1, ..., t_n) \neq 0$.

We use the sequence of Mayer–Vietoris for these data.

(a) For $k > 2$, $k \neq n$, the portion of the sequence around $H_k(\mathbb{S}^n)$

$$H_k(U) \oplus H_k(V) \;\to\; H_k(\mathbb{S}^n) \;\to\; H_{k-1}(A) \qquad (2.113)$$

has trivial groups in the left and the right term, by the inductive assumption and $A \simeq \mathbb{S}^{n-1}$. Thus the central term is also the trivial group.

(b) For $H_1(\mathbb{S}^n)$ we use the portion

$$0 \;\to\; H_1(\mathbb{S}^n) \xrightarrow{\;D_1\;} H_0(A) \xrightarrow{\;h_0\;} H_0(U) \oplus H_0(V)$$

where h_0 is mono (by 2.5.2(c)) and $D_1 = 0$, so that $H_1(\mathbb{S}^n) = 0$.

(c) For $H_n(\mathbb{S}^n)$ we use the portion

$$H_n U \oplus H_n V \xrightarrow{k_n} H_n\mathbb{S}^n \xrightarrow{D_n} H_{n-1}A \xrightarrow{h_{n-1}} H_{n-1}U \oplus H_{n-1}V.$$

The first and the last term are again trivial groups, and D_n is an isomorphism. Letting $u\colon \mathbb{S}^{n-1} \to A$ be the inclusion (a homotopy equivalence), we have the isomorphism required in (2.110)

$$D'_n = (u_{*n-1})^{-1}D_n \colon H_n(\mathbb{S}^n) \to H_{n-1}(\mathbb{S}^{n-1}), \qquad (2.114)$$

and, by the inductive assumption, $H_n(\mathbb{S}^n) \cong \mathbb{Z}$.

(d) The proof by induction is complete, and we only have to prove the naturality of D'_n with respect to a map $f\colon \mathbb{S}^n \to \mathbb{S}^n$ satisfying the conditions (2.111).

Its restrictions $f'\colon A \to A$ and $f''\colon \mathbb{S}^{n-1} \to \mathbb{S}^{n-1}$ give the following diagram

$$\begin{array}{ccccc}
H_n(\mathbb{S}^n) & \xrightarrow{D_n} & H_{n-1}(A) & \xleftarrow{u_{*n-1}} & H_{n-1}(\mathbb{S}^{n-1}) \\
\downarrow{\scriptstyle f_{*n}} & & \downarrow{\scriptstyle f'_{*n-1}} & & \downarrow{\scriptstyle f''_{*n-1}} \\
H_n(\mathbb{S}^n) & \xrightarrow[D_n]{} & H_{n-1}(A) & \xleftarrow{u_{*n-1}} & H_{n-1}(\mathbb{S}^{n-1})
\end{array} \qquad (2.115)$$

The left square commutes, by the naturality of the MV-sequence (in Theorem 2.3.7), as the conditions (2.111) imply that $f(U) \subset U$ and $f(V) \subset V$. The right square commutes, because we are applying the functor H_{n-1} to a commutative square $f'u = uf''$ in Top (since f' and f'' are both restrictions of f and u is an inclusion).

Reversing the isomorphism u_{*n-1} in the two horizontal arrows, the new square still commutes, and the naturality of D'_n is proved.

Remarks. The reader can note that $(u_{*n-1})^{-1} = p_{*n-1}$, because the maps u, p form a homotopy equivalence. We can replace (2.115) with a commutative diagram with rightward horizontal arrows.

Yet, the previous version seems preferable, because of two reasons:

- the inclusion $\mathbb{S}^{n-1} \to A$ is a canonical map, more than its retraction p,

- the commutativity of the new right square comes from the commutativity of the right square in (2.115): at the topological level, f' and f'' need not agree with the retraction $p\colon A \to \mathbb{S}^{n-1}$, or any retraction of u.

$\square$

2.4.3 Applications, I

The rest of this section investigates applications of the homology groups of the spheres, computed above. We begin by reviewing the main consequences of these results, already surveyed in 1.1.5 and 1.1.6, or obvious.

(a) If the spheres $\mathbb{S}^m$ and $\mathbb{S}^n$ are homotopy equivalent, then $m = n$.

(b) (Theorem of Topological Dimension) *If the spaces $\mathbb{R}^m$ and $\mathbb{R}^n$ are homeomorphic, then $m = n$.*

(c) The sphere $\mathbb{S}^n$ is not a retract of any contractible space.

(d) The Euler–Poincaré characteristic of $\mathbb{S}^n$ (defined in (2.63)) is

$$\chi(\mathbb{S}^n) = 1 + (-1)^n \in \mathbb{Z} \qquad (n \geqslant 0), \qquad\qquad (2.116)$$

that is, either 0 (for n odd) or 2 (for n even).

The parity of dimension of spheres will also turn up in Theorems 2.4.7 and 2.4.8.

2.4.4 Exercises and complements (Applications, II)

The following results are important. The solutions are not straightforward, except (b) and (d), and can be found below.

(a) (The Brouwer Fixed-point Theorem) *Every endomap $f\colon \mathbb{D}^n \to \mathbb{D}^n$ of the compact disc has at least a fixed point.*

(b) (Surjectivity Lemma) *Every map $f\colon \mathbb{S}^n \to \mathbb{S}^n$ homotopic to the identity is surjective.* (This point will be better analysed in 2.4.6.)

(c) (Intermediate Value Theorem on the Cube) *Let $f\colon \mathbb{I}^n \to \mathbb{I}^n$ be a continuous mapping which sends each face to itself*

$$f(\delta_i^\alpha \mathbb{I}^{n-1}) \subset \delta_i^\alpha \mathbb{I}^{n-1} \qquad (\alpha = \pm,\ i = 1, ..., n).$$

Then f is surjective and sends each face onto *itself.*

(d) For $n = 2$ this statement can be concretely visualised. We have a square cloth, and a square frame of lesser or equal dimension, whose edges are numbered (say) clockwise. We arrange the cloth on the frame (this represents the map $f\colon \mathbb{I}^2 \to \mathbb{I}^2$), so that each edge is arranged on the corresponding edge of the frame; we can make any folds we want in the cloth (f is not assumed to be injective).

However we proceed, each vertex of the cloth (belonging to two edges) goes to the corresponding vertex of the frame; therefore each edge of the cloth covers the corresponding edge of the frame, and the cloth itself covers the whole frame.

(e) (Intermediate Value Theorem on the Disc) *Let $f\colon \mathbb{D}^n \to \mathbb{D}^n$ be a continuous mapping which sends the border $\mathbb{S}^{n-1}$ into itself, so that the restriction $g\colon \mathbb{S}^{n-1} \to \mathbb{S}^{n-1}$ is homotopic to the identity. Then f is surjective.*

Solutions. (a) We suppose, for a contradiction, that f has no fixed point, and construct a retraction $p\colon \mathbb{D}^n \to \mathbb{S}^{n-1}$ of $\mathbb{S}^{n-1}$ which cannot exist.

For every $x \in \mathbb{D}^n$ we are assuming that $f(x) \neq x$, and define $p(x)$ as the intersection of $\mathbb{S}^{n-1}$ with the half-line from $f(x)$ to x

$$\tag{2.117}$$

so that $p(x) = x$ on the sphere. The continuity of p is geometrically obvious. It can be verified analytically, determining the coordinates of $p(x)$ in $\mathbb{R}^n$ — an exercise of Analytic Geometry.

(b) If $n = 0$, $\mathbb{S}^0$ is discrete and two homotopic maps $\mathbb{S}^0 \to \mathbb{S}^0$ coincide.

For $n > 0$, suppose that $f(\mathbb{S}^n) \subset \mathbb{S}^n \setminus \{x_0\}$, a contractible space (by stereographic projection); then f is homotopic to a constant map g (by 1.1.3(c)), and $f_{*n} = g_{*n} = 0\colon H_n(\mathbb{S}^n) \to H_n(\mathbb{S}^n)$, as g takes any n-cube to a degenerate one. But $H_n(\mathbb{S}^n) \neq 0$, and f cannot be homotopic to $\mathrm{id}\,\mathbb{S}^n$.

(c) This statement is trivial for $n = 0$, and amounts to the classical Intermediate Value Theorem for $n = 1$.

We suppose that the statement holds for $n - 1$, and prove it for $n \geqslant 1$. The inductive hypothesis can be applied to each face (because each face of the latter is the intersection of two faces of $\mathbb{I}^n$); it follows that $f(\delta_i^\alpha \mathbb{I}^{n-1}) = \delta_i^\alpha \mathbb{I}^{n-1}$ and $f(\partial \mathbb{I}^n) = \underline{\partial}\mathbb{I}^n$: we register that f covers the border of $\mathbb{I}^n$, and we have to prove that it also covers the interior points.

Now, $f\colon \mathbb{I}^n \to \mathbb{I}^n$ is homotopic to the identity by an affine homotopy (see (1.3)) $h\colon \mathbb{I}^n \times \mathbb{I} \to \mathbb{I}$; moreover h stays inside each face (because f and $\mathrm{id}\,\mathbb{I}^n$ both do, and each face is convex). Therefore also h stays inside the border of $\mathbb{I}^n$, in the sense that: $h(\underline{\partial}\mathbb{I}^n \times \mathbb{I}) \subset \underline{\partial}\mathbb{I}^n$.

Collapsing $\underline{\partial}\mathbb{I}^n$ to a point, f induces a map $g\colon \mathbb{S}^n \to \mathbb{S}^n$, and h induces a homotopy $h'\colon g \simeq \mathrm{id}\,\mathbb{S}^n$

$$
\begin{array}{ccc}
\mathbb{I}^n \xrightarrow{\ f\ } \mathbb{I}^n & \qquad & \mathbb{I}^n \times \mathbb{I} \xrightarrow{\ h\ } \mathbb{I}^n \times \mathbb{I} \\[2pt]
\ \downarrow{\scriptstyle p} \qquad \downarrow{\scriptstyle p} & & \ \downarrow{\scriptstyle p \times \mathbb{I}} \qquad\quad \downarrow{\scriptstyle p \times \mathbb{I}} \\[2pt]
\mathbb{S}^n \xrightarrow[\ g\ ]{} \mathbb{S}^n & & \mathbb{S}^n \times \mathbb{I} \xrightarrow[\ h'\ ]{} \mathbb{S}^n \times \mathbb{I}
\end{array}
\tag{2.118}
$$

In fact, the cartesian product $- \times \mathbb{I}$ (by a locally compact space) preserves topological quotients, as proved in 7.2.6. (Alternatively, we can remark that $\mathbb{I}^n \times \mathbb{I}$ is compact and $\mathbb{S}^n \times \mathbb{I}$ is Hausdorff, whence the surjective map $p \times \mathbb{I}$ is necessarily a topological projection.)

Therefore g is surjective (by Exercise (b)), which implies that f covers the interior part $\mathbb{I}^n \setminus \underline{\partial}\mathbb{I}^n$ (on which p is injective). Finally, f is surjective.

(d) Again, the statement is trivial for $n = 0$, and amounts to the Intermediate Value Theorem for $n = 1$. We suppose that $n \geqslant 2$.

In our hypotheses, the restriction $g \colon \mathbb{S}^{n-1} \to \mathbb{S}^{n-1}$ is surjective (by Exercise (b)). If f is not surjective, then $\operatorname{Im} f \subset \mathbb{D}^n \setminus \{x_0\}$, for an *internal* point x_0 of $\mathbb{D}^n$. Then $\mathbb{S}^{n-1}$ is a deformation retract of $\mathbb{D}^n \setminus \{x_0\}$ (which is homeomorphic to $\mathbb{D}^n \setminus \{0\}$), and we have a commutative diagram

$$
\begin{array}{ccc}
\mathbb{S}^{n-1} & \xrightarrow{\ g\ } & \mathbb{S}^{n-1} \\
{\scriptstyle m}\downarrow & & {\scriptstyle m'}\downarrow \ \ \searrow{\scriptstyle \mathrm{id}} \\
\mathbb{D}^n & \xrightarrow[\ f\ ]{} & \mathbb{D}^n \setminus \{x_0\} \xrightarrow[\ p\ ]{} \mathbb{S}^{n-1}
\end{array}
\tag{2.119}
$$

Applying H_{n-1} we get a contradiction: the homomorphism $g_{*\,n-1} \colon H_{n-1}(\mathbb{S}^{n-1}) \to H_{n-1}(\mathbb{S}^{n-1})$ is the identity of a non-trivial group, but factorises through $H_{n-1}(\mathbb{D}^n) = 0$.

2.4.5 The suspension

For a non-empty space X, the *suspension* ΣX is defined as the topological quotient

$$
\Sigma X = (X \times \mathbb{I})/R,
\tag{2.120}
$$

which collapses, independently, the bases of the cylinder IX to a lower and an upper *vertex* (or *pole*), v^- and v^+.

Plainly, the suspension of $\mathbb{S}^n$ is canonically homeomorphic to $\mathbb{S}^{n+1}$, and we shall identify them (for all $n \geqslant 0$). We have thus

$$
\mathbb{S}^n = \Sigma^n \mathbb{S}^0
\tag{2.121}
$$

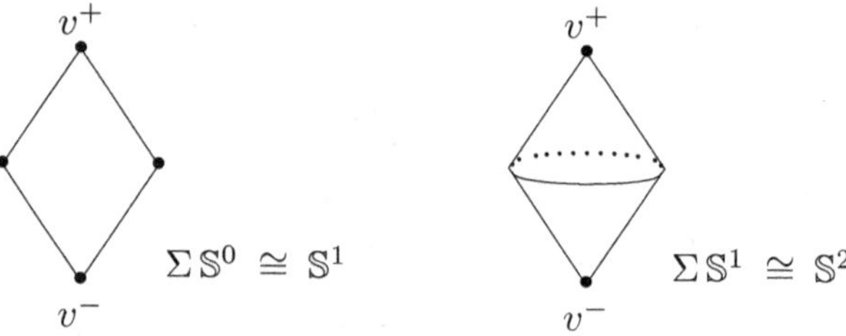

Exercises and complements. (a) Coming back to the general case, there is a natural isomorphism, for $n \geqslant 1$

$$
\delta_n X \colon H_{n+1}(\Sigma X) \to H_n(X),
\tag{2.122}
$$

given by the connecting morphism of the MV-sequence of ΣX, for the open cover

$$U = \Sigma X \setminus \{v^-\} \simeq \{*\}, \qquad V = \Sigma X \setminus \{v^+\} \simeq \{*\}, \tag{2.123}$$

with $A = U \cap V \simeq X$.

(b) If X is path connected, non-empty

$$H_1(\Sigma X) = 0 \tag{2.124}$$

is not isomorphic to $H_0(X)$. Reduced homology will give a better result, in (2.194).

(c) The endofunctor $\Sigma \colon \mathsf{Top} \to \mathsf{Top}$ plays an important role in homotopy theory. It has to be defined for all spaces, adding

$$\Sigma\emptyset = \{v^-, v^+\} \cong \mathbb{S}^0. \tag{2.125}$$

The general definition can be given by a colimit in Top, a sort of generalised pushout: see [G5], Subsection 5.2.5. Letting $\mathbb{S}^{-1} = \emptyset$, as in 1.4.1, we also have $\Sigma \mathbb{S}^{-1} \cong \mathbb{S}^0$.

2.4.6 The winding number

Let $f \colon \mathbb{S}^n \to \mathbb{S}^n$ be an endomap of the n-sphere, for $n > 0$. The associated homology homomorphism

$$f_{*n} \colon H_n(\mathbb{S}^n) \to H_n(\mathbb{S}^n) \tag{2.126}$$

is an endomorphism of an infinite cyclic group. It is thus the multiplication by an integral number $k \in \mathbb{Z}$, called the (topological) *degree*, or *winding number* of f, and written as $\deg f$.

Its properties are investigated in the (easy) exercises below.

Notes. Loosely speaking, the winding number represents how many times the sphere is wound-up on itself by f, preserving or reversing an arbitrary orientation. This is easily visualised on $\mathbb{S}^1$, as shown in Exercise (e).

We are exploiting the fact that the domain and codomain of f are the *same* sphere, and the domain and codomain of f_{*n} are the same group. If we are only given a map $f \colon X \to Y$ between two spaces homeomorphic to $\mathbb{S}^n$, we can only define the degree of f 'up to sign change' (or as a natural number). One cannot prove Theorem 2.4.8 in this way.

Exercises and complements. Let $f, g \colon \mathbb{S}^n \to \mathbb{S}^n$.

(a) The degree is invariant up to homotopy: $f \simeq g$ implies $\deg f = \deg g$.

The converse is also true: the degree classifies the homotopy classes of maps $f \colon \mathbb{S}^n \to \mathbb{S}^n$. This important result is harder to prove; here, it will only be proved for $n = 1$, in Theorem 63.7.

(b) We have:

$$\deg(fg) = (\deg f)(\deg g), \qquad \deg(\mathrm{id}\,\mathbb{S}^n) = 1. \tag{2.127}$$

The degree is a homomorphism of monoids, from $\mathsf{Top}(\mathbb{S}^n, \mathbb{S}^n)$ to the multiplicative monoid $\mathbb{Z}$.

(c) A homotopy equivalence $f\colon \mathbb{S}^n \to \mathbb{S}^n$ has degree ± 1, and any weak inverse of f has the same degree.

(d) Extending 2.4.4(b), every map $f\colon \mathbb{S}^n \to \mathbb{S}^n$ with $\deg f \neq 0$ is surjective. (One can similarly extend 2.4.4(e), assuming that $\deg g \neq 0$.)

(e) For $n = 1$, the standard map that winds-up k times the circle on itself

$$f_k\colon \mathbb{S}^1 \to \mathbb{S}^1, \qquad f_k(z) = z^k, \tag{2.128}$$

(in complex variable) has topological degree k; all the integers are thus covered by the degree.

(f) (*Degree of a suspension*) For every map $f\colon \mathbb{S}^n \to \mathbb{S}^n$, the suspension $\Sigma f\colon \mathbb{S}^{n+1} \to \mathbb{S}^{n+1}$ has the same degree

$$\deg(\Sigma f) = \deg f. \tag{2.129}$$

Applying (e), there is thus a map $\Sigma^{n-1} f_k\colon \mathbb{S}^n \to \mathbb{S}^n$ of arbitrary degree k, for every $n \geqslant 1$.

Formula (2.129) is easily visualised for $n = 1$: a map $f\colon \mathbb{S}^1 \to \mathbb{S}^1$ is extended to a map $\Sigma f\colon \mathbb{S}^2 \to \mathbb{S}^2$, which coincides with f on the equator and moves each meridian according to f; the poles v^-, v^+ stay fixed.

*(g) (The Fundamental Theorem of Algebra) *Every non-constant polynomial with coefficients in the complex field $\mathbb{C}$ has a root in $\mathbb{C}$.*

An elegant proof based on $H_2(\mathbb{S}^2)$ is given in [ES], Section XI.5. Essentially, a polynomial function $f\colon \mathbb{C} \to \mathbb{C}$ of (algebraic) degree k can be extended to a map $f^{\bullet}\colon \mathbb{S}^2 \to \mathbb{S}^2$ on the one-point compactification of $\mathbb{C}$; this map is proved to have topological degree k. If $k > 0$, $f^{\bullet}$ is surjective (by (d)) and f is too.

Taking out a point, instead of adding it, one can give a proof based on $H_1(\mathbb{S}^1)$, or equivalently on the fundamental group of the circle: see [Ha], Theorem 1.8 or [Mu], Chapter 9, Theorem 5.6.1.

2.4.7 Theorem (The antipodal map)

(a) *The* antipodal map *of the n-sphere*

$$T\colon \mathbb{S}^n \to \mathbb{S}^n, \qquad T(x) = -x, \tag{2.130}$$

has winding number $\deg T = (-1)^{n+1}$.

(b) *Suppose we have two maps* $f, g\colon \mathbb{S}^n \to \mathbb{S}^n$ *such that* $f(x) \neq Tg(x)$, *for all* $x \in \mathbb{S}^n$. *Then* $f \simeq g$ *and* $\deg f = \deg g$.

Proof (a) A reader will find it useful, and hopefully interesting, to write the proof following this outline (made explicit below). First, the map T is written as a composite of sign changes, in each variable. We prove that these $n + 1$ maps have the same degree. On $\mathbb{S}^1$ this degree is proved to be -1; by induction, this result is extended to arbitrary dimension, making use of the natural isomorphism $D'_n \colon H_n(\mathbb{S}^n) \to H_{n-1}(\mathbb{S}^{n-1})$ of (2.110). Finally $\deg T = (-1)^{n+1}$.

In fact, T is the composite of the following $n + 1$ maps (in any order)

$$i_j \colon \mathbb{S}^n \to \mathbb{S}^n, \quad i_j(t_1, ..., t_{n+1}) = (t_1, ..., -t_j, ..., t_{n+1}), \tag{2.131}$$

where $j = 1, ..., n+1$. To prove that all i_j have the same degree, we remark that the involution h that interchanges the coordinates t_1 and t_j gives a commutative diagram

$$\begin{array}{ccc} \mathbb{S}^n & \xrightarrow{\ h\ } & \mathbb{S}^n \\ {\scriptstyle i_1}\downarrow & & \downarrow{\scriptstyle i_j} \\ \mathbb{S}^n & \xrightarrow[\ h\ ]{} & \mathbb{S}^n \end{array}$$

and the relation $(\deg i_1)(\deg h) = (\deg h)(\deg i_j)$. But $\deg h = \pm 1$ (by 2.4.6(c)) can be cancelled, and $\deg i_1 = \deg i_j$.

For $n = 1$, the degree of $i_1 \colon \mathbb{S}^1 \to \mathbb{S}^1$ is easily computed using the generator $\zeta = [c + c'] \in H_1(\mathbb{S}^1)$ of (2.108). In fact, i_1 reverses the paths c, c', so that $i_{1*}[c + c'] = [i_1 c + i_1 c'] = -\zeta$ (by Exercise 2.2.8(f)).

Finally, we suppose by induction that $f' = i_1 \colon \mathbb{S}^{n-1} \to \mathbb{S}^{n-1}$ has degree -1, and prove that the same holds for $f = i_1 \colon \mathbb{S}^n \to \mathbb{S}^n$.

Noting that f only acts on the first coordinate, the conditions (2.111) are satisfied

$$f(\mathbb{S}^{n-1}) \subset \mathbb{S}^{n-1}, \quad f^{-1}\{\underline{s}\} = \{\underline{s}\}, \quad f^{-1}\{\underline{n}\} = \{\underline{n}\},$$

and D'_n is natural on f, giving a commutative square

$$\begin{array}{ccc} H_n(\mathbb{S}^n) & \xrightarrow{\ D'_n\ } & H_{n-1}(\mathbb{S}^{n-1}) \\ {\scriptstyle f_{*n}}\downarrow & & \downarrow{\scriptstyle f'_{*n}} \\ H_n(\mathbb{S}^n) & \xrightarrow[\ D'_n\ ]{} & H_{n-1}(\mathbb{S}^{n-1}) \end{array}$$

On any homology class $\zeta \in H_n(\mathbb{S}^n)$ the inductive hypothesis gives:

$$D'_n(f_{*n}(\zeta)) = f'_{*n} D'_n(\zeta) = -D'_n(\zeta) = D'_n(-\zeta).$$

Cancelling the isomorphism D'_n, we get $f_{*n}(\zeta) = -\zeta$ and $\deg f = -1$. Finally, $T = i_1 i_2 ... i_{n+1}$ has degree $(-1)^{n+1}$.

(b) Suppose we have two maps $f, g \colon \mathbb{S}^n \to \mathbb{S}^n$ such that $f(x) \neq Tg(x)$, for all $x \in \mathbb{S}^n$.

The affine homotopy $\varphi \colon f \simeq g \colon \mathbb{S}^n \to \mathbb{R}^{n+1}$ (see (1.3)) describes, for every $x \in \mathbb{S}^n$, the line segment from $f(x)$ to $g(x)$; the latter does not pass through the origin, because $f(x)$ and $g(x)$ cannot be antipodal.

$$(2.132)$$

We have thus a homotopy $\varphi \colon f \simeq g \colon \mathbb{S}^n \to \mathbb{R}^{n+1} \setminus \{0\}$, to which we can apply the normalisation map $N \colon \mathbb{R}^{n+1} \setminus \{0\} \to \mathbb{S}^n$, getting a homotopy $N\varphi$ between $Nf = f$ and $Ng = g$. $\qquad\square$

2.4.8 Theorem (Vector fields on spheres)

(a) If $n > 0$ is even, every tangent vector field on $\mathbb{S}^n$ vanishes at least in a point. Equivalently, there is no field of tangent versors on $\mathbb{S}^n$.

(b) On the other hand, every odd dimensional sphere has fields of tangent versors.

Informally, one says that the even dimensional spheres 'cannot be combed': in dimension 2, for instance, one cannot coherently comb the whole surface of a hairy ball.

Proof The geometry of the sphere is so simple that Differential Geometry is not needed here: giving a field $v(x)$ of tangent versors on $\mathbb{S}^n$ is equivalent to giving a map $f \colon \mathbb{S}^n \to \mathbb{S}^n$ such that $f(x) \perp x$ (the orthogonality relation), for all $x \in \mathbb{S}^n$, as shown in the following figure (for $n = 1$)

$$(2.133)$$

In fact, the tangent versor $v(x) \in T_x\mathbb{S}^n$ (the $\mathbb{R}$-linear space of all vectors at x, tangent to the n-sphere) corresponds to a versor $f(x) \perp x$, in the vector space $T_0\mathbb{R}^{n+1}$, which is viewed as a point of $\mathbb{S}^n$.

(a) The main part of the theorem is a corollary of Theorem 2.4.7. We suppose that there is a map $f \colon \mathbb{S}^n \to \mathbb{S}^n$ with $f(x) \perp x$, everywhere. It follows that $f(x) \neq Tx$ and $f(x) \neq x = TTx$ (where $Tx = -x$ is the antipodal map of $\mathbb{S}^n$).

Applying 2.4.7(b), we deduce that $f \simeq \mathrm{id}\,(\mathbb{S}^n)$ and $f \simeq T$. Thus $T \simeq \mathrm{id}\,\mathbb{S}^n$ and $\deg T = 1$. But we know that $\deg T = (-1)^{n+1}$ (by 2.4.7), and n must be odd.

(b) On $\mathbb{S}^1$ the following map, that corresponds to figure (2.133)

$$f\colon \mathbb{S}^1 \to \mathbb{S}^1, \qquad f(t_1, t_2) = (-t_2, t_1),$$

gives a scalar product $(x \mid f(x))$ everywhere zero: $-t_1 t_2 + t_2 t_1 = 0$.

In general, provided that $n+1$ is even, the following map gives the same result

$$f\colon \mathbb{S}^n \to \mathbb{S}^n, \qquad f(t_1, ..., t_{n+1}) = (-t_2, t_1, -t_4, t_3, ..., -t_{n+1}, t_n).$$

$\square$

2.4.9 *Exercises* (A homology generator of the sphere)

We want an explicit description of the generator of the group $H_n(\mathbb{S}^n) \cong \mathbb{Z}$, for $n \geqslant 1$.

To simplify the notation, we begin by replacing (up to homeomorphism) the standard disc $\mathbb{D}^{n+1}$ with the standard cube $\mathbb{I}^{n+1}$, and its border $\mathbb{S}^n$ with the border $S^n = \underline{\partial}\mathbb{I}^{n+1}$ of the cube.

The identity map $f_n\colon \mathbb{I}^{n+1} \to \mathbb{I}^{n+1}$ has a boundary in $C_n(\mathbb{I}^{n+1})$ whose n-cubes are (parametrised) faces of $\mathbb{I}^{n+1}$, with image in S^n. We have thus an n-chain, which is a cycle in S^n (but no more a boundary)

$$z_n = \partial_{n+1} f_n \in C_n(S^n), \qquad \partial_n z_n = \partial_n \partial_{n+1} f_n = 0. \tag{2.134}$$

The following exercises prove that its homology class $[z_n]$ is a generator of $H_n(S^n)$, and transfer this result to the standard sphere.

(a) Prove that $[z_1]$ is a generator of $H_1(S^1)$.

(b) Prove that $[z_n]$ is a generator of $H_n(S^n)$, by induction on $n \geqslant 1$.
Hints: use the MV-sequence of S^n with respect to the open subsets

$$U = S^n \setminus \{\underline{s}\}, \qquad \underline{s} = (1/2, ..., 1/2, -1),$$
$$V = S^n \setminus \{\underline{n}\}, \qquad \underline{n} = (1/2, ..., 1/2, 1). \tag{2.135}$$

(c) Construct a homeomorphism, for $n > 0$:

$$h\colon \mathbb{I}^{n+1} \to \mathbb{D}^{n+1}, \qquad h(\underline{\partial}\mathbb{I}^{n+1}) = \underline{\partial}\mathbb{D}^{n+1} = \mathbb{S}^n. \tag{2.136}$$

The previous points show that $[\partial_n h]$ is a generator of $H_n(\mathbb{S}^n)$. Concretely, it is a chain formed of $2n+2$ cubes covering the n-sphere and enfolding its cavity.

2.5 Computing homology

This Section is a list of exercises with the goal of:

- developing intuition on the topological meaning of the homology groups and their generators,

- learning how to extract information from an exact sequence.

Homology always means singular homology. For a non-empty path connected space X, we know — and shall not repeat — that $H_0(X)$ is an infinite cyclic group, generated by the homology class of any point.

2.5.1 The figure-eight space

We want to compute the homology of the *figure-eight space*, the subspace X of the plane formed of two circles tangent at a point

$$U = X \setminus \{p\} \simeq \mathbb{S}^1,$$
$$V = X \setminus \{q\} \simeq \mathbb{S}^1, \tag{2.137}$$
$$A = U \cap V = X \setminus \{p, q\} \simeq \{*\}.$$

An exercise. (a) We may guess that $H_1(X)$ is the free abelian group generated by two elements, the homology classes of two simple loops around the left or the right circle; and that the higher homology groups of X are zero.

The reader is invited to write down the proof, which is not difficult and written below. One can use the open cover (U, V) described above, taking advantage of the fact that the homology groups of U, V, and A are already known.

Solution. (a) All homology groups $H_k(X)$ are zero for $k > 1$, because they occur between two trivial groups, in an exact sequence

$$H_k(U) \oplus H_k(V) \to H_k(X) \to H_{k-1}(A). \tag{2.138}$$

(b) For $H_1(X)$ we use a segment of the sequence around this group

$$H_1 A \xrightarrow{h_1} H_1 U \oplus H_1 V \xrightarrow{k_1} H_1 X \xrightarrow{D_1} H_0 A \xrightarrow{h_0} H_0 U \oplus H_0 V. \tag{2.139}$$

Here $H_1(A) = 0$, and k_1 is a monomorphism.

$H_0(A)$, $H_0(U)$ and $H_0(V)$ are free abelian groups on a single generator $[x]$, the homology class of a point $x \in A$ — for instance the meeting point of the circles. The homomorphism $h_0 \colon H_0 A \to H_0 U \oplus H_0 V$ is computed as $h_0[x]_A = ([x]_U, [x]_V)$. Thus h_0 is mono, which means that $D_1 = 0$ and k_1 is epi.

Coming back to the sequence (2.139), k_1 is an isomorphism, and

$$H_1(X) \cong \mathbb{Z}^2. \tag{2.140}$$

For a basis of $H_1(X)$, we choose two simple loops $a, b \colon \mathbb{I} \to X$ around each circle of X, as the following ones

$$a(t) = (-1 + \cos 2\pi t, \sin 2\pi t), \qquad b(t) = (1 - \cos 2\pi t, -\sin 2\pi t). \tag{2.141}$$

We are using two loops with the same basepoint (at the meeting point of the circles). This is not relevant here, but will be for the fundamental group of X, in 6.4.4(b).

Now $[a]_U$ is a generator of $H_1(U)$, and $[b]_V$ is a generator of $H_1(V)$. We choose the elements $([a]_U, 0)$ and $(0, -[b]_V)$ as a basis of $H_1(U) \oplus H_1(V)$. The classes $[a]_X, [b]_X$ are their image in $H_1(X)$, by the isomorphism k_1, and a basis of the abelian group $H_1(X)$.

2.5.2 Exercises and complements

We can abstract some points from 2.5.1(b), that will be useful applying the MV-sequence.

(a) If the map $f \colon X \to Y$ is defined on a path connected space, the homomorphism $f_{*0} \colon H_0(X) \to H_0(Y)$ is injective.

(b) If the map $f \colon X \to Y$ takes values in a path connected space, and $X \neq \emptyset$, the homomorphism $f_{*0} \colon H_0(X) \to H_0(Y)$ is surjective.

(c) If, in the MV-sequence (2.84), the space $A = U \cap V$ is path connected, then:

- the homomorphism $h_0 \colon H_0(A) \to H_0(U) \oplus H_0(V)$ is injective,
- the homomorphism $D_1 \colon H_1(X) \to H_0(A)$ is trivial,
- the homomorphism $k_1 \colon H_1(U) \oplus H_1(V) \to H_1(X)$ is surjective.

2.5.3 Bouquets of circles

Generalising the figure-eight space, the *bouquet of circles* C_n is a space formed of n circles meeting in a single point

$$\tag{2.142}$$

It can be realised as a subspace of $\mathbb{R}^2$ (or of $\mathbb{R}^3$, if we prefer circles with the same radius). But in fact there is no need of a super-space: C_n can be obtained as a pointed sum of pointed circles, as we shall see in 2.5.4.

Exercises and complements. (a) Compute the homology groups of C_n.

(b) Compute the homology groups of the subspace X of $\mathbb{R}^2$ formed of three circles tangent in two points

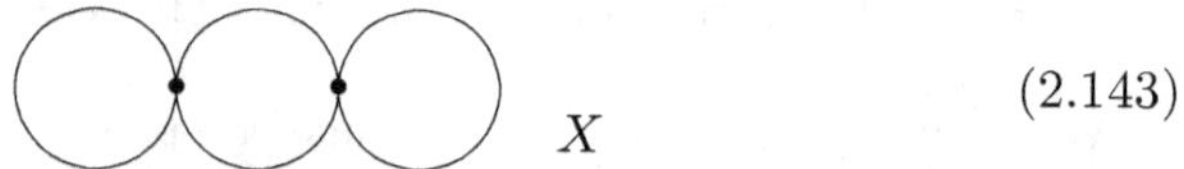

$$(2.143)$$

The reader will note that singular homology does not distinguish this space from C_3 (when directly applied to the spaces); but these spaces are not homeomorphic.

2.5.4 Sums of pointed spaces

We work in the category $\mathsf{Top}_{\bullet}$ of pointed spaces; for an object X, the basepoint is written as 0_X.

Extending a bouquet of circles, we have a family of pointed spaces $(X_i)_{i \in I}$ and want to find its categorical sum in $\mathsf{Top}_{\bullet}$, according to the general definition in 1.6.2. This is a pointed space X equipped with a family of pointed maps $u_i \colon X_i \to X$, called injections, that satisfy the universal property:

(*) for every family of pointed maps $f_i \colon X_i \to Y$ ($i \in I$), there is a unique pointed map $f \colon X \to Y$ such that $f u_i = f_i$ (for all $i \in I$).

The (rather obvious) solution is constructed below, and determined by the universal property up to pointed homeomorphism; it is denoted as $\bigvee_i X_i$, and called the *wedge*, or *pointed sum*, of the pointed spaces X_i.

To construct the solution, we begin by forming the disjoint union of the sets X_i, realised as the union of a copy $X_i \times \{i\}$ of each set of the family

$$S = \bigcup_i X_i \times \{i\}. \qquad (2.144)$$

Then we take the quotient set $X = S/R$, modulo the equivalence relation that identifies all the points $0_i = (0_{X_i}, i)$ with a single basepoint $0_X \in X$. The structural injections are defined as

$$u_i \colon X_i \to X, \qquad u_i(x) = [x, i] \quad (x \in X_i), \qquad (2.145)$$

where $[x, i]$ is the equivalence class of (x, i) in S/R; all u_i are injective.

Now we put on the pointed set X the finest topology that makes all u_i continuous: a subset $U \subset X$ is open if and only if every preimage $u_i^{-1}(U)$ is open in the space X_i.

The universal property (*) is satisfied: for a family of pointed maps $f_i \colon X_i \to Y$, we must define $f \colon X \to Y$ letting $f([x, i]) = f_i(x)$, for each $i \in I$ and $x \in X_i$. Proceeding in this way, the mapping f is well defined, because all f_i are pointed, and is pointed itself; it is also continuous: if V is open in Y, then every $u_i^{-1}(f^{-1}(V)) = f_i^{-1}(V)$ is open in X_i, and $f^{-1}(V)$ is open in X.

Exercises and complements. Let $X = \bigvee_i X_i$ be a pointed sum, with injections $u_i \colon X_i \to X$.

(a) An open subset V of X is either a nbd of the basepoint or not. The two forms are characterised as follows:

 (i) $V = \bigcup_i u_i U_i$,   each U_i is an open nbd of 0_{X_i} in X_i $(0_X \in V)$,

 (ii) $V = \bigcup_i u_i U_i$,   each U_i is open in X_i and $0_{X_i} \notin U_i$ $(0_X \notin V)$.

(b) The injections $u_i \colon X_i \to X$ of a pointed sum are topological embeddings: every X_i is homeomorphic to its image in X.

 In other words, if U is open in X_i, then $u_i(U)$ is open in $u_i(X_i)$. If, moreover, $0_{X_i} \notin U$, then $u_i(U)$ is also open in X.

(c) If every pointed space X_i is path connected (resp. connected), so is the pointed sum X.

(d) The sum of pointed sets, in $\mathsf{Set_\bullet}$, is obtained in the same way, forgetting topologies.

2.5.5 Proposition

Let $X = \bigvee_i X_i$ and $Y = \bigvee_i Y_i$ be pointed sums indexed by the same set I, with structural injections $u_i \colon X_i \to X$ and $v_i \colon Y_i \to Y$.

(a) (Functoriality) A family of pointed maps $f_i \colon X_i \to Y_i$ $(i \in I)$ gives a pointed map

$$ f = \bigvee_i f_i \colon X \to Y, \qquad f([x, i]) = [f_i(x), i] \qquad (x \in X_i), \tag{2.146} $$

characterised by the relation $f u_i = v_i f_i$ $(i \in I)$.

 The wedge is thus a functor from the cartesian power $(\mathsf{Top_\bullet})^I$ to $\mathsf{Top_\bullet}$.

(b) A family $\varphi_i \colon f_i \simeq g_i \colon X_i \to Y_i$ $(i \in I)$ of pointed homotopies gives a pointed homotopy

$$ \varphi \colon f \simeq g \colon X \to Y, \qquad \varphi([x, i], t) = [\varphi_i(x, t), i] \qquad (x \in X_i, \, t \in \mathbb{I}), \tag{2.147} $$

the unique mapping $\varphi \colon X \times \mathbb{I} \to Y$ such that $\varphi(u_i \times \mathbb{I}) = v_i \varphi_i$ (for all i).

(c) (Homotopy equivalence) If $X_i \simeq Y_i$ (for all $i \in I$), then $X \simeq Y$.

Proof (a) This is another occurrence of the functoriality of categorical sums, in 1.6.2(g). By the universal property of $\bigvee_i X_i$, f is defined as the pointed map such that $f u_i = v_i f_i$, for all $i \in I$.

(b) The mapping φ is well defined in (2.147), because $[\varphi_i(0_{X_i}, t), i] = [0_{Y_i}, i] = 0_Y$ does not depend on the index i; this also proves that $\varphi(0_X, t) = 0_Y$, for all $t \in \mathbb{I}$.

To prove the continuity of φ, we remark that:

- $\bigvee_i X_i = (\sum X_i)/R$ is a topological quotient (for the equivalence relation identifying all basepoints $(0_{X_i}, i)$),

- $(\bigvee_i X_i) \times \mathbb{I} = (\sum X_i \times \mathbb{I})/(R \times \mathbb{I})$ is also, because $\mathbb{I}$ is locally compact, and therefore the functor $- \times \mathbb{I}$ preserves topological quotients (see 7.2.6).

It is now sufficient to prove that the composite

$$\varphi p \colon \sum (X_i \times \mathbb{I}) = (\sum X_i) \times \mathbb{I} \to (\sum X_i \times \mathbb{I})/(R \times \mathbb{I}) \to Y$$

is continuous; this is true, because on every summand $X_i \times \mathbb{I}$ the mapping φp coincides with $v_i \varphi_i \colon X_i \times \mathbb{I} \to Y$.

(c) We have a family of pointed homotopy equivalences

$$f_i \colon X_i \rightleftarrows Y_i \colon g_i, \qquad \varphi_i \colon g_i f_i \simeq \mathrm{id}\, X_i, \quad \psi_i \colon f_i g_i \simeq \mathrm{id}\, Y_i. \qquad (2.148)$$

Applying (a) and (b), we get two pointed maps and two pointed homotopies:

$$f \colon X \rightleftarrows Y \colon g, \qquad \varphi \colon gf \simeq \mathrm{id}\, X, \quad \psi \colon fg \simeq \mathrm{id}\, Y.$$

$\square$

2.5.6 Theorem

Let $X = \bigvee_{i=1,\ldots,r} X_i$ be a finite sum of pointed spaces; we suppose that the basepoint of each X_i has a contractible open nbd U_i in X_i. Then

$$H_n(X) \cong \bigoplus_i H_n(X_i) \qquad (n > 0), \qquad (2.149)$$

and 0_X has a contractible open nbd in X.

Note. As to $H_0(X)$, we already know that, if all X_i are path connected, so is X.

Proof First, we note that $\bigvee_i u_i U_i$ is a contractible nbd of 0_X in X. The statement is trivially true for $r = 1$; we suppose it is for $r - 1$ and prove it for $r \geqslant 2$.

We use the MV-sequence of X, with the open cover (U, V) defined by the following wedges of pointed subspaces of X

$$U = (\bigvee_{i<r} u_i X_i) \vee u_r U_r \simeq \bigvee_{i<r} X_i,$$

$$V = (\bigvee_{i<r} u_i U_i) \vee u_r X_r \simeq X_r, \tag{2.150}$$

$$A = U \cap V = \bigvee_{i\leqslant r} u_i U_i \simeq \{*\}.$$

The homotopy equivalences are a consequence of Proposition 2.5.5(c), taking into account that all injections u_i are topological embeddings, so that each $u_i U_i$ is homeomorphic to U_i and contractible. We also note that the contractible space A is path connected (by Exercise 1.4.3(b)).

We want to prove that, for every $n \geqslant 1$, the homomorphism

$$k_n \colon H_n(U) \oplus H_n(V) \to H_n(X) \tag{2.151}$$

is an isomorphism, so that the inductive hypothesis gives:

$$H_n X \cong H_n U \oplus H_n V \cong \bigoplus_{i<r} H_n X_i \oplus H_n X_r \cong \bigoplus_i H_n X_i.$$

For $n \geqslant 2$ the claim follows from the exact sequence

$$H_n(A) \xrightarrow{h_n} H_n(U) \oplus H_n(V) \xrightarrow{k_n} H_n(X) \xrightarrow{D_n} H_{n-1}(A)$$

where $H_n(A) = H_{n-1}(A) = 0$. Finally, for $n = 1$, we have an exact sequence

$$H_1(A) \xrightarrow{h_1} H_1(U) \oplus H_1(V) \xrightarrow{k_1} H_1(X)$$

where k_1 is injective, because $H_1(A) = 0$, and is surjective by Exercise 2.5.2(c). $\qquad\square$

2.5.7 Bouquets of spheres

A *bouquet of spheres* is a finite pointed sum $X = \bigvee_{i=1,\dots,r} X_i$ where each X_i is a pointed sphere $\mathbb{S}^n$ of some dimension $n > 0$. The space X is path connected.

We recall that all standard spheres are nested, and we are using a common basepoint e_1: see 1.4.1.

Exercises and complements. (a) For $n \geqslant 1$, find a bouquet of spheres X_n with the following homology groups

$$H_k(X_n) \cong \mathbb{Z}, \text{ for } 0 \leqslant k \leqslant n, \qquad H_k(X_n) = 0, \text{ for } k > n. \tag{2.152}$$

(b) One can realise a bouquet of spheres X with arbitrary Betti numbers $\beta_1, \dots, \beta_n$ and all higher $\beta_k = 0$.

We can add disjoint open points to have an arbitrary $\beta_0 > 0$, if this has an interest.

2.5.8 Spaces with non-trivial torsion

So far we have analysed spaces whose homology groups are torsion free. We introduce now a family of path connected spaces P_n ($n \geqslant 2$) with a single torsion coefficient n, in degree 1, and trivial homology groups in higher degree

$$H_1(P_n) \cong \mathbb{Z}_n, \qquad H_k(P_n) = 0, \quad \text{for } k > 1. \tag{2.153}$$

The space P_n is the topological quotient of the standard disc $\mathbb{D}^2$

$$P_n = \mathbb{D}^2/R_n, \tag{2.154}$$

modulo the equivalence relation R_n, which is discrete in the interior of the disc and, on the border $\mathbb{S}^1$, identifies each n-tuple of points that are vertices of a regular n-sided polygon inscribed in the circle.

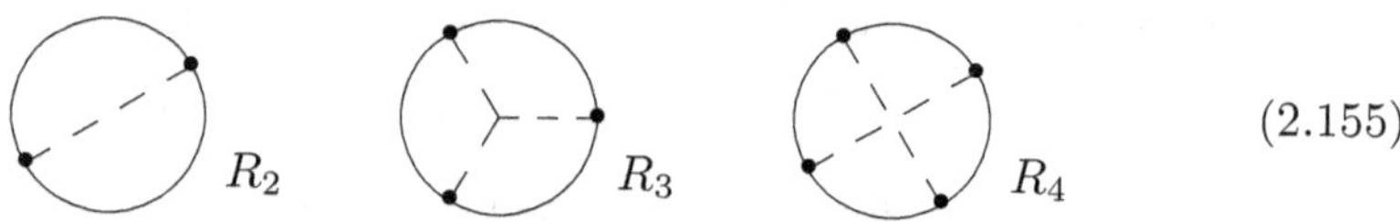

$$\tag{2.155}$$

In other words, an equivalence class $\overline{\xi}$ of R_n is either an interior point of the disc or an orbit of the action of the cyclic group $\mathbb{Z}_n$ on $\mathbb{S}^1$, generated by the rotation of $2\pi/n$ radians

$$\overline{\xi} = \{\xi,\, \xi e^{2\pi i/n},\, \xi e^{4\pi i/n},\, ...,\, \xi e^{2(n-1)\pi i/n}\} \subset \mathbb{S}^1 \subset \mathbb{C}.$$

The open disc $\operatorname{int} \mathbb{D}^2$ is topologically embedded in P_n, and will be viewed as an open subspace of the latter.

In particular the space $P_2 = \mathbb{D}^2/R_2$ is the quotient of the disc modulo the equivalence relation that identifies the antipodal points of the circle $\mathbb{S}^1$: a well-known representation of the real projective plane $\mathbb{P}^2$, that will be analysed in Section 2.6 with all (real) projective spaces $\mathbb{P}^n$.

The space $\mathbb{P}^2$ is a 2-dimensional manifold. All the other P_n are not locally euclidean: loosely speaking, a typical neighbourhood of a point $\overline{\xi}$ (with $\xi \in \mathbb{S}^1$) is homeomorphic to the union of $n \geqslant 3$ half-discs on the same diameter, and is not homeomorphic to a disc.

Exercises and complements. (a) Prove that $H_1(P_n) \cong \mathbb{Z}_n$. This requires some non-trivial work, useful and interesting.

(b) Prove that $H_k(P_n) = 0$, for $k > 1$.

2.5.9 Exercises (Torsion coefficients)

The space $P_n = \mathbb{D}^2/R_n$ defined in 2.5.8 for $n \geqslant 2$ will be pointed at the origin $0 \in \operatorname{int} \mathbb{D}^2$.

(a) Construct a path connected space X whose torsion coefficients are all of degree 1, and form an arbitrary sequence of positive integers $\tau_1, ..., \tau_k \in \mathbb{N}^*$, ordered by divisibility.

(b) Construct a path connected space whose torsion coefficients are all of a fixed degree $n \geqslant 1$, and form an arbitrary sequence as in (a).

(c) The spaces P_4 and $P_2 \vee P_2$ have different torsion coefficients.

(d) Construct two non-homeomorphic, path connected spaces with a torsion coefficient 6 in degree 1, and no other in any degree.

*(e) Putting together point (b) with Exercise 2.5.7(b) and Theorem 2.5.6, every eventually trivial sequence of finitely generated abelian groups can be realised by the homology groups of a space.

2.6 Compact surfaces and projective spaces

After the 2-sphere and the projective plane, we study some other basic compact surfaces, like the torus, the Klein bottle, and the double torus, computing their homology groups. Then we deal with the real projective spaces of any dimension.

The exposition will now be less elementary and some topological aspects will be left as understood — facts not hard to prove but requiring analytic work.

2.6.1 Quotients of the square

Many compact surfaces can be obtained as a topological quotient of the standard square $\mathbb{I}^2$, modulo an equivalence relation R which is *trivial out of the border* $\underline{\partial}\mathbb{I}^2$ (in $\mathbb{R}^2$): each equivalence class of R which is not contained in $\underline{\partial}\mathbb{I}^2$ is a singleton.

Varying the relation R, we can obtain the following instances:

(a) the sphere $\mathbb{S}^2$, if R identifies all the points of $\underline{\partial}\mathbb{I}^2$,

(b) the projective plane $\mathbb{P}^2$, if R identifies two points of $\underline{\partial}\mathbb{I}^2$ when they are symmetric with respect to the centre of the square,

(c) the torus $\mathbb{T}$, if R identifies two parallel edges 'preserving their direction', and the other two in the same way (see 2.6.2 and 2.6.3),

(d) the Klein bottle $\mathbb{K}$, if R identifies two parallel edges 'preserving direction', and the other two 'reversing direction' (see 2.6.4),

(e) the compact cylinder $\mathbb{S}^1 \times \mathbb{I}$, if R identifies two parallel edges preserving their direction,

(f) the Möbius band M, if R identifies two parallel edges reversing their direction (see 2.6.5),

(g) the square $\mathbb{I}^2$ itself, if R is the discrete relation $x = y$.

Below, we describe more precisely the torus, the Klein bottle and the Möbius band, computing their homology groups. The other non-obvious surfaces have already been considered: the sphere in 2.4.2 and the projective plane in 2.5.8.

Notes. (i) The term 'surface' is used here in an informal way: the formal definition will be given in Section 3.3.

Essentially, a *surface* (without boundary) is a connected Hausdorff space which is locally homeomorphic to the euclidean plane $\mathbb{R}^2$, while a *surface with boundary* X can also have points with an open nbd homeomorphic to the half-plane $\mathbb{R} \times [0, +\infty[$; these exceptional points form the *boundary* of X.

In the previous list, the second kind contains: the compact cylinder $\mathbb{S}^1 \times \mathbb{I}$ (with a disconnected boundary, the sum of two circles), the Möbius band M and the disc $\mathbb{D}^2$ (both with boundary a circle).

(ii) The interior $U = \,]0, 1[^2$ of the square is homeomorphic to the plane. All the spaces listed above are compactifications of U, in which U is embedded as a dense open subspace. (In fact, if V is open in U, then it is open in $\mathbb{I}^2$ and saturated for R, whence it is open in $\mathbb{I}^2/R$.)

2.6.2 The torus

We presented the torus as 'the surface of a lifebuoy', at the beginning of the general Introduction. We are now able to compute its homology groups, and make precise our intuition about its loops and 2-dimensional cavity, in Section 0.1.

(a) One can define the torus as the surface $\mathbb{T} \subset \mathbb{R}^3$ produced by the rotation around the vertical axis z of a circle C, which lies in the plane $y = 0$, with centre $(2, 0, 0)$ and radius 1

$$C = \{(x, y, z) \in \mathbb{R}^3 \mid y = 0 \text{ and } (x - 2)^2 + z^2 = 1\}, \tag{2.156}$$

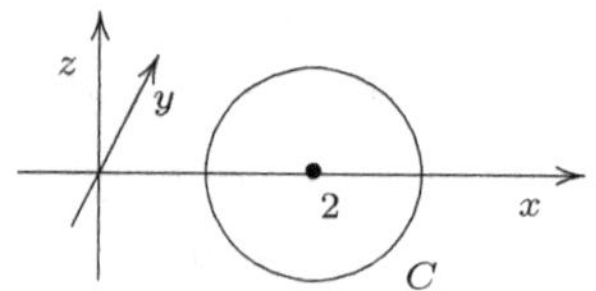

This surface is the image of the following map $f\colon \mathbb{I}^2 \to \mathbb{R}^3$

$$f(s,t) = ((2 + \cos 2\pi t)\cos 2\pi s,\; (2 + \cos 2\pi t)\sin 2\pi s,\; \sin 2\pi t) \qquad (2.157)$$

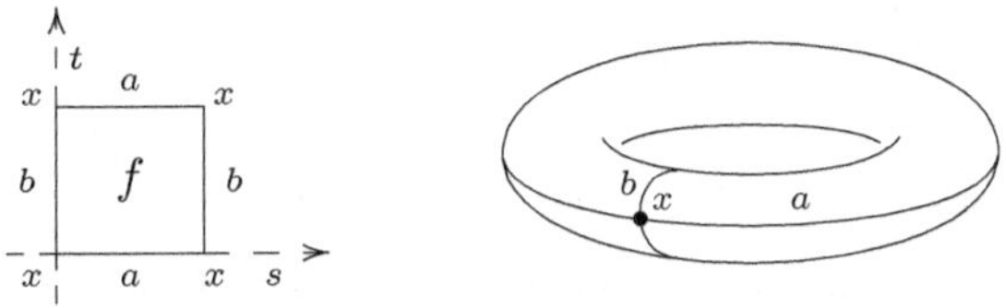

which sends the 1-indexed faces of the square to the meridian b, the 2-indexed faces to the equator a, and the four vertices of the square to the basepoint x of the loops a and b. (We are calling *equator* the 'great circle' produced by the rotation of the point $(3, 0, 0)$ of C.)

The map f induces a homeomorphism $f'\colon \mathbb{I}^2/R \to \mathbb{T}$ (from a compact to a Hausdorff space).

(b) By this homeomorphism, the torus can also be defined as the topological quotient $\mathbb{I}^2/R$. The equivalence relation R is trivial in the interior of the square, identifies the 1-indexed faces of the square 'preserving their direction', and the 2-indexed faces in the same way:

$$(0, t)\, R\, (1, t), \qquad (s, 0)\, R\, (s, 1), \qquad (s, t \in \mathbb{I}). \qquad (2.158)$$

(c) Finally, the torus can also be defined as the cartesian product $\mathbb{S}^1 \times \mathbb{S}^1$.

In fact, the relation R is the cartesian power $R_1 \times R_1$ of the equivalence relation on $\mathbb{I}$ that identifies the endpoints. It gives a bijective map $\mathbb{S}^1 \times \mathbb{S}^1 \to \mathbb{T}$, which is a homeomorphism, by the usual argument on compact Hausdorff spaces.

The torus has thus a system of coordinates in $\mathbb{S}^1 \times \mathbb{S}^1$, with latitude on the meridian and longitude on the equator.

The *n-dimensional torus* $\mathbb{T}^n$ is defined as the cartesian power $\mathbb{S}^1 \times \ldots \times \mathbb{S}^1$ of n factors. It is a compact manifold of dimension n, whose homology will be computed in Exercise 5.6.7(e).

2.6.3 *Exercises* (The homology of the torus)

(a) Prove that

$$H_1(\mathbb{T}) \cong \mathbb{Z}^2, \qquad H_2(\mathbb{T}) \cong \mathbb{Z}, \qquad (2.159)$$

and that all the higher groups are trivial. The Euler–Poincaré characteristic is 0.

Hints: use an open cover similar to that used for the projective plane, in 2.5.8.

(b) Find generators for $H_1(\mathbb{T})$ and $H_2(\mathbb{T})$. The first point is obvious, the second requires some interesting work.

2.6.4 The Klein bottle

The *Klein bottle* $\mathbb{K}$ is a (non-orientable) compact surface.

It can be defined as the quotient $\mathbb{I}^2/R$, modulo the equivalence relation which is trivial in the interior of the square, identifies the 2-indexed faces 'preserving direction', and the 1-indexed faces of the square 'reversing direction':

$$(s,0)\, R\,(s,1), \qquad (0,t)\, R\,(1,1-t) \qquad (s,t \in \mathbb{I}), \qquad (2.160)$$

as represented in the following picture

$\mathbb{K}$ cannot be realised in $\mathbb{R}^3$, but has a clear model in $\mathbb{R}^4$, that can be drawn in the plane, as above: the model should be imagined adding a fourth coordinate, to separate the 'intersection circle' where — in $\mathbb{R}^3$ — two portions of the 'tube' intersect each other. The four vertices of the square are identified with the basepoint x of the loops a and b (the images of the 2-indexed and 1-indexed faces of the square).

An exercise. (a) Prove that

$$H_1(\mathbb{K}) \cong \mathbb{Z} \oplus \mathbb{Z}_2, \qquad (2.161)$$

and that all the higher homology groups are trivial. The Euler–Poincaré characteristic is 0, and there is a torsion coefficient 2 in degree 1.

Find generators for $H_1(\mathbb{K})$.

2.6.5 The Möbius band

The *Möbius band M*, or *Möbius strip*, is a (non-orientable) compact surface with boundary.

It can be defined as the quotient $\mathbb{I}^2/R$, by the equivalence relation which is trivial in the interior of the square, and identifies two parallel faces (e.g. the 1-indexed ones) 'reversing their direction':

$$(0,t)\,R\,(1,1-t) \qquad (s\in\mathbb{I}), \qquad (2.162)$$

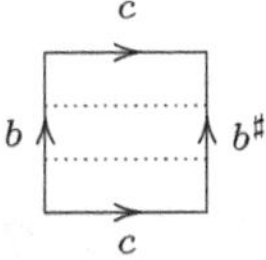

M has as a well-known model in $\mathbb{R}^3$, which can be realised with a rect-angular strip of paper, gluing the short edges as in the picture above. The 2-indexed faces a_1 and a_2 are consecutive paths, whose concatenation $a = a_1 * a_2$ is a loop in M: a parametrisation of the boundary of M, which is homeomorphic to the circle.

Due to the torsion in a strip of paper so glued, the boundary of M and the visible outline, in the picture above, do not agree everywhere.

The following path

$$c\colon \mathbb{I}\to M, \qquad c(s) = [s,1/2], \qquad (2.163)$$

is a simple loop at $[0,1/2]=[1,1/2]$, on the 'middle circle' of the strip.

The circle $\mathbb{S}^1$ is thus embedded as a deformation retract of M

$$f\colon \mathbb{S}^1\to M, \qquad f(e^{2\pi is}) = [s,1/2] \quad (s\in\mathbb{I}), \qquad (2.164)$$

and the homology groups of M are known. The isomorphism $f_*\colon H_1(\mathbb{S}^1)\to H_1(M)$ gives a generator of $H_1(M)$, the homology class $[c]$. It is also evident that the simple loop a, around the boundary of M, gives $[a] = 2[c]$.

Exercises and complements. The following exercises pertain to general topology, and their solution is left to interested readers. But their state-ment is important, here: the Möbius band allows us to better understand the Klein bottle and (even more) the projective plane — whose models in $\mathbb{R}^4$ cannot be easily visualised.

(a) Pasting two Möbius bands along their boundary we get a Klein bottle

Hints: the figure shows the Klein bottle (parametrised on the square); the middle horizontal strip is plainly a Möbius band; the two other strips, pasted along the loop c, form another Möbius band; the two bands are pasted on their (dotted) boundary.

(b) Pasting a Möbius band and a disc $\mathbb{D}^2$ along their boundary, one gets a projective plane

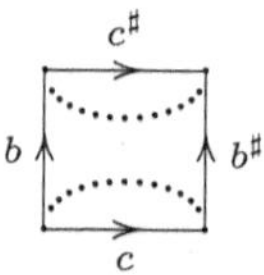

Hints: the figure shows the projective plane (parametrised on the square); the middle horizontal strip is a Möbius band; the two other parts, pasted along c, form a closed disc; the band and the disc are pasted on their (dotted) boundary.

(c) These pastings can be expressed as pushouts in Top (see 2.3.6)

$$
\begin{array}{ccc}
\mathbb{S}^1 & \xrightarrow{\ h\ } & M \\
{\scriptstyle h}\downarrow & & \downarrow{\scriptstyle u} \\
M & \xrightarrow[\ v\ ]{} & \mathbb{K}
\end{array}
\qquad\qquad
\begin{array}{ccc}
\mathbb{S}^1 & \xrightarrow{\ h\ } & M \\
{\scriptstyle k}\downarrow & & \downarrow{\scriptstyle u'} \\
\mathbb{D}^2 & \xrightarrow[\ v'\ ]{} & \mathbb{P}^2
\end{array}
\qquad\qquad (2.165)
$$

where $h\colon \mathbb{S}^1 \to M$ is induced by the loop $a\colon \mathbb{I} \to M$ around the boundary and $k\colon \mathbb{S}^1 \to \mathbb{D}^2$ is the inclusion.

2.6.6 The double torus

The 'double torus' X is the compact surface represented below

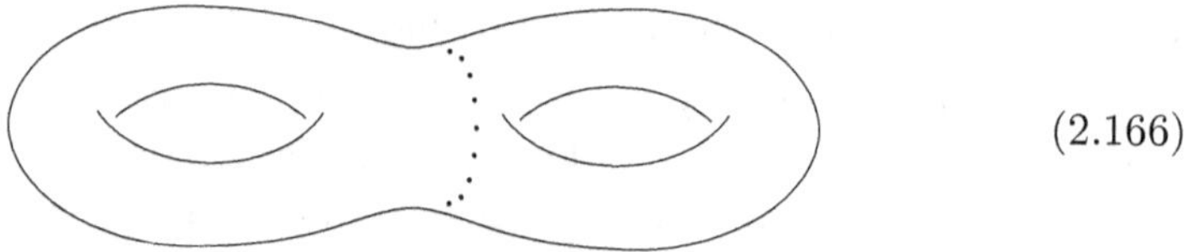

$$(2.166)$$

It can be viewed as a 'two-place lifebuoy', as here, or as a sphere with two handles. (The dotted circle at the junction is drawn at the right of the symmetry centre, for future use.)

It can also be obtained as the 'connected sum' $X = \mathbb{T} \,\sharp\, \mathbb{T}$ of two tori, as we shall see in 3.3.7.

An exercise. (a) Prove that

$$
H_1(X) \cong \mathbb{Z}^4, \qquad\qquad H_2(X) \cong \mathbb{Z}, \qquad\qquad (2.167)
$$

and that all the higher groups are trivial. The Euler–Poincaré characteristic is -2.

Hints. Use the MV-sequence for the open subspace U at the left of the dotted circle in (2.166), and a symmetric V meeting U at a cylinder.

2.6.7 The projective spaces

We end this section by computing the homology groups of the projective space $\mathbb{P}^n$. It can be defined as the topological quotient

$$\mathbb{P}^n = \mathbb{S}^n/T \qquad (n \geqslant 1), \tag{2.168}$$

of the sphere $\mathbb{S}^n$ modulo the *antipodism* relation T, which identifies each $x \in \mathbb{S}^n$ with $-x$.

We shall use the MV-sequence of $\mathbb{P}^n$, with respect to the following open cover (U_n, V_n)

$$U_n = (\mathbb{S}^n \setminus \{\underline{n}, \underline{s}\})/T \simeq \mathbb{S}^{n-1}/T \cong \mathbb{P}^{n-1},$$

$$V_n = (\mathbb{S}^n \setminus \mathbb{S}^{n-1})/T \simeq \mathbb{S}^0, \tag{2.169}$$

$$A_n = U_n \cap V_n \simeq \mathbb{S}^{n-1}.$$

We shall always omit the trivial group $H_i(V_n)$ when $i > 0$. The homomorphism $h_i \colon H_i(A_n) \to H_i(U_n)$ is thus induced by the inclusion $A_n \subset U_n$.

We have already computed the homology groups of $\mathbb{P}^1 \cong \mathbb{S}^1$ (in Section 2.4) and of $\mathbb{P}^2$ ($= P_2$, in Section 2.5).

In 2.5.8 we have used an equivalent description of $\mathbb{P}^2$, as a quotient $\mathbb{D}^2/R$ of the closed disc $\mathbb{D}^2$, where the equivalence relation R is the antipodism on the border $\mathbb{S}^1$, and discrete in the interior of the disc. This description agrees with formula (2.168), if we embed $\mathbb{D}^2$ as a closed half-sphere in $\mathbb{S}^2$. (One can proceed in the same way in any dimension.)

2.6.8 Theorem (The homology of the projective spaces)

For $n \geqslant 1$:

$$
\begin{aligned}
&H_0(\mathbb{P}^n) \cong \mathbb{Z}, && H_i(\mathbb{P}^n) = 0, \quad \text{if } i > n, \\
&H_i(\mathbb{P}^n) \cong \mathbb{Z}_2 && \text{if } i \text{ is odd and } 0 < i < n, \\
&H_i(\mathbb{P}^n) = 0 && \text{if } i \text{ is even and } 0 < i < n, \\
&H_n(\mathbb{P}^n) \cong \mathbb{Z}, \text{ if } n \text{ is odd}, && H_n(\mathbb{P}^n) = 0, \text{ if } n \text{ is even}.
\end{aligned}
\tag{2.170}
$$

Proof We know that the statement holds true for $n = 1$ and $n = 2$, and use the MV-sequence for the open cover (U_n, V_n) described in (2.169) to extend these results.

(a) To compute $H_i(\mathbb{P}^n)$, for $1 \leqslant i < n$, we suppose that (2.170) holds for $n - 1$ and prove it for $n \geqslant 3$. In the exact sequence

$$H_i(A_n) \xrightarrow{h_i} H_i(U_n) \xrightarrow{k_i} H_i(\mathbb{P}^n) \xrightarrow{D_i} H_{i-1}(A_n)$$

k_i is mono, because $H_i(A_n) = H_i(\mathbb{S}^{n-1}) = 0$; it is also epi

- by Exercise 2.5.2(c), for $i = 1$,

- or because $H_{i-1}(A_n) = H_{i-1}(\mathbb{S}^{n-1}) = 0$, for $2 \leqslant i < n$.

Therefore $H_i(\mathbb{P}^n) \cong H_i(U_n) \cong H_i(\mathbb{P}^{n-1})$, determined by the inductive hypothesis.

(b) For $H_i(\mathbb{P}^n)$ and $i > n \geqslant 3$, we suppose again that (2.170) holds for $n - 1$. Then $H_i(\mathbb{P}^n) = 0$, because of the exact sequence

$$H_i(U_n) \xrightarrow{k_i} H_i(\mathbb{P}^n) \xrightarrow{D_i} H_{i-1}(A_n)$$

with $H_i(U_n) = H_i(\mathbb{P}^{n-1}) = 0$ and $H_{i-1}(A_n) = H_{i-1}(\mathbb{S}^{n-1}) = 0$.

(c) We still have to compute $H_n(\mathbb{P}^n)$, depending on the parity of n. We prove, by induction on n odd $\geqslant 1$, that:

$$(*) \qquad H_n(\mathbb{P}^n) \cong \mathbb{Z}, \qquad H_{n+1}(\mathbb{P}^{n+1}) = 0.$$

For $n = 1$ this is known. We prove $(*)$ for n odd $\geqslant 3$, supposing that it is true for $n - 2$, that is:

$$(**) \qquad H_{n-2}(\mathbb{P}^{n-2}) \cong \mathbb{Z}, \qquad H_{n-1}(\mathbb{P}^{n-1}) = 0.$$

First, we use a portion of the exact sequence of $\mathbb{P}^n$

$$H_n(U_n) \to H_n(X_n) \xrightarrow{D_n} H_{n-1}(A_n) \to H_{n-1}(U_n)$$

where $H_n(U_n) = H_n(\mathbb{P}^{n-1}) = 0$ and $H_{n-1}(U_n) = H_{n-1}(\mathbb{P}^{n-1}) = 0$ (by hypothesis $(**)$). Therefore D_n induces an isomorphism $H_n(\mathbb{P}^n) \cong H_{n-1}(\mathbb{S}^{n-1}) \cong \mathbb{Z}$, proving the first part of $(*)$.

Secondly, we use a portion of the exact sequence of $\mathbb{P}^{n+1}$

$$0 \to H_{n+1}\mathbb{P}^{n+1} \xrightarrow{D_{n+1}} H_n A_{n+1} \xrightarrow{h_n} H_n U_{n+1} \xrightarrow{k_n} H_n \mathbb{P}^{n+1} \to 0$$

where there are isomorphisms (because n is odd):

$$H_{n+1}(U_{n+1}) = H_{n+1}(\mathbb{P}^n) = 0, \qquad H_{n-1}(A_{n+1}) = H_{n-1}(\mathbb{S}^n) = 0,$$

$$H_n(A_{n+1}) \cong H_n(\mathbb{S}^n) \cong \mathbb{Z}, \qquad H_n(U_{n+1}) \cong H_n(\mathbb{P}^n) \cong \mathbb{Z},$$

$$H_n(\mathbb{P}^{n+1}) \cong \mathbb{Z}_2.$$

Now, $\operatorname{Im} h_n = \operatorname{Ker} k_n \neq 0$ (it is a subgroup of index 2) and h_n is injective, as any non-trivial homomorphism between two infinite cyclic groups. Thus $D_{n+1} = 0$ and $H_{n+1}(\mathbb{P}^{n+1}) = 0$, proving the second part of $(*)$. $\qquad \square$

2.7 Diagram lemmas in Homological Algebra

The following issues are a basic part of Homological Algebra, and will be of frequent use in the sequel.

A reader with interests in algebra will find pleasure in proving them, by the technique of diagram chasing introduced in 2.3.3, either independently or following the hints given below. A reader more interested in topology might prefer to skip this section, and come back when necessary.

These results can be extended to any abelian category, and to various more general contexts: see [G2, G3] and references therein.

2.7.1 The Snake Lemma

The two middle squares of the diagram below are given, and supposed to be commutative, with short exact rows. The diagram is completed with the kernels and cokernels of the vertical arrows, and the induced morphisms between them (see (2.6))

$$
\begin{array}{ccccccc}
\operatorname{Ker}a & \xrightarrow{f'} & \operatorname{Ker}b & \xrightarrow{g'} & \operatorname{Ker}c & & \\
{\scriptstyle a'}\downarrow & & {\scriptstyle b'}\downarrow & & {\scriptstyle c'}\downarrow & & \\
A' & \xrightarrow{f} & B' & \xrightarrow{g} & C' & & \\
{\scriptstyle a}\downarrow & & {\scriptstyle b}\downarrow & & {\scriptstyle c}\downarrow & & \\
A'' & \xrightarrow{h} & B'' & \xrightarrow{k} & C'' & & \\
{\scriptstyle a''}\downarrow & & {\scriptstyle b''}\downarrow & & {\scriptstyle c''}\downarrow & & \\
\operatorname{Cok}a & \xrightarrow{h'} & \operatorname{Cok}b & \xrightarrow{k'} & \operatorname{Cok}c & &
\end{array}
\qquad (2.171)
$$

Then there is a connecting morphism $D\colon \operatorname{Ker}c \to \operatorname{Cok}a$ (whose dashed arrow in the diagram is the eponymous Snake), and an exact sequence, which is natural for natural transformations of the two middle squares

$$
\operatorname{Ker}a \xrightarrow{f'} \operatorname{Ker}b \xrightarrow{g'} \operatorname{Ker}c \xrightarrow{D} \operatorname{Cok}a \xrightarrow{h'} \operatorname{Cok}b \xrightarrow{k'} \operatorname{Cok}c. \qquad (2.172)
$$

Proof This is a popular lemma in Homological Algebra, closely related to Theorem 2.3.3 on the exact sequence of chain homology.

One can readily deduce the Lemma from Theorem 2.3.3, transforming the middle squares into a short exact sequence of chain complexes $A \rightarrowtail B \twoheadrightarrow C$; the complex A is trivial except for $a = \partial_1\colon A_1 \to A_0$, so that

$$
H_1(A) = \operatorname{Ker}a, \qquad H_0(A) = \operatorname{Cok}a.
$$

Similarly, B and C consist of $b\colon B_1 \to B_0$ and $c\colon C_1 \to C_0$.

The homology sequence of $A \rightarrowtail B \twoheadrightarrow C$ coincides with (2.172) (including an initial and a terminal 0) and tells us how to compute the connecting morphism $D = D_1 \colon H_1(C) \to H_0(A)$ on $z \in \operatorname{Ker} c$

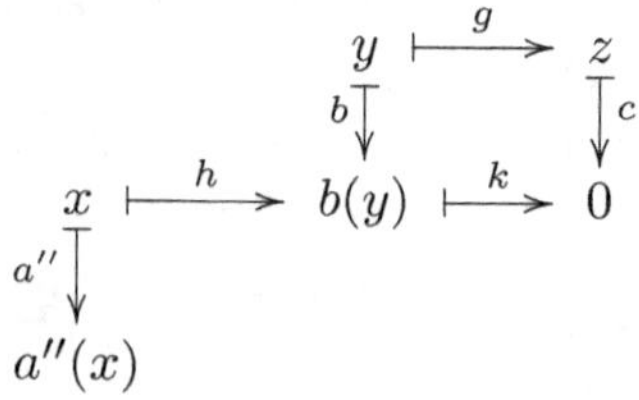

$$
\begin{array}{ccc}
& y \xmapsto{\ g\ } z \\
& b\big\uparrow \qquad \big\downarrow c \\
x \xmapsto{\ h\ } b(y) \xmapsto{\ k\ } 0 \\
a''\big\downarrow \\
a''(x)
\end{array}
\tag{2.173}
$$

One can also prove directly the Lemma (by the technique of diagram chasing used in the proof of Theorem 2.3.3), and deduce Theorem 2.3.3 from the Lemma; but this requires a longer work.

$\square$

2.7.2 The 3×3 *Lemma*

(a) In the following commutative diagram in Ab, *we suppose that the three columns and the middle row are short exact*

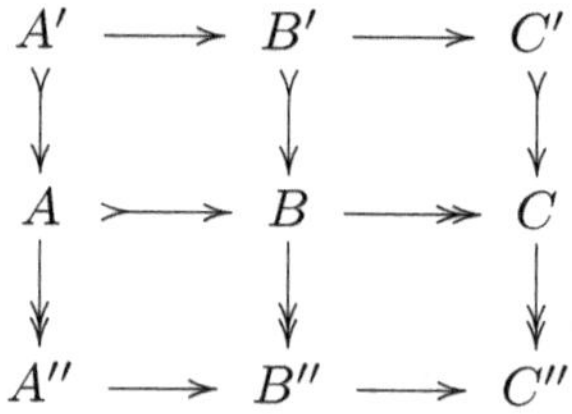

$$
\begin{array}{ccc}
A' \longrightarrow & B' \longrightarrow & C' \\
\downarrow & \downarrow & \downarrow \\
A \rightarrowtail & B \twoheadrightarrow & C \\
\downarrow & \downarrow & \downarrow \\
A'' \longrightarrow & B'' \longrightarrow & C''
\end{array}
\tag{2.174}
$$

Then the upper row is short exact if and only if the lower row is.

(b) The same holds in $\mathrm{Ch}_+\mathsf{Ab}$.

Proof (a) We assume that the upper row is short exact and prove that the lower row is also. It is sufficient to apply the Snake Lemma 2.7.1 to the two upper squares of diagram (2.174)

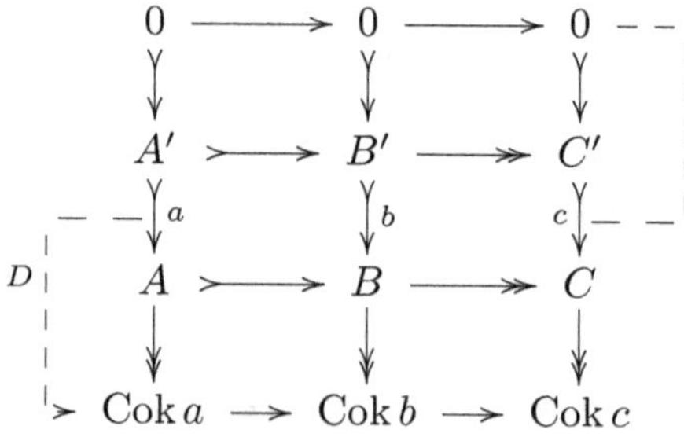

$$
\begin{array}{ccc}
0 \longrightarrow & 0 \longrightarrow & 0 \ \text{--}\,\text{--} \\
\downarrow & \downarrow & \downarrow \\
A' \rightarrowtail & B' \twoheadrightarrow & C' \\
a\big\downarrow & b\big\downarrow & c\big\downarrow \\
A \rightarrowtail & B \twoheadrightarrow & C \\
\downarrow & \downarrow & \downarrow \\
\operatorname{Cok} a \longrightarrow & \operatorname{Cok} b \longrightarrow & \operatorname{Cok} c
\end{array}
\tag{2.175}
$$

This gives the short exact sequence

$$\operatorname{Cok} a \rightarrowtail \operatorname{Cok} b \twoheadrightarrow \operatorname{Cok} c$$

corresponding to the lower row of diagram (2.174).

Similarly, if we assume that the lower row is short exact, we apply the Snake Lemma to the lower squares of (2.174), and deduce that the upper row is short exact.

(b) It is sufficient to apply (a) in each component. $\qquad\square$

2.7.3 *Exercises* (The Five Lemma)

There is given a commutative diagram in Ab, *with exact rows*

$$
\begin{array}{ccccccccc}
A & \xrightarrow{f} & B & \xrightarrow{g} & C & \xrightarrow{h} & D & \xrightarrow{k} & E \\
\downarrow{u} & & \downarrow{v} & & \downarrow{w} & & \downarrow{v'} & & \downarrow{u'} \\
A' & \xrightarrow{f'} & B' & \xrightarrow{g'} & C' & \xrightarrow{h'} & D' & \xrightarrow{k'} & E'
\end{array}
\tag{2.176}
$$

(a) If u, v, v', u' are isomorphisms, w is also.

(b) If v, v' are monomorphisms and u is epi, then w is mono.

(c) If v, v' are epimorphisms and u' is mono, then w is epi.

Hints. This is easily proved by diagram chasing. Point (a) is the main one, but it is convenient to prove (b) and (c), which imply the former.

Notes. The particular case where A, A', E, E' are trivial groups (and the rows are short exact) is called the Short Five Lemma.

In a general abelian category, (b) and (c) are dual to each other, and it would be sufficient to prove one of them.

2.7.4 *Split exact sequences*

A pair of consecutive morphisms (m, q) in Ab is said to be a *split* (short) exact sequence if the following equivalent conditions hold:

$$
A \underset{p}{\overset{m}{\rightleftarrows}} B \underset{n}{\overset{q}{\rightleftarrows}} C
\tag{2.177}
$$

(i) (m, q) is short exact and there is some n such that $qn = \operatorname{id} C$,

(ii) (m, q) is short exact and there is some p such that $pm = \operatorname{id} A$,

(iii) there are two morphisms p, n such that: $pm = \operatorname{id} A$, $qn = \operatorname{id} C$, and $mp + nq = \operatorname{id} B$.

Exercises and complements. (a) Prove the equivalence of these conditions.

(b) If these conditions hold, the morphisms p, n make B into the direct sum $A \oplus C$, completing the butterfly diagram (1.98) (with a different notation).

This decomposition of B *is not natural* on the initial data (m, q).

(c) If C is a *free* abelian group, the short exact sequence (2.177) necessarily splits. Show an example where this splitting is not determined.

(d) Let $(m, q) = (A \rightarrowtail B \twoheadrightarrow C)$ be a short exact sequence in $\mathrm{Ch}_+\mathrm{Ab}$. We say that it is *componentwise split*, or that it *splits in every degree*, if all its components (m_k, q_k) are split exact sequences of abelian groups. In particular, this certainly holds true if all the components of C are free abelian groups.

A componentwise splitting need not be a splitting in $\mathrm{Ch}_+\mathrm{Ab}$: the partial inverses $p_k \colon B_k \to A_k$ and $n_k \colon C_k \to B_k$ that we can find need not be consistent with the differentials of the complexes. The useful notion is the componentwise one, as we shall see in Chapters 4 and 5.

2.7.5 *Biproducts revisited*

Biproducts in $R\,\mathsf{Mod}$ have been examined in 1.6.4. We add now that a biproduct $C = A \oplus B$ is characterised by a diagram such that:

$$
\begin{array}{ccccc}
A & & & & B \\
& \searrow^{u} & & \swarrow^{v} & \\
1\downarrow & & C & & \downarrow 1 \\
& \swarrow_{p} & & \searrow_{q} & \\
A & & & & B
\end{array}
\tag{2.178}
$$

(*) the diagram commutes, and its diagonals are (split) exact sequences.

Exercises and complements. (a) Property (*) is indeed equivalent to property 1.6.4(ii), rewritten here as:

$$
pu = \mathrm{id}\, A, \qquad qv = \mathrm{id}\, B, \qquad up + vq = \mathrm{id}\, C.
\tag{2.179}
$$

The same characterisation holds in $\mathrm{Ch}_+\mathrm{Ab}$ (and in any abelian category).

(b) The addition $f + g$ of two homomorphisms $f, g \colon A \to B$ is determined by the categorical structure, as the following composite

$$
f + g = (A \xrightarrow{\Delta} A \oplus A \xrightarrow{f \oplus g} B \oplus B \xrightarrow{\nabla} B),
\tag{2.180}
$$

- $\Delta = (1, 1) \colon A \to A \oplus A$ is the diagonal of the product,

- $\nabla = [1, 1] \colon B \oplus B \to B$ is the codiagonal of the coproduct,

- $f \oplus g$ is categorically determined as $f \times g$, by the universal property of the categorical product (or equivalently by the universal property of the categorical sum).

2.7.6 *Exact squares*

Under the hypotheses of the Mayer–Vietoris Theorem 2.3.7, the homology functor H_n takes the left square below to a commutative square of homology groups

$$
\begin{array}{ccc}
A \xrightarrow{\ i\ } U & \qquad & H_n(A) \xrightarrow{\ i_{*n}\ } H_n(U) \\
\downarrow{\scriptstyle j} \qquad \downarrow{\scriptstyle u} & \qquad & \downarrow{\scriptstyle j_{*n}} \qquad\qquad \downarrow{\scriptstyle u_{*n}} \\
V \xrightarrow[\ v\ ]{} X & \qquad & H_n(V) \xrightarrow[\ v_{*n}\ ]{} H_n(X)
\end{array}
\qquad (2.181)
$$

Forgetting the connecting morphism, the theorem gives an exact sequence

$$
H_n(A) \xrightarrow{\ h_n\ } H_n(U) \oplus H_n(V) \xrightarrow{\ k_n\ } H_n(X)
$$

$$
h_n[z] = (i_{*n}[z], j_{*n}[z]), \qquad k_n([z],[w]) = u_{*n}[z] - v_{*n}[w],
\qquad (2.182)
$$

which means that the right square (2.181) is *exact*: cf. Hilton [Hi].

The mere commutativity of the square amounts to the condition $k_n h_n = 0$, which means that the sequence (2.181) is of order two, in the sense of 2.1.3(a).

2.8 Complements

Here we present reduced singular homology, a moderate variation of ordinary singular homology which will be useful in the next chapters.

Then we sketch the simplicial definition of singular homology, based on tetrahedra instead of cubes. The equivalence of these two forms will be proved in Section 5.6.

References are already listed in the introduction to this chapter.

2.8.1 *Reduced singular homology*

Reduced singular homology $\tilde{H}_n(X)$ only differs from ordinary homology in degrees 0 and -1. We shall see that there are sound reasons to introduce this variation.

First we introduce the category $\mathrm{Ch}_* \mathsf{Ab}$ of *unbounded* chain complexes $A = ((A_n),(\partial_n))$ of abelian groups, indexed by $n \in \mathbb{Z}$

$$
\ldots A_{n+1} \xrightarrow{\ \partial_{n+1}\ } A_n \xrightarrow{\ \partial_n\ } A_{n-1} \ldots \qquad \partial_n \partial_{n+1} = 0,
\qquad (2.183)
$$

with obvious morphisms. We have now homology functors $H_n \colon \mathrm{Ch}_* \mathsf{Ab} \to \mathsf{Ab}$, for $n \in \mathbb{Z}$.

The category $\mathrm{Ch}_+ \mathsf{Ab}$ of positive chain complexes is a full subcategory of $\mathrm{Ch}_* \mathsf{Ab}$, by adding trivial components in negative degree, as usual.

For a space X, the *augmented* (singular) *chain complex* $\tilde{C}(X)$

$$\ldots\ C_2(X) \xrightarrow{\partial_2} C_1(X) \xrightarrow{\partial_1} C_0(X) \xrightarrow{\tilde{\partial}_0} \mathbb{Z} \dashrightarrow 0 \ \ldots \qquad (2.184)$$

adds to the complex $C_+(X)$ of singular chains a single new component $\tilde{C}_{-1}(X) = \mathbb{Z}$, in degree -1, and a differential

$$\tilde{\partial}_0 \colon C_0(X) \to \tilde{C}_{-1}(X) = \mathbb{Z}, \qquad \tilde{\partial}_0(\textstyle\sum \lambda_i x_i) = \sum \lambda_i, \qquad (2.185)$$

which does satisfy the condition $\tilde{\partial}_0 \partial_1 = 0$: on each 1-cube $a \colon \mathbb{I} \to X$ we have $\partial_1(a) = a(1) - a(0)$ and $\tilde{\partial}_0 \partial_1(a) = 0$. We still write $\tilde{C}_n(X)$ as $C_n(X)$, for $n \geqslant 0$.

The augmented chain complex was used in the solution to Exercise 2.2.5(b), to compute $H_0(X)$ for a path connected space. We can think of $\tilde{C}_{-1}(X)$ as generated by the unique map $\emptyset \to X$, corresponding to an additional standard cube $\mathbb{I}^{-1} = \emptyset$.

The functor $C_+ \colon \mathsf{Top} \to \mathsf{Ch}_+\mathsf{Ab}$ is thus extended to a functor

$$\tilde{C} \colon \mathsf{Top} \to \mathsf{Ch}_*\mathsf{Ab}, \qquad (2.186)$$

and we have a *reduced singular homology functor* in each degree $n \geqslant -1$ (trivial in lower degrees)

$$\tilde{H}_n \colon \mathsf{Top} \to \mathsf{Ab}, \qquad \tilde{H}_n(X) = H_n(\tilde{C}(X)). \qquad (2.187)$$

The obvious natural transformation

$$\rho \colon \tilde{C} \to C_+ \colon \mathsf{Top} \to \mathsf{Ch}_*\mathsf{Ab}, \qquad \rho X \colon \tilde{C}(X) \twoheadrightarrow C_+(X), \qquad (2.188)$$

$$\begin{array}{ccccccccc}
\ldots\ C_2(X) & \xrightarrow{\partial_2} & C_1(X) & \xrightarrow{\partial_1} & C_0(X) & \xrightarrow{\tilde{\partial}_0} & \tilde{C}_1(X) & \dashrightarrow & 0 \ \ldots \\
\big\| & & \big\| & & \big\| & & \big\downarrow & & \big\| \\
\ldots\ C_2(X) & \xrightarrow{\partial_2} & C_1(X) & \xrightarrow{\partial_1} & C_0(X) & \dashrightarrow & 0 & \dashrightarrow & 0 \ \ldots
\end{array}$$

induces a natural transformation in each degree $n \geqslant -1$

$$\rho_{*n} \colon \tilde{H}_n \to H_n \colon \mathsf{Top} \to \mathsf{Ab}, \qquad \rho_{*n} X \colon \tilde{H}_n(X) \to H_n(X), \qquad (2.189)$$

which is the identity $\tilde{H}_n(X) = H_n(X)$ for $n > 0$, and has trivial components $\tilde{H}_{-1}(X) \to 0$ in degree -1. We are only interested in

$$\rho X = \rho_{*0} X \colon \tilde{H}_0(X) \to H_0(X), \qquad (2.190)$$

examined in the exercises below. (Note that we are using the same symbol as in $\rho X \colon \tilde{C}(X) \twoheadrightarrow C_+(X)$.)

2.8.2 Exercises and complements

We only have to explore reduced homology in degrees $0, -1$.

(a) We begin by two obvious remarks:

- for the empty space, the complex $\tilde{C}(\emptyset)$ has one non-trivial component, in degree -1; the only non-trivial group of reduced homology is $\tilde{H}_{-1}(\emptyset) = \mathbb{Z}$ (a canonical identification),

- if $X \neq \emptyset$, the differential $\tilde{\partial}_0 \colon C_0(X) \to \mathbb{Z}$ is surjective and $\tilde{H}_{-1}(X) = 0$; the group $\tilde{H}_0(X)$ is analysed below.

One may prefer to use reduced homology on non-empty spaces, to get rid of $\tilde{H}_{-1}$.

(b) There is a natural exact sequence

$$\tilde{H}_0(X) \xrightarrow{\rho X} H_0(X) \xrightarrow{DX} \mathbb{Z} \xrightarrow{pX} \tilde{H}_{-1}(X),$$

$$(DX)[\textstyle\sum_i \lambda_i x_i] = \textstyle\sum_i \lambda_i, \qquad (pX)(k) = [k]. \tag{2.191}$$

Hints. This can be proved directly, or derived from the short exact sequence of (unbounded) chain complexes $\mathrm{Ker}\,\rho \rightarrowtail \tilde{C}(X) \twoheadrightarrow C_+(X)$.

(c) For $X = \emptyset$ this exact sequence has: $\tilde{H}_0(\emptyset) = H_0(\emptyset) = 0$ and $p\emptyset = \mathrm{id}\,\mathbb{Z}$.

(d) For $X \neq \emptyset$, $\tilde{H}_{-1}(X) = 0$ and (2.183) becomes a short exact sequence

$$\tilde{H}_0(X) \rightarrowtail H_0(X) \twoheadrightarrow \mathbb{Z},$$

which necessarily splits (non canonically) in a direct sum (by 2.7.4(c)).

Concretely, we use the partition $X = \bigcup_{i \in I} X_i$ of the space X in path components (as in (2.54)). The abelian group $H_0(X)$ has a canonical basis $\xi_i = [x_i]$, where x_i is any point of the path component X_i (Exercise 2.2.5(d)). Choosing arbitrarily an index $j \in I$, we have a (non-canonical) isomorphism

$$H_0(X) \to \tilde{H}_0(X) \oplus \mathbb{Z}, \qquad \textstyle\sum_i \lambda_i [x_i] \mapsto (\textstyle\sum_{i \neq j} \lambda_i [x_i - x_j], \textstyle\sum_i \lambda_i). \tag{2.192}$$

A basis of $\tilde{H}_0(X)$ is thus given by the elements $[x_i - x_j]$, for $i \neq j$, and is in bijective correspondence with the set $I \setminus \{j\}$, namely the set of path components of X, one excluded.

(e) The space X is path connected if and only if $\tilde{H}_0(X) = 0$. If X is contractible, *all* its reduced homology groups $\tilde{H}_n(X)$ are trivial.

In several contexts, this property of reduced homology is an advantage with respect to ordinary homology.

(f) In reduced homology the only non-trivial group of the sphere $\mathbb{S}^n$ is $\tilde{H}_n(\mathbb{S}^n)$, which is infinite cyclic (also for $n = 0$): an advantage in the context of orientable manifolds.

Consider a 0-sphere $S^0 = \{p, q\} \cong \mathbb{S}^0$: then $\tilde{H}_0(S^0)$ is an infinite cyclic group, and admits two elements as generators, either $[q - p]$ or $[p - q]$.

Choosing one of them amounts to choose an orientation of S^0, from p to q or vice versa.

This will be exploited in Section 3.3: the orientation of an n-dimensional manifolds will be based on the orientation of $(n-1)$-spheres in its euclidean neighbourhoods, which leads to $\mathbb{S}^0$ for 1-dimensional manifolds. True, their connected components are lines and circles, and orienting them is not a problem; yet, a sound approach should not make exceptions.

*(g) Reduced homology can also be defined directly, without going back to chain complexes.

A space X has a unique map $f\colon X \to \{*\}$ with values in the singleton, and:

$$\tilde{H}_n(X) = \mathrm{Ker}\,(f_{*n}\colon H_n(X) \to H_n(\{*\})) \qquad (n \geqslant 0),$$
$$\tilde{H}_{-1}(X) = \mathrm{Cok}\,(f_{*0}\colon H_0(X) \to H_0(\{*\})). \tag{2.193}$$

This approach works for every homology theory: cf. [ES], Section I.7.

*(h) (*Suspension and reduced homology*) In reduced homology, the natural isomorphism (2.122) holds for all $n \geqslant 0$

$$\delta_n X\colon \tilde{H}_{n+1}(\Sigma X) \to \tilde{H}_n(X), \tag{2.194}$$

while in ordinary homology it only holds for $n > 0$. For instance, $\tilde{H}_1(\mathbb{S}^1) \cong \tilde{H}_0(\mathbb{S}^0) \cong \mathbb{Z}$.

2.8.3 *The simplicial form of singular homology*

We end this chapter showing how singular homology is constructed using the tetrahedral models Δ^n, instead of the cubical models $\mathbb{I}^n$: here a space X is explored by studying the maps $\Delta^n \to X$, where Δ^0 is the singleton, Δ^1 is a compact interval, Δ^2 a compact triangle, etc.

The simplicial construction has a formal advantage: it does not need normalisation. On the other hand, it is more complicated in many aspects we have studied or will study below. But we shall see, in Section 5.5, that it gives a simpler theory of cohomology rings.

2.8.4 *The geometry of tetrahedra*

The *standard tetrahedron* Δ^n (also called the *standard n-simplex*) is a subspace of $\mathbb{R}^{n+1}$: the convex hull of the unit points $e_0, ..., e_n$ of the cartesian axes

$$\Delta^n = \{(t_0, ..., t_n) \in \mathbb{R}^{n+1} \mid t_i \geqslant 0,\ \textstyle\sum t_i = 1\} \subset \mathbb{R}^{n+1}, \tag{2.195}$$

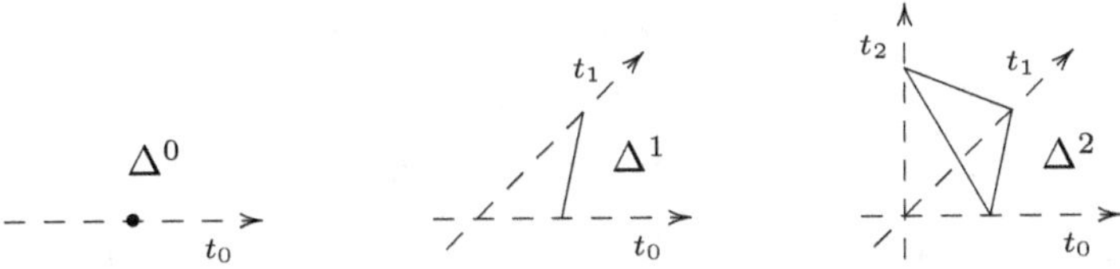

Faces and degeneracies are defined as:

$$\delta_{ni}\colon \Delta^{n-1} \to \Delta^n,$$
$$\delta_{ni}(t_0, ..., t_{n-1}) = (t_0, ..., t_{i-1}, 0, t_i, ..., t_{n-1}) \quad (0 \leqslant i \leqslant n),$$
$$\varepsilon_{ni}\colon \Delta^n \to \Delta^{n-1},$$
$$\varepsilon_{ni}(t_0, ..., t_n) = (t_0, ..., t_i + t_{i+1}, ..., t_n) \quad (0 \leqslant i \leqslant n-1).$$

$$(2.196)$$

We generally omit the index n. For singular homology we shall only use the faces, and it is important to have a concrete idea of them, in low dimension. (Degeneracies have a role in homotopy theory.)

For $n = 0$, the singleton $\Delta^0 = \{1\}$ will be written as $\{*\}$. The two faces of Δ^1 are its endpoints:

$$\delta_i\colon \{*\} \to \Delta^1, \qquad \delta_0(*) = e_1 \qquad \delta_1(*) = e_0, \qquad (2.197)$$

and the three faces of Δ^2 are parametrisations of its edges on Δ^1

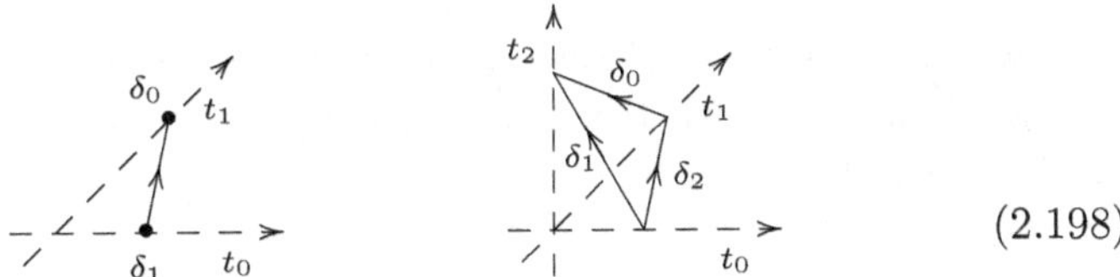

$$(2.198)$$

Faces and degeneracies satisfy the *cosimplicial relations*

$$\delta_j\delta_i = \delta_{i+1}\delta_j, \qquad \text{for } j \leqslant i,$$
$$\varepsilon_i\varepsilon_j = \varepsilon_j\varepsilon_{i+1}, \qquad \text{for } j \leqslant i,$$
$$\varepsilon_j\delta_i = \begin{cases} \delta_{i-1}\varepsilon_j, & \text{for } j < i-1, \\ \text{id}, & \text{for } j = i-1, i, \\ \delta_i\varepsilon_{j-1}, & \text{for } j > i. \end{cases} \qquad (2.199)$$

The system $((\Delta^n), (\delta_i), (\varepsilon_i))$ is called a *cosimplicial object*, in **Top**.

Notes. (a) The relation $\delta_j\delta_i = \delta_{i+1}\delta_j\colon \Delta^0 \to \Delta^2$ expresses the fact that each vertex of Δ^2 belongs to two edges. The proof of the cosimplicial relations of faces is the same as in the cubical case of Exercise 2.2.1(a), for $\alpha = \beta = 0$. The other relations are similarly checked (but are not used to define singular homology).

(b) Δ^n can also be realised in $\mathbb{R}^n$, as the convex hull of the origin and the unit points $e_1, ..., e_n$. This form, used by some authors, requires disparate formulas for faces (and degeneracies), and makes calculations more complicated.

2.8.5 The singular simplicial set

For a space X, we apply the contravariant functor

$$\mathsf{Top}(-, X)\colon \mathsf{Top}^{\mathrm{op}} \to \mathsf{Set}$$

to the system $((\Delta^n), (\delta_i), (\varepsilon_i))$, obtaining the *singular simplicial set* $\mathrm{Smp}(X)$ of the space X

$$(\mathrm{Smp}(X))_n = \mathsf{Top}(\Delta^n, X),$$

$$\partial_i(a) = a\delta_i\colon \Delta^{n-1} \to X, \qquad \text{for } a\colon \Delta^n \to X, \tag{2.200}$$

$$e_i(a) = a\varepsilon_i\colon \Delta^n \to X, \qquad \text{for } a\colon \Delta^{n-1} \to X.$$

The arrows satisfy now the *simplicial relations*, dual to the cosimplicial ones (since we are applying a contravariant functor)

$$\partial_i\partial_j = \partial_j\partial_{i+1}, \qquad \text{for } j \leqslant i,$$

$$e_j e_i = e_{i+1}e_j, \qquad \text{for } j \leqslant i,$$

$$\partial_i e_j = \begin{cases} e_j\partial_{i-1}, & \text{for } j < i - 1, \\ \mathrm{id}\,, & \text{for } j = i - 1, i, \\ e_{j-1}\partial_i, & \text{for } j > i. \end{cases} \tag{2.201}$$

In general, a *simplicial set* $K = ((K_n), (\partial_i), (e_i))$ is a sequence of sets $(K_n)_{n\geqslant 0}$ equipped with faces and degeneracies that satisfy the simplicial relations (2.201)

$$\partial_i\colon K_n \to K_{n-1} \qquad (0 \leqslant i \leqslant n),$$

$$e_i\colon K_{n-1} \to K_n \qquad (0 \leqslant i \leqslant n - 1), \tag{2.202}$$

$$K_0 \rightrightarrows K_1 \Rrightarrow K_2 \quad \ldots$$

When useful we write the faces and degeneracies as $\partial_{ni}\colon K_n \to K_{n-1}$ and $e_{ni}\colon K_{n-1} \to K_n$, specifying the degree n. The elements of K_n are called *n-simplices*, and *vertices* or *edges* when n is 0 or 1, respectively. Every n-simplex has $n + 1$ vertices.

A *morphism* of simplicial sets $f = (f_n)\colon K \to L$ is a sequence of mappings $f_n\colon K_n \to L_n$ that commute with faces and degeneracies. With componentwise composition, they form the category SmpSet.

We have now the *singular simplicial functor*, which acts on a space X as we have seen, and on a map $f\colon X \to Y$ by covariant composition

$$\mathrm{Smp}\colon \mathsf{Top} \to \mathsf{SmpSet}, \qquad X \mapsto \mathrm{Smp}(X),$$

$$f_\sharp\colon \mathrm{Smp}(X) \to \mathrm{Smp}(Y), \qquad f_{\sharp n}(a\colon \Delta^n \to X) = fa. \tag{2.203}$$

Literature. Simplicial sets can be defined as presheaves $K\colon \underline{\Delta}^{\mathrm{op}} \to \mathsf{Set}$ on the *simplicial site* $\underline{\Delta}$. The latter is the small category of positive finite ordinals and monotone maps: see [M2], Section VII.5, or [G5], Section 5.3. There is an excellent, concise book on simplicial objects, by J.P. May [May].

2.8.6 Homology

We define the *chain functor* of simplicial sets

$$\mathrm{Ch}_+\colon \mathsf{SmpSet} \to \mathrm{Ch}_+\mathsf{Ab}, \qquad \mathrm{Ch}_+(K) = ((F(K_n)), (\partial_n)), \qquad (2.204)$$

where, for a simplicial set K the component $C_n(K)$ is the free abelian group $F(K_n) = \mathbb{Z}K_n$. The differential is defined on the basis K_n as a linear combination of the faces $\partial_{ni}\colon K_n \to K_{n-1}$

$$\partial_n(a) \;=\; \Sigma_i\,(-1)^i\,\partial_{ni}(a) \in F(K_{n-1}) \qquad (i = 0, 1, ..., n). \qquad (2.205)$$

The relation $\partial_n\partial_{n+1} = 0$ is a consequence of the simplicial relations of the faces, as in the cubical case (see 2.2.3(b)).

On a morphism $f\colon K \to L$ of simplicial sets, the components of $\mathrm{Ch}_+(f)$ are the linear extensions of the components f_n

$$\mathrm{Ch}_+(f) = (C_n(f))_{n\geqslant 0},$$
$$C_n(f) = F(f_n)\colon C_n(K) \to C_n(L), \qquad \Sigma\,\lambda_i a_i \mapsto \Sigma\,\lambda_i f_n(a_i). \qquad (2.206)$$

Composing with the functor $\mathrm{Smp}\colon \mathsf{Top} \to \mathsf{SmpSet}$ of (2.203), we have the functor *of the singular simplicial chain complex*:

$$S_+\colon \mathsf{Top} \to \mathrm{Ch}_+\mathsf{Ab}, \qquad S_+(X) = \mathrm{Ch}_+(\mathrm{Smp}(X)),$$
$$f_{\sharp n}(a) = fa \qquad (\text{for } f\colon X \to Y \text{ and } a\colon \Delta^n \to X \text{ in } \mathsf{Top}). \qquad (2.207)$$

Finally, the homology functor $H_n\colon \mathrm{Ch}_+\mathsf{Ab} \to \mathsf{Ab}$ of chain complexes (in (2.24)) gives the n-th *singular homology functor*, in simplicial form

$$H_n\colon \mathsf{Top} \to \mathsf{Ab}, \quad H_n(X) = H_n(\mathrm{Ch}_+(\mathrm{Smp}(X))) = H_n(S_+(X)). \qquad (2.208)$$

For a topological space X, a (formal) linear combination of singular simplices $a_i\colon \Delta^n \to X$, with coefficients in $\mathbb{Z}$

$$c = \Sigma_i\,\lambda_i a_i \in S_n(X) = F(\mathsf{Top}(\Delta^n, X)), \qquad (2.209)$$

is called a *singular simplicial n-chain* of the space X. Again, it is an *n-cycle* if $\partial_n(c) = 0$, and an *n-boundary* if $c \in \mathrm{Im}\,\partial_{n+1}$.

Every cycle c has a *homology class* $[c] \in H_n(X)$, which vanishes if and only if c is a boundary.

Exercises and complements. (a) A 0-chain is a formal linear combination $\sum \lambda_i x_i$ of points $x_i \colon \{*\} \to X$; it is always a 0-cycle.

(b) A 1-chain is a formal linear combination of 1-simplices $a \colon \Delta^1 \to X$. The boundary of a 1-simplex

$$a \colon \Delta^1 \to X, \quad \partial_1(a) = a(0,1) - a(1,0) = a(e_1) - a(e_0), \qquad (2.210)$$

is again the formal difference of its endpoints; a is a 1-cycle if and only if it is a loop.

(c) Prove that the homology of the singleton is $\mathbb{Z}$ in degree zero and trivial elsewhere, also in the current setting.

    This is a consequence of Lemma 2.8.7, below. Nevertheless, a direct proof shows the technical difference with the corresponding cubical issue.

(d) In order to view the maps $\Delta^1 \to X$ as standard paths in the space X, we identify Δ^1 with the standard interval $\mathbb{I}$, according to the homeomorphism $\mathbb{I} \to \Delta^1$, $t \mapsto (1-t, t)$, from e_0 to e_1.

    This agrees with the differential of a 1-cube

$$a \colon \mathbb{I} \to X, \qquad \partial_1(a) = a(1) - a(0), \qquad (2.211)$$

and with the orientations marked in figure (2.198).

(e) Also here, as in 2.2.5(d), $H_0(X)$ is a free abelian group, with a canonical basis $[x_i]_H$ $(i \in I)$, where x_i is any point of the path component X_i. The proof is the same as in 2.2.5(d).

2.8.7 Lemma

For a non-empty convex euclidean space X, $H_0(X) \cong \mathbb{Z}$ and $H_k(X) = 0$ for $k > 0$.

Proof The argument only needs that X be star-shaped, for a point $x_0 \in X$.

    The chain complex $J(\mathbb{Z})$ associated to the abelian group $\mathbb{Z}$ (see (2.18)) is a retract of $S_+(X)$, in an obvious way:

$$pi \colon J(\mathbb{Z}) \to S_+(X) \to J(\mathbb{Z}),$$

$$i_0(1) = x_0, \qquad p_0(x) = 1, \qquad p_n = 0 \quad (n > 0).$$

    Proving that this is a deformation retract will show that $H_+(X)$ is the graded group $J(\mathbb{Z})$.

    For simplicity we assume that x_0 is the origin; thus, for any $x \in X$, the path a from the origin to x can be written as $a(t) = tx$, for $t \in \mathbb{I}$.

We have to build a chain homotopy $\varphi\colon ip \simeq \mathrm{id}$, i.e. a sequence

$$\varphi_n\colon S_n(X) \to S_{n+1}(X), \quad \partial_{n+1}\varphi_n(a) + \varphi_{n-1}\partial_n(a) = a - i_n p_n(a)$$

(for every n-simplex $a\colon \Delta^n \to X$).

We define $\varphi_n(a)\colon \Delta^{n+1} \to X$ letting $\varphi_n(a)(1,0,...,0) = 0$ and

$$\varphi_n(a)(t_0, ..., t_{n+1}) = (1 - t_0)a(\lambda t_1, ..., \lambda t_{n+1}), \quad \text{for } t_0 < 1, \tag{2.212}$$

where $\lambda = (1 - t_0)^{-1}$.

Now, for $n = 0$ and $a \in X$ we do have:

$$\partial_1\varphi_0(a) = \varphi_n(a)(0,1) - \varphi_n(a)(1,0) = a - 0 = a - i_0 p_0(a).$$

Finally, for $n > 0$ and $a : \Delta^n \to X$

$$\begin{aligned}
\partial_{n+1}\varphi_n(a) &= \Sigma_{i \leqslant n+1}\, (-1)^i\, \varphi_n(a)\delta_{n+1,i} \\
&= a - \varphi_{n-1}(\Sigma_{i \leqslant n}\, (-1)^i a\delta_{ni}) = a - \varphi_{n-1}\partial_n(a).
\end{aligned} \tag{2.213}$$

$\square$

2.8.8 *Homotopy invariance*

The study of the simplicial form of singular homology goes on, much in the same way as for the cubical form previously studied. The proofs and computations of the basic theory are generally more complex; other issues can be simpler (as we shall see much later, in Section 5.5).

Here we hint at diverse ways of proving the homotopy invariance of the simplicial form.

(a) (*A direct, constructive proof*) This is likely the clearest argument, if more complex than in the cubical case. Given a homotopy $\varphi\colon f \simeq g\colon X \to Y$ in Top, we build a chain homotopy

$$\Phi\colon f_\sharp \simeq g_\sharp\colon S_+(X) \to S_+(Y), \tag{2.214}$$

covering the product $\Delta^n \times \mathbb{I}$ with $n + 1$ tetrahedra $[v_0, ..., v_i, w_i, ..., w_n]$ of dimension $n+1$, which only meet at faces. (The square brackets $[...]$ denote the convex hull of the specified points, in $\mathbb{R}^{n+1}$.)

For $n = 1$, the following figure shows two triangles covering the rectangle $\Delta^1 \times \mathbb{I}$ and meeting at an edge

$$\Delta^1 \times \mathbb{I} = [v_0, w_0, w_1] \cup [v_0, v_1, w_1] = \operatorname{Im}\kappa_0 \cup \operatorname{Im}\kappa_1 \tag{2.215}$$

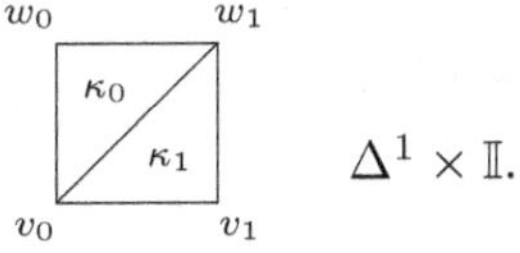

In the general case:

- $v_0, ..., v_n$ are the vertices of the n-tetrahedron $\Delta^n \times \{0\} \subset \mathbb{R}^{n+1}$,

- $w_0, ..., w_n$ are the vertices of the n-tetrahedron $\Delta^n \times \{1\} \subset \mathbb{R}^{n+1}$,

- $\kappa_i \colon \Delta^{n+1} \to \Delta^n \times \mathbb{I}$ is the affine embedding that takes the vertices $e_0, ..., e_{n+1}$ of Δ^{n+1} to the points $v_0, ..., v_i, w_i, ..., w_n$ of $\Delta^n \times \mathbb{I}$, in this order (for $i = 0, ..., n$).

Finally we let

$$\Phi_n(a) = \Sigma_{0 \leqslant i \leqslant n} \, (-1)^i \, \varphi a \kappa_i$$

(for any simplex $a \colon \Delta^n \to X$), and it is not difficult to verify that $\partial_{n+1} \Phi_n + \Phi_{n-1} \partial_n = ga - fa$.

All the details can be found in [Ha], Theorem 2.10. In this way Lemma 2.8.7 is pleonastic, since a non-empty convex euclidean space is contractible, and the homology of the singleton is already computed, in Exercise 2.8.6(c).

(b) On the other hand, one can deduce homotopy invariance from Lemma 2.8.7, using the Acyclic Model Theorem of Eilenberg–Mac Lane (see 5.6.3), as in [Sp2], Section 4.4.

Using the terminology of Section 5.6, the point is that the functor $S_+ \colon$ Top $\to$ Ch$_+$Ab is free on the models Δ^n, while the functor $S_+(- \times \mathbb{I})$ is acyclic on them, because each $\Delta^n \times \mathbb{I}$ is convex.

A similar layout is followed in [Vi], Theorem 1.10, applying the technique of acyclic models in this particular case, rather than the general theorem.

(c) We shall see in Section 5.6 that the Acyclic Model Theorem (with the help of Lemma 2.8.7) proves that the simplicial form of singular homology is naturally isomorphic to the cubical form.

As we already know that the cubical form is homotopy invariant, also the simplicial form is. This approach is followed in [HiW], Theorem 8.4.11.

3

Relative singular homology and homology theories

Singular homology is generally studied in the extended form of *relative singular homology*, defined on the category Top_2 of *relative pairs of topological spaces*: these are pairs (X, A) formed of a space X and a subspace A. This issue is dealt with in the first three sections of this chapter.

In Section 3.4 we list the Eilenberg–Steenrod axioms for homology theories, noting that they hold for (relative) singular homology. We also state the corresponding axioms for the contravariant version, a cohomology theory.

In the next two sections we introduce Alexander–Spanier cohomology and its multiplicative structure: the sequence of all cohomology groups (with coefficients in a ring) becomes a graded ring.

We end this chapter by briefly hinting at de Rham cohomology of differentiable manifolds.

3.1 Main definitions

We define relative singular homology, prove its homotopical invariance and the exact homology sequence of a pair of spaces. The term 'pair' has here a special, traditional meaning.

3.1.1 Relative pairs of spaces

We begin by introducing the category Top_2 of *pairs of topological spaces*.

An object of this category is a *pair* (X, A) formed of a space X and a *subspace* A; more explicitly, it can be called a *relative pair* of topological spaces. A (*relative*) *map* $f \colon (X, A) \to (Y, B)$ is a continuous mapping $f \colon X \to Y$ such that $f(A) \subset B$; their composition is as in Top.

Top is embedded as a full subcategory of Top_2, identifying the space X with the pair $(X, \emptyset)$.

137

The relative pair (X, A) is usually read as 'X modulo A' and viewed as a surrogate of a quotient, more 'flexible' than collapsing the subspace A to a point.

A (*relative*) *homotopy* $\varphi \colon f_0 \simeq f_1 \colon (X, A) \to (Y, B)$ of relative maps is a map of Top_2 such that:

$$\varphi \colon (X \times \mathbb{I}, A \times \mathbb{I}) \to (Y, B), \qquad \varphi(x, \alpha) = f_\alpha(x) \tag{3.1}$$

(for $x \in X$, $\alpha = 0, 1$). It is the same as an ordinary homotopy $\varphi \colon f_0 \simeq f_1 \colon X \to Y$ such that $\varphi(A \times \mathbb{I}) \subset B$.

In this context, all issues related to Top (objects, maps, homotopies, homology groups...) are called *absolute*, while those related to Top_2 are called *relative*.

Exercises and complements. (a) The homotopy relation in Top_2 is a congruence of categories: reflexive, symmetric, transitive and consistent with composition.

(b) *Homotopy equivalence* in Top_2 is defined in the usual way, and amounts to isomorphism in the quotient category $\mathsf{Top}_2/\!\simeq$.

(c) Relative homotopies derive from the *relative cylinder functor*

$$I \colon \mathsf{Top}_2 \to \mathsf{Top}_2, \qquad I(X, A) = (X \times \mathbb{I}, A \times \mathbb{I}), \tag{3.2}$$

with a structure (faces, degeneracy, etc.) similar to that of the absolute cylinder functor, in (1.85).

(d) The category $\mathsf{Top}_\bullet$ is also embedded as a full subcategory of Top_2, identifying the pointed space (X, x_0) with the pair $(X, \{x_0\})$. In this embedding, pointed homotopies are the same as relative homotopies.

(e) The expression 'relative homeomorphism' has a traditional meaning, defined in 3.2.5. An invertible arrow $f \colon (X, A) \to (Y, B)$ in Top_2 will simply be called a *homeomorphism* (of relative pairs). It is the same as a homeomorphism $f \colon X \to Y$ such that $f(A) = B$.

3.1.2 Relative chains

For a pair (X, A) in Top_2, we consider the complex $C_+(X) = \mathrm{Ch}_+(\mathrm{Cub}(X))$ of singular chains of the space X, and its *subcomplex* $C_+(A)$: we are following the usual abuse of notation, identifying a cube $\mathbb{I}^n \to A$ with a cube of X whose image is contained in A (as in 2.3.4).

The quotient complex, with induced differentials (cf. (2.69))

$$C_+(X, A) = C_+(X)/C_+(A),$$
$$C_n(X, A) = C_n(X)/C_n(A), \qquad \partial_n(\bar{c}) = (\partial_n(c))^-, \tag{3.3}$$

is called the *complex of relative singular chains* of the pair (X, A), or *of X modulo A*. We write as $\bar{c}$ the equivalence class $c + C_n(A)$ of a chain $c \in C_n(X)$.

We have thus a short exact sequence in the category $\mathsf{Ch}_+\mathsf{Ab}$

$$C_+(A) \xrightarrow{\;m\;} C_+(X) \xrightarrow{\;p\;} C_+(X, A), \tag{3.4}$$

where m is the inclusion and p is the projection on the quotient.

Applying the Induction Lemma for chain complexes (in Exercise 2.3.1(c)), a relative map $f \colon (X, A) \to (Y, B)$ gives a commutative diagram with short exact rows

$$
\begin{array}{ccccc}
C_+(A) & \rightarrowtail & C_+(X) & \twoheadrightarrow & C_+(X, A) \\
\big\downarrow{\scriptstyle f_\sharp} & & \big\downarrow{\scriptstyle f_\sharp} & & \big\downarrow{\scriptstyle f_\sharp} \\
C_+(B) & \rightarrowtail & C_+(Y) & \twoheadrightarrow & C_+(Y, B)
\end{array}
\tag{3.5}
$$

We are using the same notation $f_\sharp$ for three chain morphisms:

- $f_\sharp \colon C_+(X) \to C_+(Y)$, associated to the absolute map $f \colon X \to Y$,

- $f_\sharp \colon C_+(A) \to C_+(B)$, the restricted chain morphism (associated to the restricted map $A \to B$),

- the induced morphism on quotients, associated to the relative map f

$$f_\sharp \colon C_+(X, A) \to C_+(Y, B), \qquad \bar{c} \mapsto (f_\sharp(c))^-. \tag{3.6}$$

All this defines the functor of *relative* (cubical) *chains*

$$C_+ \colon \mathsf{Top}_2 \to \mathsf{Ch}_+\mathsf{Ab}, \qquad C_+(X, A) = C_+(X)/C_+(A), \tag{3.7}$$

which extends the absolute form, as we identify $C_+(X, \emptyset)$ with $C_+(X)$.

Exercises and complements. (a) Each component $C_n(X, A)$ is a free abelian group, with a canonical basis $c_n(X, A)$ formed of the non-degenerate n-cubes $a \colon \mathbb{I}^n \to X$ whose image is not contained in A.

(b) The short exact sequence (3.4) splits in every degree (see 2.7.4(d)).

As usual, the splitting need not be consistent with differentials. In each component there is a canonical splitting, given by the basis described in (a), but this splitting is not respected by relative maps, and is not natural on Top_2.

3.1.3 Relative singular homology

Composing the functor $C_+ \colon \mathsf{Top}_2 \to \mathsf{Ch}_+\mathsf{Ab}$ of relative chains, with the chain homology functor $H_n \colon \mathsf{Ch}_+\mathsf{Ab} \to \mathsf{Ab}$, we have the *relative singular homology functor* of degree $n \geqslant 0$:

$$H_n \colon \mathsf{Top}_2 \to \mathsf{Ab}, \qquad H_n(X, A) = H_n(C_+(X, A)),$$
$$H_n(f \colon (X, A) \to (Y, B)) = f_{*n} \colon H_n(X, A) \to H_n(Y, B). \tag{3.8}$$

Relative homology extends the absolute form: $H_n(X) = H_n(X, \emptyset)$ (and similarly on maps). $H_n(X, A)$ is called the *n-th group of singular homology of X modulo A.*

In particular, a *trivial pair* (X, X) has a trivial chain complex $C_+(X, X)$, and trivial homology $H_n(X, X) = 0$ in every degree. More generally, we shall see that $H_n(X, A) = 0$ (for all n) whenever the inclusion $A \to X$ is a homotopy equivalence.

Exercises and complements. We are given a chain $c \in C_n(X)$ and the associated relative chain $\bar{c} \in C_n(X, A)$. The following (easy) exercises are left to the reader.

(a) $\bar{c}$ is a cycle of $C_n(X, A)$ if and only if: $\partial_n(c) \in C_{n-1}(A)$.

(b) $\bar{c}$ is a boundary of $C_n(X, A)$ if and only if: $c \in B_n(X) + C_n(A)$.

3.1.4 Theorem (Homotopy invariance)

The relative singular homology functor $H_n \colon \mathsf{Top}_2 \to \mathsf{Ab}$ is homotopy invariant. It takes homotopy equivalent pairs to isomorphic groups.

Proof It is an easy consequence of the corresponding Theorem 2.2.6 for the absolute case, and can be written as an exercise. As in the proof of 2.2.6, it is convenient to write the cylinder functor in the form $I(X, A) = (\mathbb{I} \times X, \mathbb{I} \times A)$, to avoid a sign change depending on degree.

We start from a relative homotopy $\varphi \colon f \simeq g \colon (X, A) \to (Y, B)$, as defined in (3.1). The absolute homotopy $\varphi \colon f \simeq g \colon X \to Y$ has an associated chain homotopy

$$\Phi \colon f_\sharp \simeq g_\sharp \colon C_+(X) \to C_+(Y),$$

$$\Phi_n \colon C_n X \to C_{n+1} Y, \quad \Phi_n(a \colon \mathbb{I}^n \to X) = \varphi(\mathbb{I} \times a) \colon \mathbb{I}^{n+1} \to Y, \qquad (3.9)$$

$$(\partial_{n+1} \Phi_n + \Phi_{n-1} \partial_n)(a) = ga - fa.$$

Since φ is a relative homotopy, Φ_n restricts to a homomorphism $\Phi'_n \colon C_n(A) \to C_{n+1}(B)$. Applying the Induction Lemma 2.1.4 we have a commutative diagram with short exact rows (for $n \geqslant 0$)

$$\begin{array}{ccccc}
C_n(A) & \rightarrowtail & C_n(X) & \twoheadrightarrow & C_n(X, A) \\
\Phi'_n \downarrow & & \Phi_n \downarrow & & \Phi''_n \downarrow \\
C_{n+1}(A) & \rightarrowtail & C_{n+1}(X) & \twoheadrightarrow & C_{n+1}(X, A)
\end{array} \qquad (3.10)$$

The induced homomorphisms $\Phi''_n \colon C_n(X, A) \to C_{n+1}(Y, B)$ form a chain homotopy $f_\sharp \simeq g_\sharp \colon C_+(X, A) \to C_+(Y, B)$, so that, for every $n \geqslant 0$, $f_{*n} = g_{*n} \colon H_n(X, A) \to H_n(Y, B)$. $\qquad \square$

3.1.5 Theorem (The homology sequence of a pair)

For every pair of topological spaces (X, A), the following sequence is exact and natural (for maps $(X, A) \to (Y, B)$ of Top_2)

$$
\begin{array}{c}
\ldots \longrightarrow H_2(X, A) \\
H_1(A) \xrightarrow{u_{*1}} H_1(X) \xrightarrow{v_{*1}} H_1(X, A) \\
H_0(A) \xrightarrow{u_{*1}} H_0(X) \xrightarrow{v_{*1}} H_0(X, A) \longrightarrow 0.
\end{array}
\qquad (3.11)
$$

Here $u\colon A \to X$ is the inclusion and $v\colon (X, \emptyset) \to (X, A)$ is defined by the identity of X. The connecting homomorphism D_n is calculated as:

$$
D_n\colon H_n(X, A) \to H_{n-1}(A), \qquad D_n[\bar{c}]_{(X,A)} = [\partial_n(c)]_A, \qquad (3.12)
$$

where $c \in C_n(X)$ and $\partial_n(c) \in C_{n-1}(A)$ (a cycle in A, which need not be a boundary in A).

Proof In the short exact sequence of chain complexes (3.4)

$$
C_+(A) \xrightarrow{\;m\;} C_+(X) \xrightarrow{\;p\;} C_+(X, A), \qquad m = u_\sharp, \; p = v_\sharp, \qquad (3.13)
$$

the morphisms are induced by the relative maps $u\colon (A, \emptyset) \to (X, \emptyset)$ and $v\colon (X, \emptyset) \to (X, A)$.

The associated homology sequence (in 2.3.3) gives the exact sequence (3.11). Moreover the sequence (3.13) is natural for morphisms of Top_2, and therefore the sequence (3.11) is too.

We only have to check that the connecting morphisms $D_n\colon H_n(X, A) \to H_{n-1}(A)$ is indeed calculated as in (3.12).

Using the definition in 2.3.4, on a cycle $\bar{c} \in C_n(X, A)$

$$
\begin{array}{ccc}
c \longmapsto \bar{c} & \quad & (n) \\
\downarrow \qquad \downarrow & & \\
\partial_n c \longmapsto \partial_n c \longmapsto 0 & \quad & (n-1)
\end{array}
\qquad (3.14)
$$

we lift $\bar{c}$ to a representative chain $c \in C_n(X)$; its differential $\partial_n(c)$ belongs to $C_{n-1}(A)$ (as remarked in 3.1.3(a)), and $D_n[\bar{c}] = [\partial_n(c)]$. $\qquad \square$

3.1.6 Exercises and complements

(a) If A is a deformation retract of X, then $H_n(X, A) = 0$ for every n. More generally, the same holds if the inclusion $A \to X$ is a homotopy equivalence.

(b) $H_0(X, A)$ is the free abelian group generated by the path components of X which do not meet A.

(c) Consider the relative pair $(\mathbb{D}^{n+1}, \mathbb{S}^n)$, for $n \geqslant 0$. The connecting morphism of the sequence of relative homology gives a canonical isomorphism

$$D_k \colon H_k(\mathbb{D}^{n+1}, \mathbb{S}^n) \cong \tilde{H}_k(\mathbb{S}^n) \qquad \text{(for all } k\text{).} \qquad (3.15)$$

All the relative homology groups $H_k(\mathbb{D}^{n+1}, \mathbb{S}^n)$ are thus trivial, except $H_n(\mathbb{D}^{n+1}, \mathbb{S}^n) \cong \mathbb{Z}$.

(d) (*Pointed homology*) For a point $x \in X$, we have identified the pointed space (X, x) with the relative pair $(X, \{x\})$, in 3.1.1(d). The relative homology group $H_n(X, x)$ will be called the *pointed homology* of X at x. Prove that it is isomorphic to the reduced homology of X

$$H_n(X, x) \cong \tilde{H}_n(X) \qquad (n \geqslant 0), \qquad (3.16)$$

in a canonical way, depending on the pair (X, x). (Here $\tilde{H}_{-1}(X) = 0$.)

In particular $H_0(X, x)$ is canonically isomorphic to the free abelian group generated by the set of the path components of X not containing x.

3.1.7 Exercises and complements (Relative and reduced homology)

Reduced homology is well related to relative homology, and better related than ordinary homology (in degree zero, where there is a difference).

(a) First, we remark that relative homology is 'already reduced': for a relative pair (X, A), the quotient chain complex $\tilde{C}(X)/\tilde{C}(A)$ can (and will) be identified with $C_+(X, A) = C(X)/C(A)$.

(b) For a relative pair (X, A), there is an exact sequence in reduced homology which only differs from the ordinary one in degree $\leqslant 0$, and is related to the former by the natural transformation ρ, in (2.189), forming a commutative diagram with exact rows

$$
\begin{array}{ccccccccc}
\ldots H_1(X, A) & \to & \tilde{H}_0 A & \twoheadrightarrow & \tilde{H}_0 X & \to & H_0(X, A) & \to & \tilde{H}_{-1} A & \twoheadrightarrow & \tilde{H}_{-1} X \\
\| & & \downarrow{\scriptstyle \rho A} & & \downarrow{\scriptstyle \rho X} & & \| & & \downarrow & & \downarrow \\
\ldots H_1(X, A) & \twoheadrightarrow & H_0 A & \twoheadrightarrow & H_0 X & \twoheadrightarrow & H_0(X, A) & \longrightarrow & 0 & \longrightarrow & 0
\end{array}
\qquad (3.17)
$$

(c) If $A \neq \emptyset$, then $\tilde{H}_{-1}(A) = \tilde{H}_{-1}(X) = 0$ and we have a commutative diagram with exact rows

$$
\begin{array}{ccccccccc}
\ldots & H_1(X, A) & \xrightarrow{D_1} & \tilde{H}_0 A & \xrightarrow{u_{*0}} & \tilde{H}_0 X & \xrightarrow{v_{*0}} & H_0(X, A) & \longrightarrow & 0 \\
& \| & & \downarrow{\scriptstyle \rho A} & & \downarrow{\scriptstyle \rho X} & & \| & & \\
\ldots & H_1(X, A) & \xrightarrow[D_1]{} & H_0 A & \xrightarrow[u_{*0}]{} & H_0 X & \xrightarrow[v_{*0}]{} & H_0(X, A) & \longrightarrow & 0
\end{array}
\qquad (3.18)
$$

(d) In the pair (X, A), the space A has trivial reduced homology if and only if the canonical map $v\colon X \to (X, A)$ gives a family of isomorphisms $v_{*n}\colon \tilde{H}_n(X) \cong H_n(X, A)$, for $n \geqslant 0$.

(e) In the pair (X, A), the space X has trivial reduced homology if and only if all the connecting homomorphisms $D_n\colon H_n(X, A) \cong \tilde{H}_{n-1}(A)$ of reduced homology are isomorphisms. (This point extends 3.1.6(c).)

(f) The pair (X, A) has $H_n(X, A) = 0$, for all n, if and only if the inclusion $u\colon A \to X$ induces isomorphisms $u_{*n}\colon H_n(A) \cong H_n(X)$ (for $n \geqslant 0$), if and only if it induces isomorphisms $u_{*n}\colon \tilde{H}_n(A) \cong \tilde{H}_n(X)$ (for $n \geqslant -1$).

3.1.8 Theorem (The homology sequence of a triple)

For a space X and subspaces $B \subset A \subset X$, we have two canonical morphisms in Top_2

$$(A, B) \xrightarrow{u} (X, B) \xrightarrow{v} (X, A), \tag{3.19}$$

given by the inclusion $A \subset X$ and id X, *and a long exact homology sequence in* Ab, *called the exact sequence of the (relative) triple* (X, A, B)

$$\tag{3.20}$$

$$\begin{array}{l} \ldots \longrightarrow H_2(X, A) \\ \quad\quad D_2 \\ H_1(A, B) \xrightarrow{u_{*1}} H_1(X, B) \xrightarrow{v_{*1}} H_1(X, A) \\ \quad\quad D_1 \\ H_0(A, B) \xrightarrow{u_{*1}} H_0(X, B) \xrightarrow{v_{*1}} H_0(X, A) \to 0 \end{array}$$

where $D_n[\bar{c}] = [\partial_n(c)]$.

This sequence is natural for morphisms $f\colon (X, A, B) \to (X', A', B')$ of relative triples of spaces; this obviously means a map $f\colon X \to X'$ such that $f(A) \subset A'$ and $f(B) \subset B'$.

Note. *Theorem 3.1.5 corresponds to the case $B = \emptyset$.*

Proof The 3×3 Lemma, in 2.7.2, proves that the lower row of the following diagram of chain complexes is short exact

$$\tag{3.21}$$

$$\begin{array}{ccc} C_+(B) \rightarrowtail C_+(B) \longrightarrow 0 \\ \downarrow \quad\quad\quad \downarrow \quad\quad\quad \downarrow \\ C_+(A) \rightarrowtail C_+(X) \twoheadrightarrow C_+(X, A) \\ \downarrow \quad\quad\quad \downarrow \quad\quad\quad \downarrow \\ C_+(A, B) \to C_+(X, B) \to C_+(X, A) \end{array}$$

as we know that the three columns and the other two rows are short exact.

The exact homology sequence of the lower row gives the exact sequence (3.20). Its naturality is obvious.

The connecting morphisms $D_n \colon H_n(X, A) \to H_{n-1}(A, B)$ is indeed computed as stated above: the proof is the same as at the end of 3.1.5. $\square$

3.1.9 *Comments

Theorem 3.1.8 says that relative singular homology, with the connecting homomorphisms D_n, forms an *exact connected sequence* of functors from Top_2 to Ab, provided we can interpret the sequence (3.19)

$$(A, B) \xrightarrow{\;u\;} (X, B) \xrightarrow{\;v\;} (X, A),$$

as a short exact sequence in Top_2, and actually as *the* general short exact sequence of topological pairs (up to isomorphism).

This layout is developed in [G3], and briefly sketched in [G5], Section 6.7.

Essentially, the category Top_2 is equipped with a class of 'null objects', those of the form (X, X), that is X *modulo* X; a morphism $f \colon (X, A) \to (Y, B)$ is *null* if it factorises through a null object, or equivalently if $f(X) \subset B$. Kernels and cokernels are defined by the usual universal property, with respect to the ideal of null morphisms, and Top_2 satisfies axioms making it a 'homological category' (when equipped with the ideal previously defined).

Exact sequences are defined accordingly, and (3.19) is proved to be the general short exact sequence in Top_2. The particular case $A \rightarrowtail X \twoheadrightarrow (X, A)$ shows that (X, A) is indeed the quotient of X modulo A (with respect to our ideal).

3.2 Excision and compact pairs

We prove the Excision Theorem for singular homology, one of the main properties of relative homology. Loosely speaking, in a relative pair (X, A) we can cut out of each space a subspace U which is 'strongly included' in A, in the sense that $\mathrm{cl}_X U \subset \mathrm{int}_X A$.

Then we begin exploring some of its consequences: for a 'strong compact pair' (X, A), the relative homology $H_n(X, A)$ is canonically isomorphic to the reduced homology $\tilde{H}_n(X/A)$ of the quotient of the space X that collapses the subspace A to a point: a case were a relative pair behaves, in homology, as a topological quotient.

3.2.1 Theorem (Excision in singular homology)

If X is a topological space and $U \subset A \subset X$ are subspaces with $\mathrm{cl}_X U \subset \mathrm{int}_X A$, the inclusion map $i\colon (X \setminus U, A \setminus U) \to (X, A)$ of relative pairs induces an isomorphism in homology, in each degree

$$i_{*n}\colon H_n(X \setminus U, A \setminus U) \cong H_n(X, A), \qquad (3.22)$$

called the excision isomorphism.

Proof It is another consequence of the Subdivision Theorem 2.3.5. By hypothesis, the pair $\mathcal{U} = (X \setminus U, A)$ is a generalised open cover of X:

$$\mathrm{int}(X \setminus U) \cup \mathrm{int}\,A = (X \setminus \mathrm{cl}\,U) \cup \mathrm{int}\,A = X,$$

and the inclusion $j\colon C_+(X;\mathcal{U}) \to C_+(X)$ induces isomorphisms in homology. It also gives rise to a commutative diagram in $\mathrm{Ch}_+\mathrm{Ab}$, with short exact rows

$$
\begin{array}{ccccc}
C_+(A) & \rightarrowtail & C_+(X;\mathcal{U}) & \twoheadrightarrow & E_+ \\
\| & & \downarrow{\scriptstyle j} & & \downarrow{\scriptstyle i} \\
C_+(A) & \rightarrowtail & C_+(X) & \twoheadrightarrow & C_+(X, A)
\end{array}
\qquad (3.23)
$$

where $E_+ = C_+(X;\mathcal{U})/C_+(A)$.

Applying Theorem 2.3.3 (on the exact sequence of chain homology, and its naturality), we have a commutative diagram with exact rows

$$
\begin{array}{ccccccccc}
\ldots H_n A & \to & H_n(X;\mathcal{U}) & \to & H_n(E_+) & \to & H_{n-1}A & \to & H_{n-1}(X;\mathcal{U}) \\
\| & & \downarrow{\scriptstyle j_{*n}} & & \downarrow{\scriptstyle i_{*n}} & & \| & & \downarrow{\scriptstyle j_{*n-1}} \\
\ldots H_n A & \to & H_n X & \to & H_n(X, A) & \to & H_{n-1}A & \to & H_{n-1}X
\end{array}
\qquad (3.24)
$$

Applying the Five Lemma 2.7.3, each $i_{*n}\colon H_n(E_+) \to H_n(X, A)$ is an isomorphism.

It is now sufficient to prove that the chain complex E_+ is canonically isomorphic to $C_+(X \setminus U, A \setminus U)$. This follows from a Noether isomorphism in each component

$$
\begin{aligned}
E_n &= C_n(X;\mathcal{U})/C_n(A) = (C_n(X \setminus U) + C_n(A))/C_n(A) \\
&\cong C_n(X \setminus U)/(C_n(X \setminus U) \cap C_n(A)) \\
&= C_n(X \setminus U)/(C_n(A \setminus U) = C_n(X \setminus U, A \setminus U).
\end{aligned}
\qquad (3.25)
$$

More formally, we are applying a Noether isomorphism in the category of chain complexes. $\qquad\square$

3.2.2 Strong compact pairs

A *compact Hausdorff pair* (X, A) will be a relative pair of compact Hausdorff spaces; equivalently, X is compact Hausdorff and A is a closed subspace.

We say that (X, A) is a *strong compact pair* if, moreover, A is a strong deformation retract of a compact neighbourhood of A in X (see 1.1.3).

We write as X/A the quotient of X where A is collapsed to a single point x_0 (often viewed as a pointed space, with basepoint x_0). We write as $p: X \to X/A$ the canonical projection, that sends all points of A to x_0 and leaves the other points unchanged.

The case $A = \emptyset$ requires some attention: here it is convenient to assume that $X/\emptyset$ is the topological sum $X + \{x_0\}$, adding a point $x_0 \notin X$, as discussed below, in (b)–(d); note that in this case p is not surjective. (One can avoid considering this case, but we prefer to avoid exceptions.)

Examples and complements. (a) The relative pair $(\mathbb{D}^n, \mathbb{S}^{n-1})$ is a strong compact pair.

(b) For a topological space X, the meaning of $X/\emptyset$ can depend on the context. Here we want X/A to be always a pointed space.

In other contexts we may want to form a mere space and take $X/\emptyset = X$. Indeed, we have already seen that the relative pair $(X, \emptyset)$ is identified with the space X, and read as X modulo $\emptyset$.

*(c) Adjoint functors (see Section 7.1) make the present definition of X/A clearer. The embedding $U: \mathsf{Top}_\bullet \to \mathsf{Top}_2$ considered in 3.1.1(d) has a left adjoint F

$$U: \mathsf{Top}_\bullet \to \mathsf{Top}_2, \qquad U(Y, y_0) = (Y, \{y_0\}),$$
$$F: \mathsf{Top}_2 \to \mathsf{Top}_\bullet, \qquad F(X, A) = X/A, \tag{3.26}$$

characterised by the natural bijection

$$\mathsf{Top}_2((X, A), (Y, \{y_0\})) \to \mathsf{Top}_\bullet(X/A, (Y, y_0)), \tag{3.27}$$

that takes a relative map $(X, A) \to (Y, \{y_0\})$ to the induced pointed map $X/A \to (Y, y_0)$. If $A = \emptyset$, we must take $F(X, \emptyset) = X + \{x_0\}$.

*(d) The space X/A can be obtained as the pushout of the inclusion i and the (unique) map $A \to \{*\}$ (see 2.3.6)

$$\begin{array}{ccc}
A & \xrightarrow{\ i\ } & X \\
\downarrow & & \downarrow{\scriptstyle p} \\
\{*\} & \xrightarrow[\ v\ ]{} & X/A
\end{array} \tag{3.28}$$

This construction gives indeed $X/\emptyset = X + \{*\}$.

3.2.3 Lemma

(a) If (X, A) is a compact Hausdorff pair, then $(X/A, x_0)$ is also and the projection $p\colon X \to X/A$ is closed.

(b) If (X, A) is a strong compact pair, then $(X/A, x_0)$ is also. More precisely, if A has a compact nbd U in X, of which A is a strong deformation retract, then x_0 has a compact nbd U/A in X/A, of which $\{x_0\}$ is a strong deformation retract.

Proof For $A = \emptyset$ all this is obvious, and we can suppose that $A \neq \emptyset$.

(a) Letting (X, A) be a compact Hausdorff pair, we have to prove that X/A is Hausdorff. Two points $x \neq y$ in the open subspace $X \setminus A$ have disjoint open nbds contained in $X \setminus A$ and saturated for the equivalence relation that collapses A. If $x \in A$ and $y \notin A$, there exist two disjoint open sets U, V such that $A \subset U$ and $y \in V$ (because every compact Hausdorff space is regular); again, these open sets are saturated for the equivalence relation.

The projection $p\colon X \to X/A$ is closed, because X is compact and X/A is Hausdorff.

(b) We suppose that A is a strong deformation retract of U (a compact nbd of A in X), by a homotopy φ (see (1.7))

$$\begin{aligned}
&\varphi\colon X \times \mathbb{I} \to X, \quad \varphi(x, 0) \in A, \\
&\varphi(x, 1) = x, \qquad \varphi(a, t) = a \qquad (\text{for } x \in X, a \in A, t \in \mathbb{I}).
\end{aligned} \tag{3.29}$$

The projection $q\colon U \to U/A$ is closed, as above. There is a (unique) mapping $\psi\colon U/A \times \mathbb{I} \to U/A$ making the following diagram commutative

$$\begin{array}{ccc}
U \times \mathbb{I} & \xrightarrow{\ \varphi\ } & U \\
{\scriptstyle q \times \mathbb{I}}\downarrow & & \downarrow{\scriptstyle q} \\
U/A \times \mathbb{I} & \xrightarrow[\ \psi\ ]{} & U/A
\end{array} \tag{3.30}$$

because two distinct pairs $(x, t) \neq (x', t')$ are identified by $q \times \mathbb{I}$ if and only if $x, x' \in A$ and $t = t'$; but then $\varphi(x, t) = x$ and $\varphi(x', t') = x'$ are identified by q. Moreover ψ is continuous, because so is the composite $\psi(q \times \mathbb{I}) = q\varphi$, and $q \times \mathbb{I}$ is a topological projection (as any closed surjective map).

The map ψ satisfies

$$\begin{aligned}
&\psi\colon U/A \times \mathbb{I} \to U/A, \quad \psi(x, 0) = x_0, \\
&\psi(x, 1) = x, \qquad \psi(x_0, t) = x_0 \quad (\text{for } x \in U/A, t \in \mathbb{I}),
\end{aligned} \tag{3.31}$$

so that $\{x_0\}$ is a strong deformation retract of U/A, which is indeed a compact nbd of $\{x_0\}$ in X/A. $\qquad\square$

3.2.4 Theorem

Let (X, A) be a strong compact pair. The canonical projection $p\colon (X, A)$ $\to (X/A, \{x_0\})$ induces a canonical isomorphism, for every $n \geqslant 0$

$$p_{*n}\colon H_n(X, A) \to H_n(X/A, x_0) \cong \tilde{H}_n(X/A), \qquad (3.32)$$

where the second isomorphism is already known: see (3.16).

Proof We let $X' = X/A$. By hypothesis, A is a strong deformation retract of some compact nbd U (of A in X). By Lemma 3.2.3, $\{x_0\}$ is a strong deformation retract of its compact nbd $U' = U/A$.

The compact Hausdorff space X is normal. Since A is a closed subset contained in $\operatorname{int} U$, there is an open set V of X such that

$$A \subset V \subset \operatorname{cl} V \subset \operatorname{int} U. \qquad (3.33)$$

We shall form a commutative diagram, where the solid arrows are isomorphisms, and therefore the dashed ones are also

$$
\begin{array}{ccccc}
H_n(X, A) & \xrightarrow{\ v_*\ } & H_n(X, U) & \xleftarrow{\ i_*\ } & H_n(X \setminus V, U \setminus V) \\
{\scriptstyle p_*}\big\downarrow & & {\scriptstyle p_*}\big\downarrow & & \big\downarrow{\scriptstyle p_*} \\
H_n(X', x_0) & \xrightarrow{\ v_*\ } & H_n(X', U') & \xleftarrow{\ i_*\ } & H_n(X' \setminus pV, U' \setminus pV)
\end{array}
\qquad (3.34)
$$

(a) First, in the upper row, A is a strong deformation retract of U, and the exact sequence of the triple (X, U, A) gives an isomorphism $v_{*n}\colon H_n(X, A) \to H_n(X, U)$ associated to the canonical map $v\colon (X, A) \to (X, U)$.

Similarly, in the lower row, $\{x_0\}$ is a strong deformation retract of U', and we have a canonical isomorphism $v_{*n}\colon H_n(X', x_0) \to H_n(X', U')$.

(b) Secondly, we have two excision isomorphisms

$$i_{*n}\colon H_n(X \setminus V, U \setminus V) \to H_n(X, U),$$
$$i_{*n}\colon H_n(X' \setminus pV, U' \setminus pV) \to H_n(X', U'),$$

by excision of V (since $\operatorname{cl} V \subset \operatorname{int} U$) and of pV.

Let us recall that $p\colon X \to X/A$ is closed, and preserves the open sets containing A (which are saturated for p).

(c) Thirdly, the projection $p\colon X \to X/A$ restricts to the identity of $X \setminus A$; it gives an identity of relative pairs $(X \setminus V, U \setminus V) \to (X' \setminus pV, U' \setminus pV)$ and the identity $p_{*n}\colon H_n(X \setminus V, U \setminus V) \to H_n(X' \setminus pV, U' \setminus pV)$.

(d) We have thus proved that $p_{*n}\colon H_n(X, A) \to H_n(X', x_0)$ is an isomorphism. $\qquad \square$

3.2.5 Relative homeomorphisms

A relative map $f\colon (X, A) \to (Y, B)$ between strong compact pairs is said to be a *relative homeomorphism* if it has a bijective restriction $X \setminus A \to Y \setminus B$.

This is the traditional terminology: f is not supposed to be an invertible morphism of Top_2.

In this situation f induces a homeomorphism $f'\colon X/A \to X/B$. In fact, we have a commutative diagram in Top, at the left

$$
\begin{array}{ccc}
X \xrightarrow{\ f\ } Y & \qquad & (X, A) \xrightarrow{\ f\ } (Y, B) \\
\ \ \downarrow{\scriptstyle p} \qquad \downarrow{\scriptstyle q} & & \ \ \downarrow{\scriptstyle p} \qquad\qquad\quad \downarrow{\scriptstyle q} \\
X/A \xrightarrow[f']{} Y/B & & (X/A, x_0) \xrightarrow[f']{} (Y/B, y_0)
\end{array}
\tag{3.35}
$$

where f' is a bijective map between compact Hausdorff spaces (by 3.2.3(a)), and therefore a homeomorphism.

As a consequence, we also have a commutative diagram in Top_2, at the right in (3.35), where f' is a pointed homeomorphism.

3.2.6 Relative homeomorphism Theorem

If $f\colon (X, A) \to (Y, B)$ is a relative homeomorphism between strong compact pairs, then $f_{*n}\colon H_n(X, A) \to H_n(Y, B)$ is an isomorphism, for every n.

Proof Applying H_n to the right diagram (3.35), the homeomorphism of relative pairs f' induces an isomorphism; the same is true of the relative maps p and q, by Theorem 3.2.4.

It follows that $f_{*n}\colon H_n(X, A) \to H_n(Y, B)$ is an isomorphism. $\qquad\square$

3.3 Local homology and orientable manifolds

The reader likely knows that the orientability of a differentiable manifold is based on the Jacobian determinant of the transition morphisms of an atlas: if these determinants are positive everywhere, the current atlas is oriented, and gives an orientation of the manifold.

For a topological n-manifold M the transition morphisms cannot be derived, and we need a different approach. A 'local orientation' at a point $x \in M$ is defined by the choice of a generator ζ_x of the group $H_n(M, M \setminus \{x\})$: an infinite cyclic group called the *local homology group* of M at x; concretely, this amounts to orienting the 'little spheres' centred at x. An orientation of M is a consistent choice of local orientations at all points — if such a choice exists.

We only hint here at this classical, non-elementary subject of homology theory. The classification of compact surfaces is briefly described in 3.3.7 and 3.3.8, without proofs. Poincaré duality will be stated in 5.3.8.

For the sake of simplicity, all spaces are here assumed to be Hausdorff; in particular, their points are closed. On the other hand, manifolds need not be assumed to be paracompact.

Literature. Topological manifolds and their homology are studied in many books on Topology and Algebraic Topology, like [Mu], Chapter 12, [Mas3], Chapter IX, and [Ha], Section 3.3. The classification of compact surfaces is dealt with in [Mas1] and [Mu]. Differentiable manifolds are the basic topic of Differential Geometry [Sv]. A general theory of 'local structures', from manifolds to fibre bundles, was established by C. Ehresmann, since the 1960's; a new version can be found in [G4].

Other approaches to the orientation of a manifold M are based on cohomology ([ES], Section XI.6), or a single relative cohomology class of the product $M \times M$ ([Sp2], Sections 6.2 and 6.3).

Caution. Isomorphisms should be managed with special care, here: an infinite cyclic group is oriented by the choice of a generator, say ζ or $-\zeta$; an arbitrary isomorphism need not preserve it.

3.3.1 Orienting the spheres

We begin by considering the orientation of spheres; this will be extended to the local orientation of the euclidean spaces, and then to the local and global orientation of manifolds.

Let C be a space homeomorphic to $\mathbb{S}^1$; for instance, a circle in some $\mathbb{R}^k$. Then the group $H_1(C)$ is infinite cyclic.

By definition, an *orientation* of C is the choice of a generator $\zeta \in H_1(C)$, or equivalently the choice of an isomorphism $\mathbb{Z} \to H_1(C)$. Concretely, we have seen in 2.4.1(f) that this amounts to choose a way of turning around C. The other generator, namely $-\zeta$, gives the *opposite* orientation.

A circle in $\mathbb{R}^2$ has a standard orientation, given by the homology class of the counterclockwise simple loop; but a circle in any higher $\mathbb{R}^k$ has no canonical orientation.

More generally, let S be a space homeomorphic to a standard sphere $\mathbb{S}^n$, for $n \geqslant 1$. Again, the group $H_n(S)$ is infinite cyclic. An *orientation* of S is the choice of a generator $\zeta \in H_n(S)$, or equivalently the choice of an isomorphism $\mathbb{Z} \to H_n(S)$.

For a 0-sphere $S = \{p, q\} \cong \mathbb{S}^0$ we can follow the same pattern, using the reduced homology (as already discussed in 2.8.2(f)): the infinite cyclic group $\tilde{H}_0(\{p, q\})$ is generated by $[q - p]$, and also by $[p - q]$. Choosing one of them amounts to choose an orientation, from p to q or vice versa.

3.3.2 Local homology

To extend this notion, we introduce the *local homology of a space X at a point $x \in X$*, as the family of groups

$$H_k(X \mid x) = H_k(X, X \setminus \{x\}). \tag{3.36}$$

Its local character is made explicit by the excision isomorphism

$$i_{*k} \colon H_k(U, U \setminus \{x\}) \to H_k(X, X \setminus \{x\}),$$

for any open nbd U of x in X (induced by the inclusion $U \subset X$). In fact, we are cutting out the complement $X \setminus U$, which is closed in X and contained in the open subset $X \setminus \{x\}$ (as we are supposing all spaces to be Hausdorff).

The notation $H_k(X \mid x)$ is borrowed from [Ha], Section 3.3. Local homology at x should not be confused with pointed homology at x, defined in 3.1.6(d).

More generally we are interested in the *local homology groups* of the space X at a subspace A

$$H_k(X \mid A) = H_k(X, X \setminus A), \qquad H_k(X \mid A) \cong H_k(U \mid A), \tag{3.37}$$

where U is any open nbd of $\mathrm{cl}\, A$ in X: we are using again an excision isomorphism $H_k(U, U \setminus A) \to H_k(X, X \setminus A)$.

3.3.3 Local orientations

We now examine the local properties of the euclidean space $X = \mathbb{R}^n$, at a point $x \in X$.

We shall ignore its canonical orientation, because these properties will be used for spaces equipped with a homeomorphism onto X.

There are canonical isomorphisms

$$H_k(X, X \setminus \{x\}) \xrightarrow{D_k} \tilde{H}_{k-1}(X \setminus \{x\}) \xrightarrow{N_*} \tilde{H}_{k-1}(S^{n-1}), \tag{3.38}$$

where

- S^{n-1} is a sphere of X centred at x,

- D_k is the connecting (iso)morphism of the reduced homology sequence of the pair $(X, X \setminus \{x\})$ (with X contractible, as in Exercise 3.1.7(e)),

- the isomorphism N_* is induced by the normalisation map $N \colon X \setminus \{x\} \to S^{n-1}$, for the deformation retract $S^{n-1} \subset X \setminus \{x\}$.

A (local) *orientation* of X *at a point* x is the choice of a generator ζ of the infinite cyclic group

$$H_n(X \mid x) = H_n(X, X \setminus \{x\}) \cong \tilde{H}_{n-1}(S^{n-1}), \qquad (3.39)$$

and amounts to orienting all the $(n-1)$-spheres of X centred at x, in a consistent way.

An orientation of X at the point x gives an orientation of X at any point y, via the canonical isomorphisms associated to inclusions

$$H_n(X, X \setminus \{x\}) \leftarrow H_n(X, X \setminus B) \rightarrow H_n(X, X \setminus \{y\}), \qquad (3.40)$$

where B is any open ball of X containing x and y. (The left isomorphism comes from the homology sequence of the triple $(X, X \setminus \{x\}, X \setminus B)$, in 3.1.8, where $X \setminus B$ is a deformation retract of $X \setminus \{x\}$.)

Globally, this coherent system of local orientations forms an *orientation* of X.

Remarks. The only non-trivial local homology group $H_k(\mathbb{R}^n \mid x)$ has degree n. This is again an intrinsic characterisation of the topological dimension of $\mathbb{R}^n$.

3.3.4 *Topological manifolds and surfaces*

We briefly review the basic facts of this topic, developed in the literature cited above.

(a) A (topological) *n-manifold,* or *manifold of dimension n,* is a Hausdorff topological space X locally homeomorphic to $\mathbb{R}^n$, in the sense that every point has an open nbd homeomorphic to $\mathbb{R}^n$.

By definition, a manifold X is locally path connected. Therefore every path connected component of X is open; as they are disjoint, the space X is connected if and only if it is path connected.

There is, up to homeomorphism, one connected 0-manifold, the singleton. *There is one compact connected 1-manifold, the circle.*

A *surface* is a connected 2-manifold. We have examined in Chapter 2 various compact surfaces, like the sphere $\mathbb{S}^2$, the torus, the projective plane $\mathbb{P}^2$, the Klein bottle, etc. A celebrated classification theorem (stated in 3.3.8) says that — up to homeomorphism — they form a countable family.

The higher dimensional manifolds are much more complex.

(b) More generally, an *n-manifold with boundary* is a Hausdorff topological space X where every point has an open nbd homeomorphic to $\mathbb{R}^n$ or to the euclidean half-space $H^n = \{x \in \mathbb{R}^n \mid x_1 \geqslant 0\}$.

Then the *intrinsic interior* of X is the set of points which have an open nbd homeomorphic to $\mathbb{R}^n$, and the (intrinsic) *boundary* of X is the complement of the intrinsic interior. These subspaces of X are intrinsically defined, and should not be confused with relative notions, like the interior or border of X *in* some super-space X'. The boundary of X is a manifold of dimension $n-1$, usually denoted as ∂X.

In particular, a *surface with boundary* is a connected 2-manifold with boundary; the latter is a 1-dimensional manifold, possibly disconnected. Various compact examples were described in 2.6.1.

(c) The cartesian product of two manifolds of dimensions m, n is a manifold of dimension $m + n$. The product $X \times Y$ of a manifold X and a manifold with boundary Y is a manifold with boundary.

*(d) A connected manifold X (without boundary) is a *homogeneous space*: for every pair of points $x, y \in X$ there is a homeomorphism $f \colon X \to X$ that takes x to y.

This is obvious for $\mathbb{R}^n$, using translations. Therefore, for every $x \in X$, the set U_x of all the points $y \in X$ related in this way to x is open. All U_x form a partition of X in disjoint open sets; by connectedness, there is only one of them.

3.3.5 Orientable manifolds

We can now define orientation for an n-manifold X (of positive dimension).

At any point $x \in X$, the local homology group $H_n(X \mid x) = H_n(X, X \setminus x)$ is isomorphic to $H_n(U \mid x)$, where U is any open nbd of x; it is an infinite cyclic group, as shown in (3.38) (taking U homeomorphic to $\mathbb{R}^n$).

A *local orientation* of the n-manifold X at a point x is a generator ζ_x of $H_n(X \mid x)$.

An *orientation* of X is a family $(\zeta_x)_{x \in X}$ of local orientations at each point, which is coherent. This means that any $x \in X$ has a euclidean nbd (homeomorphic to $\mathbb{R}^n$) containing an open n-ball B containing x, such that there exists a local orientation $\zeta_B \in H_n(X \mid B)$ whose image by the canonical isomorphism $H_n(X \mid B) \to H_n(X \mid y)$ is ζ_y, for all $y \in B$.

Remarks and complements. (a) Recalling that $H_k(X \mid x) = 0$ for $k \neq n$ (by (3.38), again), the local homology of X at any point *structurally determines* the dimension of X.

(b) If the manifold X has an orientation $(\zeta_x)_{x \in X}$, the family $(-\zeta_x)_{x \in X}$ is also an orientation, the *opposite* one.

(c) The manifold X is orientable if and only if all its connected components are. A connected orientable manifold has precisely two orientations, opposite to each other.

3.3.6 *Complements

Further developing the theory of topological manifolds is beyond the bounds
of this book. The rest of this section will give further information, without
proofs, to awaken interest in reading, or studying, the sources cited above.

The situation is particularly clear in the case of a (non-empty) compact
connected n-manifold M.

(a) M is orientable if and only if $H_n(M) \cong \mathbb{Z}$. In this case, for every $x \in M$,
the identity of M induces an isomorphism $i_x \colon H_n(M) \to H_n(M, M \setminus \{x\})$,
and a generator $\zeta \in H_n(M)$ produces an orientation of M, the family
$\zeta_x = i_x(\zeta)$ $(x \in X)$.

(b) M is non-orientable if and only if $H_n(M) = 0$.

(c) $H_k(M) = 0$, for $k > 0$.

(d) Poincaré duality binds the singular homology and cohomology groups
of M. It will be stated in 5.3.8.

3.3.7 *The connected sum of manifolds and surfaces

There is an enticing operation acting on two connected manifolds X, Y of
the same dimension $n > 0$. Loosely speaking, the *connected sum* $X \sharp Y$ is
obtained by taking out of each of them a 'little' open n-disc, and pasting the
remaining subspaces X', Y' along their boundary S (a sphere of dimension
$n - 1$).

The result is a connected manifold *independent* of these choices (up to
homeomorphism, of course), essentially because of the homogeneity property of connected manifolds (in 3.3.4(d)). One proves that the operation
is associative (up to homeomorphism); the sphere $\mathbb{S}^n$ is a unit element for
this operation: we take out an open disc and replace it.

Here we are interested in the connected sum $X \sharp Y$ of connected surfaces;
the junction around the (dotted) circle S looks as follows

$$(3.41)$$

The connected sum $\mathbb{T}^{\sharp 2} = \mathbb{T} \sharp \mathbb{T}$ of two copies of the torus is the double
torus drawn and studied in 2.6.6.

Repeating this operation we get an orientable surface $\mathbb{T}^{\sharp p}$ which can be represented as a lifebuoy with p places, or a sphere with p handles. Its homology groups are easily computed, by induction:

$$H_1(\mathbb{T}^{\sharp p}) \cong \mathbb{Z}^{2p}, \qquad H_2(\mathbb{T}^{\sharp p}) \cong \mathbb{Z}. \tag{3.42}$$

3.3.8 *The Classification Theorem of Compact Surfaces*

The compact surfaces form, up to homeomorphism, a countable set with an amazingly simple classification. The proof is complex, and we only give the statement. It can be seen in Massey [Mas1], or in [Mu], Chapter 12.

(i) Every orientable compact surface is homeomorphic to a connected sum $\mathbb{T}^{\sharp p}$ of $p \geqslant 0$ tori (where $\mathbb{T}^{\sharp 0}$ stands for the 2-sphere).

(ii) Every non-orientable compact surface is homeomorphic to a connected sum $(\mathbb{P}^2)^{\sharp p}$ of $p \geqslant 1$ projective planes.

Remarks. The non-orientable compact surfaces are better viewed considering the following points (not hard to prove).

(a) The connected sum $\mathbb{P}^2 \sharp \mathbb{P}^2$ of two projective planes is the pasting of two Möbius bands along their boundary, that is a Klein bottle $\mathbb{K}$ (as we have already seen in 2.6.5(b) and 2.6.5(a)).

(b) The connected sums $\mathbb{T} \sharp \mathbb{P}^2$ and $\mathbb{K} \sharp \mathbb{P}^2$ are homeomorphic. *Hints:* it is easy to see that $\mathbb{T} \sharp M \cong \mathbb{K} \sharp M$, where M is a Möbius band contained in $\mathbb{P}^2$.

(c) We can now represent any non-orientable compact surface as the connected sum of an orientable surface $\mathbb{T}^{\sharp n}$ with either a projective plane or a Klein bottle:

$$\mathbb{P}^{\sharp 2n+1} \cong \mathbb{T}^{\sharp n} \sharp \mathbb{P}, \qquad \mathbb{P}^{\sharp 2n+2} \cong \mathbb{T}^{\sharp n} \sharp \mathbb{K} \qquad (n \geqslant 0). \tag{3.43}$$

3.4 Eilenberg–Steenrod axioms for homology and cohomology

The book of Eilenberg and Steenrod [ES] was a landmark in Algebraic Topology, establishing a list of axioms sufficiently general to cover most of the theories in use, and sufficiently strong to determine the theory on 'good spaces', for a fixed abelian group 'of coefficients'.

In particular, relative singular homology is a homology theory in this sense, with coefficient group $\mathbb{Z}$. Singular homology and cohomology with an arbitrary coefficient group will be built in Sections 4.3 and 4.4.

3.4.1 The axioms

A *homology theory* $H = ((H_n), (D_n))$ on Top_2 consists of the following data.

(a) For every relative pair (X, A) and every $n \geqslant 0$ we have an abelian group, called the *n-th homology group of X modulo A*

$$H_n(X, A). \tag{3.44}$$

The embedding of Top in Top_2 gives the absolute group $H_n(X) = H_n(X, \emptyset)$.

(b) For every relative map $f \colon (X, A) \to (Y, B)$ and every $n \geqslant 0$ we have an *induced homomorphism*

$$H_n(f) = f_{*n} \colon H_n(X, A) \to H_n(Y, B). \tag{3.45}$$

(c) For every relative pair (X, A) and every $n \geqslant 1$ we have a *connecting homomorphism*

$$D_n(X, A) \colon H_n(X, A) \to H_{n-1}(A). \tag{3.46}$$

The following axioms have to be satisfied.

(HT.1) (*Functoriality*) The data (a) and (b) form a functor $H_n \colon \mathsf{Top}_2 \to \mathsf{Ab}$, called the *n-th homology functor* of the theory, for every $n \geqslant 0$.

(HT.2) (*Naturality*) The family of connecting homomorphisms $D_n(X, A)$ forms a natural transformation

$$D_n \colon H_n \to H_{n-1} P'' \colon \mathsf{Top}_2 \to \mathsf{Ab} \qquad (n \geqslant 1), \tag{3.47}$$

where $P'' \colon \mathsf{Top}_2 \to \mathsf{Top}$ is the second projection: $P''(X, A) = A$.

(HT.3) (*Homotopy invariance*) The functors $H_n \colon \mathsf{Top}_2 \to \mathsf{Ab}$ are invariant under relative homotopies.

(HT.4) (*Exactness*) For every relative pair (X, A), the sequence (3.11) is exact. It is called the *homology sequence* of the pair.

(HT.5) (*Excision*) If $U \subset A \subset X$ are subspaces and $\mathrm{cl}_X U \subset \mathrm{int}_X A$, the inclusion map $i \colon (X \setminus U, A \setminus U) \to (X, A)$ has an associated isomorphism, for every $n \geqslant 0$, called the *excision isomorphism*

$$i_{*n} \colon H_n(X \setminus U, A \setminus U) \cong H_n(X, A). \tag{3.48}$$

(HT.6) (*Dimension*) The singleton space $\{*\}$ has trivial homology groups in positive degree:

$$H_n(\{*\}) = 0, \quad \text{for } n > 0. \tag{3.49}$$

The group $G = H_0(\{*\})$ of the standard point is called the *coefficient group* of the theory.

Notes. The original version of the axioms, in [ES], is more general: it is formulated for degrees $n \in \mathbb{Z}$ and an 'allowable' subcategory of Top_2. For instance, the full subcategory formed by the relative pairs of compact Hausdorff spaces is allowable.

The name of 'Dimension Axiom' is explained by the fact that, without it, one could translate the theory, letting $H'_n = H_{n+h}$ (for a fixed $h \in \mathbb{N}^*$).

Of course, there are diverse singleton spaces $\{x\}$, but all of them are related by a coherent system of (uniquely determined) homeomorphisms, and the groups $H_0(\{x\})$ are related by a coherent system of (uniquely determined) isomorphisms.

3.4.2 The singular theory

We have already proved that relative singular homology is a homology theory in this sense:

- the data are defined in 3.1.3 and 3.1.5, where the properties of functoriality, exactness and naturality are also established,

- homotopy invariance is proved in Theorem 3.1.4, the excision property in Theorem 3.2.1, and the dimension axiom in (2.56), where we have also seen that $H_0(\{*\}) \cong \mathbb{Z}$, in a canonical way.

Because of this, we say that singular homology is a homology theory *with integral coefficients*, or *with coefficients in* $\mathbb{Z}$.

For a given group G, we shall construct, in Chapter 4, the *singular theory with coefficient group* G (up to canonical isomorphism).

Essentially, we construct the singular chain complex $C_+(X; G)$ with coefficients in G, letting

$$C_n(X; G) = F_G(\mathrm{Cub}_n(X))/F_G(\mathrm{Deg}_n(X)), \qquad (3.50)$$

where, for a set S, the direct sum $F_G(S) = \bigoplus_{i \in S} G$ replaces the free abelian group $F(S) = \mathbb{Z}S = \bigoplus_{i \in S} \mathbb{Z}$. An n-chain is now an essentially finite formal sum $\sum \lambda_i a_i$ with coefficients $\lambda_i \in G$ and $a_i \colon \mathbb{I}^n \to X$ (up to degenerate chains).

Proceeding in this elementary way would soon become complicated, and a more structural way is usually followed: the chain groups $C_n(X, A; G)$ will be obtained as $C_n(X, A) \otimes G$, using tensor products of abelian groups.

3.4.3 Exercises and complements

A homology theory H is given, with coefficients in G. Several issues we have established for the singular theory are a consequence of the previous axioms, and hold for every homology theory, adapting some points related to G. We only mention a few cases, but an interested reader can state and work out many more (or look for them in [ES]).

(a) (*Trivial pairs*) Every pair (X, X) has trivial groups $H_n(X, X) = 0$. More generally, if the inclusion $A \to X$ is a homotopy equivalence, then $H_n(X, A) = 0$, for every n.

(b) (*The empty space*) In particular, all $H_n(\emptyset)$ are trivial groups.

(c) (*The associated reduced theory*) The theory H has an associated *reduced homology theory* $\tilde{H}$, defined by extending the procedure used in Exercise 2.8.2(g) for the singular theory.

*(d) The exact sequence of a triple, stated and proved for singular homology in Theorem 3.1.8, can be extended to any homology theory. The proof is purely within Homological Algebra, and can be found in [ES], Section I.10, or in [Sp2], Section 4.8.

*(e) The Mayer–Vietoris Theorem 2.3.7 holds for any homology theory, if restricted to a 'proper triad' (X; U, V): see [ES], Section I.15.

3.4.4 Cohomology theories

Cohomology theories are the contravariant version of homology theories; traditionally, all degrees are written as superscripts.

Formally, a cohomology theory can be defined as a homology theory with values in $\mathsf{Ab}^{\mathrm{op}}$, the abelian category opposite to Ab.

A *cohomology theory* $H = ((H^n), (D^n))$ consists of the following data.

(a) For every relative pair (X, A) and every $n \geqslant 0$ we have an abelian group, called the *n-th cohomology group of* X *modulo* A

$$H^n(X, A). \tag{3.51}$$

(b) For every relative map $f \colon (X, A) \to (Y, B)$ and every $n \geqslant 0$ we have an *induced homomorphism*

$$H^n(f) = f^{*n} \colon H^n(Y, B) \to H^n(X, A). \tag{3.52}$$

(c) For every relative pair (X, A) and every $n \geqslant 0$ we have a *connecting homomorphism*

$$D^n(X, A) \colon H^n(A) \to H^{n+1}(X, A). \tag{3.53}$$

These data are subject to the following axioms.

(HT*.1) (*Functoriality*) The data (a) and (b) form a contravariant functor $H^n \colon \mathsf{Top}_2 \dashrightarrow \mathsf{Ab}$, called the *n-th cohomology functor* of the theory, for every $n \geqslant 0$.

(HT*.2) (*Naturality*) The family of connecting homomorphisms $D^n(X, A)$ forms a natural transformation of contravariant functors

$$D^n \colon H^n P'' \to H^{n+1} \colon \mathsf{Top}_2 \dashrightarrow \mathsf{Ab} \quad (n \geqslant 0), \tag{3.54}$$

where $P'' \colon \mathsf{Top}_2 \to \mathsf{Top}$ is the second projection: $P''(X, A) = A$.

(HT*.3) (*Homotopy invariance*) The contravariant functors $H^n \colon \mathsf{Top}_2 \dashrightarrow \mathsf{Ab}$ are invariant under relative homotopies.

(HT*.4) (*Exactness*) For every relative pair (X, A), the *cohomology sequence*

$$
\begin{array}{l}
0 \longrightarrow H^0(X, A) \xrightarrow{\;-v^{*0}\;} H^0(X) \xrightarrow{\;-u^{*0}\;} H^0(A) \\
\qquad\qquad \overset{D^0}{\longleftarrow} \\
H^1(X, A) \xrightarrow{\;-v^{*1}\;} H^1(X) \xrightarrow{\;-u^{*1}\;} H^1(A) \\
\qquad\qquad \overset{D^1}{\longleftarrow} \\
H^2(X, A) \xrightarrow{\;-v^{*2}\;} \ldots
\end{array}
\tag{3.55}
$$

is exact (for the canonical maps $u \colon A \to X$ and $v \colon (X, \emptyset) \to (X, A)$).

(HT*.5) (*Excision*) If $U \subset A \subset X$ are subspaces and $\mathrm{cl}\, U \subset \mathrm{int}\, A$, the inclusion map $i \colon (X \setminus U, A \setminus U) \to (X, A)$ has an associated *excision isomorphism*, for every $n \geqslant 0$

$$i^{*n} \colon H^n(X, A) \cong H^n(X \setminus U, A \setminus U). \tag{3.56}$$

(HT*.6) (*Dimension*) The singleton space $\{*\}$ has trivial cohomology groups in positive degree:

$$H^n(\{*\}) = 0, \qquad \text{for } n > 0. \tag{3.57}$$

The theory H has *coefficient group* $G = H^0(\{*\})$. Also here we speak of a theory *with integral coefficients* when $G \cong \mathbb{Z}$.

Singular cohomology with an arbitrary coefficient group G will be defined in Chapter 4, using a cochain complex $C^+(X, A; G)$ derived from the singular chain complex

$$C^n(X, A; G) = \mathrm{Hom}(C_n(X, A), G). \tag{3.58}$$

3.4.5 Comments

There are different homology theories with the same coefficient group, and different cohomology theories.

In this book we shall construct Alexander–Spanier cohomology and singular cohomology, and see that, in degree zero, they give different results on a space X, related:

- either to the set $\mathrm{Cc}(X)$ of connected components, in the first case (see 3.5.8(d)),

- or to the set $\Pi_0(X)$ of path components, in the second (see 4.4.3(d)).

However, all homology theories with the same coefficient group coincide on 'good spaces'. We shall now review, without proof, the original uniqueness theorem of Eilenberg–Steenrod, stated for 'triangulated pairs'.

3.4.6 Simplicial complexes

Triangulation of spaces, namely the study of spaces which can be obtained by a 'regular pasting' of tetrahedra, is an old topic, going back to the beginning of Combinatorial and Algebraic Topology.

Following the terminology of [ES], in Sections II.2–3, triangulation is based on simplicial complexes – not to be confused with simplicial sets, defined in 2.8.5.

Remarks. As in [ES] we avoid considering the empty simplex. Including it, in dimension -1, would require some minor modifications. Of course, triangulated pairs of kind $(X, \emptyset)$ cannot be avoided.

(a) Loosely speaking, an *n-simplex* s is a copy of the standard tetrahedron Δ^n non-embedded in $\mathbb{R}^{n+1}$. Precisely, s is a (non-empty) finite set of $n+1$ elements, called the *vertices* of s. A *point* of s is a real function

$$x\colon s \to \mathbb{R}, \qquad x(v) \geqslant 0, \qquad \Sigma_v\, x(v) = 1 \qquad (\text{for } v \in s), \tag{3.59}$$

whose values are called the *barycentric coordinates* of the point x. The *geometric realisation* $|s|$ of s is the space of these points, with the topology defined by the *euclidean distance*

$$d(x, y) = (\Sigma_v\, (x(v) - y(v))^2)^{1/2}.$$

A total ordering $v_0 < v_1 < ... < v_n$ of the set s gives a homeomorphism

$$f\colon |s| \to \Delta^n, \qquad f(x) = (x(v_0), x(v_1), ..., x(v_n)), \tag{3.60}$$

between the space $|s|$ and the n-simplex Δ^n, proving that $|s|$ is compact and path connected.

The topology of $|s|$ can also be defined by means of this bijection (independently of the chosen ordering).

A non-empty subset $s' \subset s$ is also called a *face* of s, and a *p-face* if it has $p + 1$ elements; note that every face of a face of s is a face of s. Plainly, there is a topological embedding $|s'| \to |s|$, where the point $x\colon s' \to \mathbb{R}$ is extended to s, letting $x(v) = 0$ for all vertices $v \in s \setminus s'$.

(b) Loosely speaking, a *simplicial complex* K is union of 'faces' (of any dimension) of the standard tetrahedron Δ^n (forgetting again any embedding in $\mathbb{R}^{n+1}$).

Precisely, K is a simplex s equipped with a set of faces of s (non-empty subsets), which are called *faces* of K. The assigned faces are assumed to satisfy the condition:

(*) every face of a face of K is a face of K.

The *geometric realisation* $|K|$ is defined to be the subspace of $|s|$ consisting of the points which belong to the realisation of some face of K. In other words, $|K| = \bigcup |s'|$, where s' varies in the set of faces of K. The *dimension* of K is the maximal dimension of its faces.

A *subcomplex* K' of K is a subset $s' \subset s$ equipped with the faces of K which are contained in s'. Plainly, there is a topological embedding $|K'| \to |K|$.

Remarks and examples. (a) Every n-simplex s can be viewed as a simplicial complex K of dimension n, admitting all its faces. Its geometric realisation is that of s.

(b) An n-simplex s also determines a simplicial complex K', admitting all the faces of s of dimension $< n$ (if any). Its geometric realisation is the border of Δ^n in the n-plane $\Sigma t_i = 0$ of $\mathbb{R}^{n+1}$. K' is a subcomplex of the previous K.

(c) A simplicial complex with no faces is realised as the empty space.

*(d) Simplicial complexes form a specific category, where a morphism $f \colon K \to K'$ is a mapping $s \to s'$ between the underlying sets, that takes every face of K to a face of K'. Such a morphism has a geometric realisation, as a map $|K| \to |K'|$. (This category will not be used here.)

3.4.7 Triangulated pairs

Finally, a *triangulated pair* (X, A) is a relative pair of topological spaces which has a homeomorphism (i.e. an isomorphism in Top_2)

$$f \colon (X, A) \to (|K|, |K'|), \tag{3.61}$$

where K is a simplicial complex and K' is a simplicial subcomplex of K.

We are interested in the full subcategory TrTop_2 of Top_2 formed of triangulated pairs. The following Uniqueness Theorem has a straightforward extension to the full subcategory of Top_2 consisting of all relative pairs which are homotopically equivalent to triangulated pairs.

3.4.8 *Uniqueness Theorem* [ES]

Let H and H' be two homology theories (satisfying Eilenberg–Steenrod's axioms), with coefficient groups G and G', respectively.

Restricting these theories to the category TrTop_2 of triangulated pairs, every homomorphism $h\colon G \to G'$ can be uniquely extended to a family of natural transformations

$$h_n\colon H_n \to H'_n\colon \mathsf{TrTop}_2 \to \mathsf{Ab} \tag{3.62}$$

(with $h_0(\{*\}) = h$), that commute with the connecting morphisms of H and H', forming commutative diagrams

$$
\begin{array}{ccc}
H_n(X, A) & \xrightarrow{\ h\ } & H'_n(X, A) \\
\partial \downarrow & & \downarrow \partial' \\
H_{n-1}(A) & \xrightarrow[\ h\ ]{} & H'_{n-1}(A)
\end{array}
\tag{3.63}
$$

If $h\colon G \to G'$ is an isomorphism, all the natural transformations h_n are natural isomorphisms.

Cohomology theories have a dually-similar uniqueness property.

Proof See [ES], Theorem III.10.1. $\qquad\qquad\square$

3.4.9 *Simplicial homology*

Homology (with a fixed coefficient group) is thus unique on triangulated pairs. It is called *simplicial homology*; for this subject, one can see [ES], Chapter 3.

However, if one is interested in its concrete use to compute homology groups, it would be preferable to replace simplicial complexes with 'face-simplicial sets' (introduced by Eilenberg–Zilber [EZ1] as 'semisimplicial complexes', a name which was later used in a different sense). For this subject we refer to [Ha], Section 2.1, where face-simplicial sets are called Δ-*complexes*.

3.5 Alexander–Spanier cohomology

We present now a cohomology theory with an arbitrary coefficient group G, which comes out naturally in contravariant form.

This theory was introduced by Alexander for compact metric spaces, in 1935. It was extended to general spaces by Spanier [Sp1], in 1948, and briefly exposed in his book [Sp2], Section 6.4, as 'Alexander cohomology'; it is usually named after both authors.

3.5.1 Cochain complexes

The theory of cochain complexes is quite similar to that of chain complexes, and only needs some adaptation: the degree is written as a superscript, a differential $d^n \colon A^n \to A^{n+1}$ rises the degree by one, while a component of a homotopy lowers it.

A (positive) *cochain complex* of abelian groups $A = ((A^n)_{n \geqslant 0}, (d^n)_{n \geqslant 0}$ is thus a diagram in Ab

$$\ldots\, 0 \dashrightarrow A^0 \xrightarrow{d_0} A^1 \,\ldots\, A^{n-1} \xrightarrow{d_{n-1}} A^n \xrightarrow{d_n} A^{n+1} \,\ldots \qquad (3.64)$$

with *differentials* $d^n \colon A^n \to A^{n+1}$, so that all composites $d^n d^{n-1}$ are trivial. The complex can be extended at the left with trivial groups and differentials.

A morphism $f \colon A \to B$ has components $f^n \colon A^n \to B^n$ that commute with the differentials. They compose componentwise, forming the category $\mathrm{Ch}^+\mathsf{Ab}$ of cochain complexes of abelian groups.

In each component A^n of A we have the subgroups of *n-cocycles* and *n-coboundaries*

$$Z^n(A) = \operatorname{Ker}(d^n \colon A^n \to A^{n+1}) \subset A^n,$$
$$B^n(A) = \operatorname{Im}(d^{n-1} \colon A^{n-1} \to A^n) \subset A^n. \qquad (3.65)$$

In particular $B^0(A) = \operatorname{Im}(d^{-1} \colon 0 \to A) = 0$. The condition $d^n d^{n-1} = 0$ is equivalent to saying that $B^n(A) \subset Z^n(A)$ (for $n \geqslant 0$): *every coboundary is a cocycle.*

The *n-th cohomology group* of the complex A is

$$H^n(A) = Z^n(A)/B^n(A) = \operatorname{Ker}(d^n)/\operatorname{Im}(d^{n-1}) \qquad (n \geqslant 0). \qquad (3.66)$$

Every cocycle $z \in Z^n(A)$ has a *cohomology class* $[z] \in H^n(A)$, written as $[z]^H$ when useful. Two cocycles $z, z' \in A^n$ are *cohomologous* if $z - z'$ is a coboundary, so that $[z] = [z']$.

This gives a sequence of (covariant!) *cochain cohomology functors*

$$H^n \colon \mathrm{Ch}^+\mathsf{Ab} \to \mathsf{Ab}, \qquad A \mapsto H^n(A),$$
$$H^n(f \colon A \to B) = (f^{*n} \colon H^n(A) \to H^n(B)), \quad f^{*n}[z] = [f^n(z)]. \qquad (3.67)$$

A *homotopy* $\varphi = (\varphi^n) \colon f \simeq g \colon A \to B$ of cochain morphisms is a sequence of homomorphisms $\varphi^n \colon A^n \to B^{n-1}$ such that

$$d^{n-1}\varphi^n + \varphi^{n+1}d^n = g^n - f^n. \qquad (3.68)$$

The cochain cohomology functors are homotopy invariant.

All issues about exactness can be lifted from Ab to $\mathrm{Ch}^+\mathsf{Ab}$, using the sequence of forgetful functors $U^n \colon \mathrm{Ch}^+\mathsf{Ab} \to \mathsf{Ab}$, $A \mapsto A^n$. The arrows of monomorphisms and epimorphisms of cochain complexes are still written as $\rightarrowtail$ and $\twoheadrightarrow$.

3.5.2 Complements

(a) The relationship between chain and cochain complexes is clearer in the unbounded case: reversing degrees gives an obvious isomorphism that interchanges the two theories

$$\mathrm{Ch}_* \mathsf{Ab} \to \mathrm{Ch}^* \mathsf{Ab},$$

$$((A_n), (\partial_n \colon A_n \to A_{n-1})) \mapsto ((A^n), (d^n \colon A^n \to A^{n+1})), \qquad (3.69)$$

where $A^n = A_{-n}$ and $d^n = \partial_{-n}$, for $n \in \mathbb{Z}$.

   This layout is internal to abelian groups, and makes no use of duality in abelian categories.

(b) Theorem 2.3.3 has a version for cochain complexes: a short exact sequence $A \rightarrowtail B \twoheadrightarrow C$ in $\mathrm{Ch}^+ \mathsf{Ab}$ gives a *long exact cohomology sequence*

$$0 \to H^0(A) \to H^0(B) \to H^0(C) \xrightarrow{D_0} H^1(A) \to \ldots \qquad (3.70)$$

The connecting homomorphism is computed with the following diagram, an obvious modification of (2.74) for differentials that rise the degree

$$D^n \colon H^n(C) \to H^{n+1}(A), \qquad D^n[c] = [a], \qquad (3.71)$$

$$
\begin{array}{ccccc}
 & b & \xrightarrow{\ g^n\ } & c & \qquad (n) \\
 & \ \ \downarrow{\scriptstyle d} & & \ \ \downarrow{\scriptstyle d} & \\
a & \xrightarrow{\ f^{n+1}\ } db & \xrightarrow{\ g^{n+1}\ } & 0 & \qquad (n+1)
\end{array}
$$

(c) A contravariant functor $F \colon \mathsf{Ab} \dashrightarrow \mathsf{Ab}$ that preserves zero homomorphisms takes chain complexes to cochain complexes, and vice versa. The contravariant functor $\mathrm{Hom}(-, G)$ will be used in Section 4.2 to transform the singular chain complex of a space into the singular cochain complex with coefficients in G.

(d) Kernels, cokernels, exact sequences and (co)chain complexes in the category $R\,\mathsf{Mod}$ of R-modules are dealt with as in Ab.

   The categories $\mathrm{Ch}_+(R\,\mathsf{Mod})$ and $\mathrm{Ch}^+(R\,\mathsf{Mod})$ of chain and cochain complexes of R-modules have homology and cohomology functors

$$H_n \colon \mathrm{Ch}_+(R\,\mathsf{Mod}) \to R\,\mathsf{Mod},$$

$$H^n \colon \mathrm{Ch}^+(R\,\mathsf{Mod}) \to R\,\mathsf{Mod}. \qquad (3.72)$$

3.5.3 A preliminary cochain complex

To define Alexander–Spanier cohomology (with coefficients in G), we begin by considering, for a space X, a cochain complex $C^+(X;G)$ of abelian groups, which does not depend on the topology of X

$$C^n(X;G) = \mathsf{Set}(X^{n+1}, G), \qquad d^n \colon C^n(X;G) \to C^{n+1}(X;G),$$
$$(d^n\lambda)(x_1, ..., x_{n+2}) = \Sigma_{1\leqslant i\leqslant n+2}\,(-1)^i\,\lambda(x_1, ..., \hat{x}_i, ..., x_{n+2}),$$

$$(3.73)$$

where $C^n(X;G)$ is an abelian group for the pointwise sum: $(\lambda + \mu)(x) = \lambda(x) + \mu(x)$. The symbol $\hat{x}_i$ means that the i-th coordinate is omitted.

For a map $f \colon Y \to X$, we have a cartesian power $f^\bullet \colon Y^{n+1} \to X^{n+1}$ and a cochain morphism

$$f^\sharp \colon C^+(X;G) \to C^+(Y;G),$$
$$f^{\sharp n}(\lambda \colon X^{n+1} \to G) = \lambda f^\bullet \colon Y^{n+1} \to G.$$

$$(3.74)$$

Exercises and complements. These definitions require some simple verifications.

(a) The composite $d^n d^{n-1} \colon C^{n-1}(X;G) \to C^{n+1}(X;G)$ is zero. *Hints:* one can express the differential using cubical cofaces $\varepsilon_i \colon X^n \to X^{n-1}$, defined as in (2.36).

(b) In (3.74), $f^\sharp$ is indeed a morphism of cochain complexes: $d^n f^{\sharp n} = f^{\sharp n+1} d^n$.

(c) We have defined a functor $C^+(-;G) \colon \mathsf{Top}^{\mathrm{op}} \to \mathsf{Ch}^+\mathsf{Ab}$, contravariant on Top.

*(d) A homomorphism $h \colon G \to G'$ gives an obvious morphism

$$h_\sharp \colon C^+(X;G) \to C^+(X;G'), \qquad h_\sharp(\lambda \colon X^{n+1} \to G) = h\lambda. \qquad (3.75)$$

Putting everything together, we have a functor in two variables

$$C^+ \colon \mathsf{Top}^{\mathrm{op}} \times \mathsf{Ab} \to \mathsf{Ch}^+\mathsf{Ab}, \qquad (X,G) \mapsto C^+(X;G), \qquad (3.76)$$

defined on the product category $\mathsf{Top}^{\mathrm{op}} \times \mathsf{Ab}$.

3.5.4 Alexander–Spanier cochains

Now we take the topology of X into account, and define the *complex of Alexander–Spanier cochains of X with coefficient group G*, as the quotient

$$\overline{C}^+(X;G) = C^+(X;G)/C_0^+(X;G), \qquad (3.77)$$

where $C_0^+(X;G)$ is the subcomplex of $C^+(X;G)$ of *locally zero* functions: a function $\lambda \colon X^{n+1} \to G$ is said to be locally zero (at the diagonal) if it vanishes at a nbd of any point of the diagonal of X^{n+1}.

Equivalently, there exists an open cover $\mathcal{U}$ of X such that λ vanishes on U^{n+1}, for each $U \in \mathcal{U}$. $C_0^n(X; G)$ is the subgroup of $C^n(X; G)$ of locally zero functions; these components are plainly closed under the differential.

For a map $f \colon Y \to X$, the homomorphism $f^{\sharp n} \colon C^n(X; G) \to C^n(Y; G)$ restricts to a homomorphism $f^{\sharp n} \colon C_0^n(X; G) \to C_0^n(Y; G)$.

In fact, if $\lambda \colon X^{n+1} \to G$ is locally zero with respect to the open cover $\mathcal{U}$ of X, then $f^{\sharp n}(\lambda) = \lambda f^\bullet \colon Y^{n+1} \to G$ is locally zero with respect to the obvious open cover $\mathcal{V}$ of Y, formed of the open sets $f^{-1}(U) \subset Y$, for $U \in \mathcal{U}$.

The sequence $f^\sharp = (f^{\sharp n})$ is consistent with the differentials, and forms a morphism of cochain complexes.

We have thus two functors:

$$\begin{aligned} C_0^+(-; G) \colon \mathsf{Top}^{\mathrm{op}} &\to \mathrm{Ch}^+\mathsf{Ab}, & X &\mapsto C_0^+(X; G), \\ \overline{C}^+(-; G) \colon \mathsf{Top}^{\mathrm{op}} &\to \mathrm{Ch}^+\mathsf{Ab}, & X &\mapsto \overline{C}^+(X; G), \end{aligned} \tag{3.78}$$

whose (contravariant) action on morphisms will still be written as $f \mapsto f^\sharp$. Again, the coefficient group can be allowed to vary.

3.5.5 Alexander–Spanier cohomology

Composing the functor $\overline{C}^+(-; G) \colon \mathsf{Top}^{\mathrm{op}} \to \mathrm{Ch}^+\mathsf{Ab}$ with the (covariant) functors H^n of cochain cohomology, we get the functors of *Alexander–Spanier cohomology*, with coefficient group G

$$\begin{aligned} \overline{H}^n(-; G) \colon \mathsf{Top}^{\mathrm{op}} &\to \mathsf{Ab}, & \overline{H}^n(X; G) &= H^n(\overline{C}^+(X; G)), \\ (f \colon Y \to X) &\mapsto (f^{*n} \colon \overline{H}^n(X; G) \to \overline{H}^n(Y; G)). \end{aligned} \tag{3.79}$$

An exercise. (a) (*The dimension axiom*) The singleton $X = \{*\}$ gives:

$$\overline{H}^n(X; G) = 0, \text{ for } n > 0, \qquad \overline{H}^0(X; G) \cong G. \tag{3.80}$$

The isomorphism in degree 0 essentially takes a function $\lambda \colon \{*\} \to G$ to its value $\lambda(*)$. *Hints:* compute the cochain complexes $C^+(X; G)$ and $\overline{C}^+(X; G)$.

3.5.6 Relative Alexander–Spanier cohomology

For a pair (X, A), the inclusion $u \colon A \to X$ gives a morphism of $\mathrm{Ch}^+\mathsf{Ab}$

$$u^\sharp \colon \overline{C}^+(X; G) \to \overline{C}^+(A; G), \qquad u^{\sharp n}(\overline{\lambda}) = (\lambda u^\bullet)^- \tag{3.81}$$

(for $\lambda \colon X^{n+1} \to G$) which is easily seen to be epi.

Here $\lambda u^\bullet \colon A^{n+1} \to G$ is the restriction of $\lambda \colon X^{n+1} \to G$. As to surjectivity, a class $\overline{\mu} \in \overline{C}^n(A; G))$ is represented by a function $\mu \colon A^{n+1} \to G$; we extend the latter to the function $\lambda \colon X^{n+1} \to G$ which vanishes out of A^{n+1}, getting $\lambda u^\bullet = \mu$ and $(\lambda u^\bullet)^- = \overline{\mu}$.

We define the *complex of relative Alexander–Spanier cochains* of (X, A), with coefficient group G

$$\overline{C}^+(X, A; G) = \mathrm{Ker}\,(u^\sharp \colon \overline{C}^+(X; G) \to \overline{C}^+(A; G)), \qquad (3.82)$$

where $\overline{C}^n(X, A; G)$ is the subgroup of all classes $\overline{\lambda} \in \overline{C}^n(X; G)$ such that $\lambda \colon X^{n+1} \to G$ vanishes on A^{n+1}.

We have defined a functor

$$\overline{C}^+(-; G) \colon \mathsf{Top_2}^{\mathrm{op}} \to \mathsf{Ch}^+\mathsf{Ab}, \quad (X, A) \mapsto \overline{C}^+(X, A; G). \qquad (3.83)$$

Applying cochain cohomology, we get the functors of *relative Alexander–Spanier cohomology*, with coefficients in G

$$\begin{aligned}
&\overline{H}^n(-; G) \colon \mathsf{Top_2}^{\mathrm{op}} \to \mathsf{Ab}, \quad \overline{H}^n(X, A; G) = H^n(\overline{C}^+(X, A; G)), \\
&(f \colon (Y, B) \to (X, A)) \mapsto (f^{*n} \colon \overline{H}^n(X, A; G) \to \overline{H}^n(Y, B; G)).
\end{aligned} \qquad (3.84)$$

3.5.7 The exact cohomology sequence

For a relative pair (X, A), the previous definition of $\overline{C}^+(X, A; G)$ gives a (natural) short exact sequence in $\mathsf{Ch}^+\mathsf{Ab}$

$$\overline{C}^+(X, A; G) \;\overset{v^\sharp}{\rightarrowtail}\; \overline{C}^+(X; G) \;\overset{u^\sharp}{\twoheadrightarrow}\; \overline{C}^+(A; G), \qquad (3.85)$$

where the embedding $v^\sharp$ of $\mathrm{Ker}\,u^\sharp$ is associated to the relative map $v \colon (X, \emptyset) \to (X, A)$ given by $\mathrm{id}\,X$.

The associated exact sequence of relative Alexander–Spanier cohomology

$$\begin{aligned}
0 \;\longrightarrow\; &\overline{H}^0(X, A; G) \;\overset{v^{*0}}{\longrightarrow}\; \overline{H}^0(X; G) \;\overset{u^{*0}}{\longrightarrow}\; \overline{H}^0(A; G) \\
&\overline{H}^1(X, A; G) \;\overset{v^{*1}}{\longrightarrow}\; \overline{H}^1(X; G) \;\overset{u^{*1}}{\longrightarrow}\; \overline{H}^1(A; G) \\
&\overline{H}^2(X, A; G) \;\overset{v^{*2}}{\longrightarrow}\; \dots
\end{aligned} \qquad (3.86)$$

is natural, with connecting morphism $D^n \colon \overline{H}^n(A; G) \to \overline{H}^{n+1}(X, A; G)$

$$D^n[(u^{\sharp n}\lambda)^-] = [(d^n\lambda)^-] \qquad (\text{for } \lambda \colon X^{n+1} \to G). \qquad (3.87)$$

In fact, the restriction $u^{\sharp n}(\lambda) = \lambda u^\bullet$ is a cocycle of $\overline{C}^n(A; G)$; the function $d^n\lambda \colon X^{n+2} \to G$ vanishes on A^{n+2} and is a cocycle in $\overline{C}^{n+1}(X, A; G)$.

3.5.8 *Exercises and complements* (Degree 0)

Alexander–Spanier cohomology in degree zero is related to the ring $\mathcal{A}(X)$ introduced in (1.15), and studied in 1.2.3.

(a) For a relative pair of spaces (X, A) there is a canonical isomorphism

$$\mathcal{A}(X, A; G) \to \overline{H}^0(X, A; G), \qquad \lambda \mapsto [\overline{\lambda}], \qquad (3.88)$$

where $\mathcal{A}(X, A; G)$ is the abelian group of locally constant functions $\lambda \colon X \to G$ which vanish on A, with pointwise addition.

Hints. Build a suitable augmentation $\mathcal{A}(X, A; G) \to \overline{C}^0(X, A; G)$. Note that a function $X \to G$ is locally constant if and only if it is continuous with respect to the discrete topology of G.

(b) In particular, for a space X, there is a canonical isomorphism

$$\mathcal{A}(X; G) \to \overline{H}^0(X; G), \qquad \lambda \mapsto [\overline{\lambda}], \qquad (3.89)$$

where $\mathcal{A}(X; G)$ is the abelian group of locally constant functions $\lambda \colon X \to G$.

(c) With coefficients in a ring R, the abelian group $\mathcal{A}(X; R)$ is naturally a ring. In particular, we have the ring $\mathcal{A}(X; \mathbb{Z}) = \mathcal{A}(X)$ previously introduced.

The whole graded abelian group $\overline{H}^*(X; R)$ will be made into a graded ring, in the next section.

(d) If the connected components of the space X are open, there is a canonical isomorphism

$$\overline{H}^0(X; G) \cong G^I, \qquad (3.90)$$

where $I = \mathrm{Cc}(X)$ is the set of connected components of X (as in (1.14)).

(e) On the other hand, we shall see in 4.4.3(d) that singular cohomology has an isomorphism $H^0(X; G) \cong G^{\Pi_0(X)}$, where $\Pi_0(X)$ is the set of *path components* of X.

3.5.9 *The axioms*

Alexander–Spanier cohomology satisfies the Eilenberg–Steenrod axioms: it is a cohomology theory with coefficient group (isomorphic to) G.

We have verified the functoriality axioms in 3.5.5, the naturality and the exactness axioms in 3.5.7, the dimension axiom in Exercise 3.5.5(a).

We refer to Spanier's book [Sp2] for the remaining two. The excision axiom is verified in Section 6.4, Lemma 4. The proof of homotopy invariance is difficult, and takes the whole Section 6.5.

3.6 The product in Alexander–Spanier cohomology

Let R be a commutative (unital) ring. For every space X, the graded abelian group $\overline{H}^*(X; R)$ of Alexander–Spanier cohomology has a natural structure of graded R-algebra, where the product of two elements of degrees p, q has degree $p + q$.

3.6.1 Algebras on a commutative ring

An R-*algebra* A is an R-module which is also equipped with an (internal) product xy, so that:

(i) the underlying abelian group becomes a ring with this product: the product is associative and distributes over the sum, on both sides,

(ii) the scalar multiplication and the product are consistent:

$$(\lambda x)y = \lambda(xy) = x(\lambda y) \qquad \text{(for } \lambda \in R \text{ and } x, y \in A), \tag{3.91}$$

and we can write λxy, without parentheses.

We reserve the name of *multiplication* to the external operation, defined on $R \times A$, and the name of *product* to the internal one, defined on $A \times A$. An R-algebra is said to be unital, or commutative, if it is as a ring.

A *homomorphism* of R-algebras $f \colon A \to B$ has to preserve all operations: it is both a homomorphism or R-modules and of rings. They form the category $R\mathsf{Alg}$ of R-algebras.

A *subalgebra* of A is a subset which is both a submodule and a subring. An R-*ideal* of A is a sub-R-module of A which is an ideal for the ring structure.

Remarks. A ring is 'the same' as a $\mathbb{Z}$-algebra, with multiplication given by multiples. A commutative ring is an algebra on itself.

3.6.2 Graded algebras

A *graded R-algebra* $A = (A^n)_{n \geqslant 0}$ is a graded R-module equipped with a (graded) product

$$A^p \times A^q \to A^{p+q}, \qquad (x, y) \mapsto xy \qquad (p, q \geqslant 0), \tag{3.92}$$

satisfying the following axioms (for $x, x' \in A^p$, $y, y' \in A^q$, $z \in A^r$, $\lambda \in R$)

(i) (*Consistency*) $(\lambda x)y = \lambda(xy) = x(\lambda y)$,

(ii) (*Distributivity*) $(x + x')y = xy + x'y$, $x(y + y') = xy + xy'$,

(iii) (*Associativity*) $x(yz) = (xy)z$.

The first two properties say that the mapping (3.92) is R-bilinear. The graded R-algebra A is said to be *unital* if it has a unit $\underline{1} \in A^0$ (with $\underline{1}x = x = x\underline{1}$, for all x). It is said to be *skew-commutative* if

$$xy = (-1)^{pq}\, yx \qquad (\text{for } x \in A^p,\, y \in A^q). \tag{3.93}$$

The 'Koszul sign rule' takes place here: interchanging an element of degree p with an element of degree q comes about with a sign change $(-1)^{pq}$. This behaviour is typical of differential graded structures.

A *differential graded R-algebra* $A = ((A^n)_{n \geqslant 0}, (d^n)_{n \geqslant 0})$ is also equipped with R-linear differentials, which make it into a cochain complex of R-modules and agree with the product:

$$d^n \colon A^n \to A^{n+1}, \qquad d^n d^{n-1} = 0,$$
$$d^n(xy) = (d^p x)y + (-1)^p\, x(d^q y) \qquad (\text{for } x \in A^p,\, y \in A^q). \tag{3.94}$$

It is (and will often be) understood that $n = p+q$. The Koszul rule is applied, viewing d as a generalised morphism of graded modules, of degree 1.

Morphisms of graded R-algebras $f = (f^n)\colon A \to B$ are morphisms of graded modules which preserve the product: $f^n(xy) = (f^p x)(f^q y)$. In the differential case we further require that the components f^n commute with differentials.

We have thus the categories RGra and RDga of *graded R-algebras* and *differential graded R-algebras*. The latter will also be called dg-algebras, leaving the ring R as understood.

From now on, all R-algebras are assumed to be unital, with unit $\underline{1}$ (including the graded and differential graded algebras.)

Remarks. (a) In a graded R-algebra A, the component A^0 is an R-algebra. The ring of scalars R is mapped into A^0, by $\lambda \mapsto \lambda\underline{1}$, but this map need not be injective. We also note that $d^0(\underline{1}) = 0$ (writing $\underline{1} = \underline{1}\,\underline{1}$), and therefore $d^0(\lambda\underline{1}) = 0$ for all scalars $\lambda \in R$.

(b) A graded R-algebra A (resp. a dg-algebra on R) can be made into an ordinary R-algebra $\hat{A}$ (resp. an ordinary dg-algebra on R), letting

$$\hat{A} = \oplus A^n, \tag{3.95}$$

and extending sums and products in the obvious way: writing an essentially finite family $(a_n) \in \hat{A}$ as an essentially finite sum $\Sigma_n a_n$, the product is

$$(\Sigma_p a_p)(\Sigma_q b_q) = \Sigma_n (\Sigma_{p+q=n} a_p b_q). \tag{3.96}$$

In the graded algebras associated to cohomology theories (as Alexander–Spanier's, or de Rham's, or the singular theory) this 'transformation' is practically useless: for instance, adding differential forms of different degrees has little use and little sense. But this is not true for a polynomial ring $R[X]$, which has an obvious graduation given by its submodules of monomials.

3.6.3 The cohomology algebra

Let $A = ((A^n)_{n \geqslant 0}, (d^n)_{n \geqslant 0})$ be a differential graded R-algebra.

The cohomology graded R-module $H^*(A) = (H^n(A))$ inherits a graded product

$$H^p(A) \times H^q(A) \to H^{p+q}(A), \qquad ([x], [y]) \mapsto [xy] \qquad (p, q \geqslant 0), \quad (3.97)$$

that makes it into a graded algebra, with unit $[\underline{1}]$. The following exercises are nearly obvious.

Exercises and complements. (a) The product of cohomology classes is well defined.

(b) Cohomology gives a covariant functor

$$H^* \colon \mathsf{RDga} \to \mathsf{RGra}. \tag{3.98}$$

3.6.4 The Alexander–Spanier graded algebra

For a space X, the cochain complex $C^*(X; R)$ has a natural structure of differential graded R-algebra

$$C^p(X; R) \times C^q(X; R) \to C^{p+q}(X; R),$$
$$(\lambda \bullet \mu)(x_1, ..., x_{n+1}) = \lambda(x_1, ..., x_{p+1})\, \mu(x_{p+1}, ..., x_{n+1}), \tag{3.99}$$

as verified in Exercise (a).

Moreover the cochain subcomplex $C_0^+(X; R)$ of locally zero functions is an ideal of this graded algebra, in the obvious sense: if λ or μ is locally zero, then $\lambda \bullet \mu$ is also. Therefore, the quotient $\overline{C}^*(X; R) = C^*(X; R)/C_0^+(X; R)$ inherits a structure of dg-algebra.

The Alexander–Spanier cohomology graded R-module is thus a graded R-algebra, and we have a functor

$$\overline{H}^* \colon \mathsf{Top} \to \mathsf{RGra}. \tag{3.100}$$

Exercises and complements. (a) The product (3.99) is consistent with the differential, as in (3.94).

(b) (*Functoriality*) Verify that, for a map $f \colon Y \to X$, the cochain morphism $f^\sharp \colon \overline{C}^*(X; R) \to \overline{C}^*(Y; R)$ is a morphism of dg-algebras; therefore it induces a homomorphism of graded algebras, in cohomology.

3.7 Hints at de Rham cohomology

In this section we sketch de Rham cohomology for open euclidean spaces. A reader familiar with basic differential geometry will know how this part can be extended to differentiable manifolds. (An automatic extension, from local to global, is also possible, following [G4].)

Literature. Differentiable manifolds, differential forms and de Rham cohomology can be studied in Spivak's book [Sv]. De Rham's theorem can be found in the Appendix of [Mas3].

3.7.1 An introduction

We work in the category $C^\infty \text{Top}$ of open euclidean spaces (i.e. open subspaces of some $\mathbb{R}^n$) and C^∞-maps, that is maps that have all partial derivatives. The coordinates of $\mathbb{R}^3$ are written as x, y, z, the coordinates of $\mathbb{R}^n$ in the usual form $x^1, x^2, ..., x^n$.

The reader is likely acquainted with differential forms on an open subspace U of $\mathbb{R}^3$. This is an expression of the form

$$\omega = \lambda \qquad \text{(in degree 0)},$$

$$\omega = \lambda\, dx + \mu\, dy + \nu\, dz \qquad \text{(in degree 1)},$$

$$\omega = \lambda\, dydz + \mu\, dxdz + \nu\, dxdy \qquad \text{(in degree 2)},$$

$$\omega = \lambda\, dxdydz \qquad \text{(in degree 3)},$$

where $\lambda, \mu, \nu \colon U \to \mathbb{R}$ are C^∞-functions on U.

A C^∞-function $\lambda \colon U \to \mathbb{R}$ is a form of degree 0, and has a differential, which is a form of degree 1, whose coefficients are the three partial derivatives of λ

$$d\lambda = \frac{\partial \lambda}{\partial x}\, dx + \frac{\partial \lambda}{\partial y}\, dy + \frac{\partial \lambda}{\partial z}\, dz. \tag{3.101}$$

A form $\omega = \lambda\, dx + \mu\, dy + \nu\, dz$ of degree 1 has a differential of degree 2

$$d\omega = \left(\frac{\partial \nu}{\partial y} - \frac{\partial \mu}{\partial z}\right) dydz + \left(\frac{\partial \nu}{\partial x} - \frac{\partial \lambda}{\partial z}\right) dxdz + \left(\frac{\partial \mu}{\partial x} - \frac{\partial \lambda}{\partial y}\right) dxdy. \tag{3.102}$$

A form $\omega = \lambda\, dydz + \mu\, dxdz + \nu\, dxdy$ of degree 2 has a differential of degree 3

$$d\omega = \left(\frac{\partial \lambda}{\partial x} - \frac{\partial \mu}{\partial y} + \frac{\partial \nu}{\partial z}\right) dxdydz. \tag{3.103}$$

We shall now extend differential forms and their calculus to an open subspace U of $\mathbb{R}^n$. The partial derivatives of the C^∞-function $\lambda \colon U \to \mathbb{R}$ will also be written as $D_1\lambda, ..., D_n\lambda$.

3.7.2 Main definitions

For an open euclidean space $U \subset \mathbb{R}^n$, we write as

$$\Omega^0(U) = C^\infty \mathsf{Top}(U, \mathbb{R}), \tag{3.104}$$

the commutative algebra of real-valued C^∞-functions on U, with the point-wise operations in $\mathbb{R}$. Its unit $\underline{1}$ is the constant function at $1 \in \mathbb{R}$.

A *p-index* $\mathbf{i}$ of dimension n will be a subset of the set $\{1, ..., n\}$ containing p elements; it is empty if $p = 0$ or $p > n$. There are $\binom{n}{p}$ of them, which we rewrite in the form

$$dx^{\mathbf{i}} = dx^{i_1} dx^{i_2} ... dx^{i_p}, \qquad dx^{\emptyset} = \underline{1}, \tag{3.105}$$

where it is understood that $\mathbf{i} = \{i_1, i_2, ..., i_p\}$ and $i_1 < i_2 < ... < i_p$, when $\mathbf{i} \neq \emptyset$. Each $dx^{\mathbf{i}}$ will be called a *basic p*-form. (As usual, we assume that $\binom{n}{p} = 0$ for $p > n$.)

We write as $\Omega^p(U)$ the free $\Omega^0(U)$-module of dimension $\binom{n}{p}$ generated by the basic p-forms. An element of $\Omega^p(U)$, called a *differential form* of degree p on U, or a *p-form*, is a formal linear combination of basic forms

$$\omega = \sum_{\mathbf{i}} \lambda_{\mathbf{i}}\, dx^{\mathbf{i}} = \sum_{\mathbf{i}} \lambda_{\mathbf{i}}\, dx^{i_1} dx^{i_2} ... dx^{i_p} \tag{3.106}$$

where $\mathbf{i}$ varies in the set of p-indices of dimension n, and all $\lambda_{\mathbf{i}}$ belong to the algebra $\Omega^0(U)$. $\Omega^p(U)$ will also be considered as a vector space on $\mathbb{R}$.

Thus $\Omega^p(U) = 0$ for $p > n$. For $p = 0$ we find again the algebra of C^∞-functions on U, introduced in (3.104).

It is convenient to extend the meaning of $dx^{i_1} dx^{i_2} ... dx^{i_p}$ to any p-uple $i = (i_1, i_2, ..., i_p)$ of integers in $\{1, ..., n\}$, letting

$$
\begin{aligned}
dx^{i_1} dx^{i_2} ... dx^{i_p} &= 0 && \text{if these indices are not distinct,} \\
dx^{i_1} dx^{i_2} ... dx^{i_p} &= \sigma(i)\, dx^{\mathbf{i}} && \text{otherwise,}
\end{aligned}
\tag{3.107}
$$

where $\sigma(i)$ is the sign of the permutation $(i_1, ..., i_p)$ of the ordered set $\mathbf{i}$.

Remarks. (a) For $0 \leqslant p \leqslant n$, the module $\Omega^p(U)$ is canonically isomorphic to $\Omega^{n-p}(U)$, sending a p-index $\mathbf{i}$ to its complement in $\{1, ..., n\}$.

(b) For the empty set $U = \emptyset$, everything becomes trivial: $\Omega^0(U)$ is the trivial algebra, and $\Omega^p(U) = 0$ for all p.

3.7.3 The product of differential forms

We have defined a graded module $\Omega^+(U)$ on the commutative algebra $\Omega^0(U)$; now we make it into a graded algebra.

The *exterior product* of a basic p-form and a basic q-form is defined as

$$dx^{\mathbf{i}} \wedge dx^{\mathbf{j}} = 0 \qquad\qquad \text{if } \mathbf{i} \cap \mathbf{j} \neq \emptyset,$$
$$dx^{\mathbf{i}} \wedge dx^{\mathbf{j}} = \sigma(i,j)\, dx^{\mathbf{k}} \qquad \text{otherwise,} \tag{3.108}$$

where $\mathbf{k} = \mathbf{i} \cup \mathbf{j}$ (a disjoint union) and $\sigma(i,j)$ is the sign of the permutation $(i_1, ..., i_p, j_1, ..., j_q)$ of the ordered set $\mathbf{k}$. Using the extended symbols introduced in (3.107), the formulas (3.108) can be rewritten as:

$$(dx^{i_1} ... dx^{i_p}) \wedge (dx^{j_1} ... dx^{j_q}) = dx^{i_1} ... dx^{i_p} dx^{j_1} ... dx^{j_q}. \tag{3.109}$$

The product is extended to all differential forms, by bilinearity on $\Omega^0(U)$

$$\Omega^p(U) \times \Omega^q(U) \to \Omega^{p+q}(U),$$
$$\left(\textstyle\sum_{\mathbf{i}} \lambda_{\mathbf{i}}\, dx^{\mathbf{i}}\right) \wedge \left(\textstyle\sum_{\mathbf{j}} \mu_{\mathbf{j}}\, dx^{\mathbf{j}}\right) = \textstyle\sum_{\mathbf{i},\mathbf{j}} \lambda_{\mathbf{i}} \mu_{\mathbf{j}} (dx^{\mathbf{i}} \wedge dx^{\mathbf{j}}),$$
$$\left(\textstyle\sum_i \lambda_i\, dx^{i_1} ... dx^{i_p}\right) \wedge \left(\textstyle\sum_j \mu_j\, dx^{j_1} ... dx^{j_q}\right)$$
$$= \textstyle\sum_{i,j} \lambda_i \mu_j\, dx^{i_1} ... dx^{i_p} dx^{j_1} ... dx^{j_q}. \tag{3.110}$$

Exercises. All this is simpler than it might initially seem, and does produce a graded algebra.

(a) For an open subspace U of $\mathbb{R}^3$, compute the exterior product of the forms $\omega = \lambda_1\, dx + \lambda_2\, dy + \lambda_3\, dz$ and $\omega' = \mu_1\, dydz + \mu_2\, dxdz + \mu_3\, dxdy$.

(b) The product makes $\Omega^+(U)$ into a skew-commutative graded algebra, as defined in 3.6.2.

3.7.4 *The differential*

We complete the construction, making $\Omega^+(U)$ into a dg-algebra on the real field, called the *exterior algebra on* $\Omega^0(U)$, or the *exterior algebra of differential forms* on U.

The differential of C^∞-functions is well known

$$d^0 \colon \Omega^0(U) \to \Omega^1(U), \qquad d^0(\lambda) = \textstyle\sum_{1 \leqslant k \leqslant n} D_k \lambda\, dx^k. \tag{3.111}$$

We extend it to all degrees, letting

$$d^p \colon \Omega^p(U) \to \Omega^{p+1}(U), \qquad d^p\left(\textstyle\sum_{\mathbf{i}} \lambda_{\mathbf{i}}\, dx^{\mathbf{i}}\right) = \textstyle\sum_{\mathbf{i}} d^0(\lambda_{\mathbf{i}}) \wedge dx^{\mathbf{i}},$$
$$d^p\left(\textstyle\sum_i \lambda_i\, dx^{i_1} ... dx^{i_p}\right) = \textstyle\sum_{i,k} D_k \lambda_i\, dx^k dx^{i_1} ... dx^{i_p}, \tag{3.112}$$

where $i = (i_1, ..., i_p)$ ranges over the p-uples of integers in $\{1, ..., n\}$ and $k = 1, ..., n$. Let us note that this mapping is only $\mathbb{R}$-linear: a homomorphism of real vector spaces.

Exercises and complements. (a) For $n = 3$ this definition gives the formulas (3.101)–(3.103).

(b) Prove that the differential acts on a product $\omega \wedge \omega'$ of differential forms (of degrees p and q) as in (3.94)

$$d^{p+q}(\omega \wedge \omega') = (d^p\omega) \wedge \omega' + (-1)^p \omega \wedge (d^q\omega'). \tag{3.113}$$

Hints: begin with C^∞-functions.

(c) The differential makes $\Omega^+(U)$ into a dg-algebra on the real field.

3.7.5 Cohomology

Cocycles and coboundaries form two $\mathbb{R}$-linear subspaces of $\Omega^p(U)$

$$\begin{aligned}
Z^p(U) &= \mathrm{Ker}\,(d^p : \Omega^p(U) \to \Omega^{p+1}(U)) \subset \Omega^p(U), \\
B^p(U) &= \mathrm{Im}\,(d^{p-1} : \Omega^{p-1}(U) \to \Omega^p(U)) \subset Z^p(U).
\end{aligned} \tag{3.114}$$

A differential p-form $\omega = \sum_i \lambda_i\, dx^i$ is said to *closed* if it is a cocycle, i.e. $d^p\omega = 0$. It is said to be *exact* if it is a coboundary: $\omega = d^{p+1}\omega'$, for some form ω' of degree $p+1$.

Every exact form is closed; the converse need not be true, and this failure is measured by the cohomology vector space of U

$$H_R^p(U) = Z^p(U)/B^p(U), \tag{3.115}$$

whose elements $[\omega]$ are the de Rham p-cohomology classes of U.

Exercises and complements. (a) Compute $H_R^0(U)$ and prove that the vector space $H_R^0(U)$ has dimension 1 if and only if U is non-empty and connected.

(b) A 1-form $\omega = \sum_i \lambda_i\, dx^i \in \Omega^1(U)$ is closed if and only if

$$\frac{\partial \lambda_i}{\partial x^j} - \frac{\partial \lambda_j}{\partial x^i} = 0, \qquad \text{for all } i, j. \tag{3.116}$$

(c) In Calculus, differential forms on open euclidean subspaces of $\mathbb{R}^3$ are also viewed as vector fields; the differential of a function is then called the *gradient;* saying that $H_R^1(U) = 0$ means that every closed vector field $X = (\lambda_1, \lambda_2, \lambda_3)$ (satisfying the conditions (3.116)) is the gradient of some C^∞-function on U. The reader can similarly interpret the conditions $H_R^2(U) = 0$ and $H_R^3(U) = 0$.

3.7.6 Smooth maps

A C^∞-map $f = (f^1, ..., f^n)\colon V \to U$ has associated $\mathbb{R}$-linear morphisms

$$f^{\sharp p}\colon \Omega^p(U) \to \Omega^p(V),$$

$$f^{\sharp p}(\textstyle\sum_i \lambda_i \, dx^{i_1} ... \, dx^{i_p}) = \sum_i \lambda_i {\circ} f \, d^0 f^{i_1} \wedge ... \wedge d^0 f^{i_p}, \tag{3.117}$$

where the composite $\lambda_i {\circ} f\colon V \to U \to \mathbb{R}$ is not written by juxtaposition, to avoid confusion with pointwise multiplication of real valued functions. In particular, $f^{\sharp 0}(\lambda) = \lambda {\circ} f$.

The following exercises show that $f^\sharp = (f^{\sharp p})\colon \Omega^+(U) \to \Omega^+(V)$ is a morphism of dg-algebras.

Exercises and complements. (a) $f^\sharp$ preserves the exterior product of differential forms. *Hints:* begin with C^∞-functions.

(b) The homomorphisms $f^{\sharp p}$ commute with the differentials.

3.7.7 Functoriality

We have so forth defined a contravariant functor

$$\Omega^+\colon C^\infty\mathsf{Top} \dashrightarrow \mathbb{R}\mathsf{Dga}. \tag{3.118}$$

The proof of the functorial property $(gf)^\sharp = f^\sharp g^\sharp$ is similar to the previous ones, and is left to an interested reader.

Composing with the cohomology functor $H^*\colon \mathbb{R}\mathsf{Dga} \to \mathbb{R}\mathsf{Gra}$ of (3.98) (for real coefficients), we get the contravariant functor of de Rham cohomology

$$H_R^*\colon C^\infty\mathsf{Top} \dashrightarrow \mathbb{R}\mathsf{Dga}, \qquad H_R^*(U) = H^*(\Omega^+(U)),$$

$$f^* = H_R^*(f)\colon H_R^*(U) \to H_R^*(V), \tag{3.119}$$

on open euclidean spaces (for $f\colon V \to U$ in $C^\infty\mathsf{Top}$).

3.7.8 *Complements

As we said from the beginning, the natural setting of de Rham cohomology is the category of differentiable manifolds and differentiable maps.

De Rham's theorem establishes a natural isomorphism of graded algebras

$$H_R^*(M) \to H^*(M; \mathbb{R}), \tag{3.120}$$

where M is a differentiable manifold and $H^*(M; \mathbb{R})$ is the graded algebra of singular cohomology of M with coefficients in the real field, defined in Section 4.4 (see also Section 5.5). The proof of this theorem can be found in [Mas3], Appendix, Section 3.

4

Singular homology with coefficients

We have already seen cohomology theories that are naturally born with an arbitrary abelian group of coefficients (as Alexander–Spanier's), or a precise group of coefficients (as de Rham cohomology, with real coefficients).

In this chapter we introduce singular homology with coefficients in an abelian group G, using the tensor product $A \otimes B$ of abelian groups, and the action of the functor $- \otimes G$ on chain complexes.

Similarly, we introduce singular cohomology with coefficients in G, using the hom-group $\mathrm{Hom}(A, B)$ and the action of the contravariant functor $\mathrm{Hom}(-, G)$ on chain complexes.

We shall see, in 4.3.8, that an appropriate group of coefficients can detect facts that the integral coefficients do not distinguish. Even more importantly, we shall see in Section 5.5 that singular cohomology with coefficients in a ring R forms a graded R-algebra.

In particular, it determines de Rham cohomology when R is the real field, as recalled in 3.7.8.

4.1 Tensor product of modules and Hom

Although we really need the tensor product and Hom functors for abelian groups, we prefer to work in a more natural context: the category $R\,\mathsf{Mod}$ of modules on a fixed commutative ring R. The category Ab is canonically isomorphic to the category $\mathbb{Z}\,\mathsf{Mod}$ of modules on the ring of integers (as recalled in 1.3.6), and we make no distinction between them.

We review the definition and main properties of the tensor product and Hom-functor of R-modules, a classical topic dealt with in many books on Homological Algebra, like [M1, CE], or Linear Algebra, as [Bou1]. The exactness properties of these functors are deferred to Section 4.2.

R is a fixed commutative (unital) ring; commutativity is essential, here.

One may prefer to exclude the trivial ring $\{0\}$, on which all modules are trivial. Otherwise, some caution is required in a few points: cf. Remark 4.1.1(a).

4.1.1 Linear and multilinear mappings

The hom-set $R\,\mathrm{Mod}(A, B)$ of R-homomorphisms $A \to B$, or R-linear mappings, has a canonical structure of R-module, written as $\mathrm{Hom}_R(A, B)$, or $\mathrm{Hom}(A, B)$ when no confusion may arise.

Its operations are the pointwise addition and pointwise multiplication by scalars $\lambda \in R$

$$(f + g)(a) = f(a) + g(a), \qquad (\lambda f)(a) = \lambda f(a). \tag{4.1}$$

These operations are preserved by compositions: for $h\colon A' \to A$ and $k\colon B \to B'$ we have:

$$k(f + g)h = kfh + kgh, \qquad k(\lambda f)h = \lambda(kfh). \tag{4.2}$$

Now, for a product module $A = \prod_i A_i$ $(i = 1, ..., n)$, a mapping $\varphi\colon A \dashrightarrow B$ with values in a module B is said to be *multilinear*, or *n-linear*, on R if it is R-linear in each variable:

$$\begin{aligned}
\varphi(a_1, &..., \lambda a_i' + \mu a_i'', ..., a_n) \\
&= \lambda\varphi(a_1, ..., a_i', ..., a_n) + \mu\varphi(a_1, ..., a_i'', ..., a_n).
\end{aligned} \tag{4.3}$$

Multilinear mappings $\varphi\colon A \dashrightarrow B$ will be denoted by dot-marked arrows. They have a pointwise addition and pointwise scalar multiplication, that extend those of the linear ones (for $n = 1$).

Composing these mappings is less obvious: given a *family* of homomorphisms $h_i\colon A_i' \to A_i$ $(i = 1, ..., n)$ and *one* homomorphism $k\colon B \to B'$, the obvious composite is multilinear

$$k\,\varphi\,(\textstyle\prod_i f_i)\colon \textstyle\prod_i A_i' \to B', \qquad (x_1, ..., x_n) \mapsto k(\varphi(f_1(x_1), ..., f_n(x_n))). \tag{4.4}$$

For abelian groups, a mapping $\varphi\colon \prod_i A_i \dashrightarrow B$ is multilinear on $\mathbb{Z}$ if (and only if) it is *multiadditive*, i.e. additive in each variable.

We have reviewed in 1.2.5 the free abelian group $F(I) = \mathbb{Z}I$ on a set I. Here we shall use the *free R-module* on a set I

$$F(I) = RI = \bigoplus_{i \in I} R, \qquad \eta\colon I \to |RI|, \tag{4.5}$$

with the usual mapping $\eta(i) = e_i = (\delta_{ij})_{j \in I}$, the canonical basis. It is characterised by the *universal property*:

(i) for every mapping $f\colon I \to |A|$ with values in (the underlying set of) an R-module A, there is a unique homomorphism $g\colon RI \to A$ such that $g(e_i) = f(i)$, for all $i \in I$.

By abuse of notation, the index i will also stay for $e_i \in RI$, and we simply say that every mapping $f\colon I \to |A|$ has a unique extension $g\colon RI \to A$ in $R\,\mathsf{Mod}$, with $g(i) = f(i)$, for all $i \in I$. (This is commented in Remark (a), below.)

We have thus the *free R-module functor*

$$F\colon \mathsf{Set} \to R\,\mathsf{Mod}, \qquad F(I) = RI = \bigoplus_{i\in I} R,$$
$$F(f)\colon RI \to RJ, \qquad F(f)(i) = f(i) \qquad (\text{for } f\colon I \to J), \tag{4.6}$$

which can be written as F_R when useful.

Remarks. (a) For the trivial ring $R = \{0\}$, all R-modules are trivial, and free on any set I: the unique map $\eta\colon I \to \{0\}$ trivially satisfies the universal property.

Outside of this case, $0 \neq 1$ in R and the mapping η in (4.5) is injective: we can identify i with e_i and I with its image in RI. Our abuse of notation is then fully justified.

(b) Being a free object is not a property of the mere category $R\,\mathsf{Mod}$, but of the forgetful functor $U\colon R\,\mathsf{Mod} \to \mathsf{Set}$, as made clear by the universal property (i).

In fact, the free R-module functor is left adjoint to U: see Section 7.1.

4.1.2 Tensor product of modules

The *tensor product* of two R-modules A, B is defined by a universal property whose goal is to transform bilinear mappings into homomorphism.

Precisely, we look for a module $A \otimes_R B$ *equipped with a bilinear mapping* $\varphi_0\colon A \times B \dashrightarrow A \otimes_R B$ such that, for every bilinear mapping $\varphi\colon A \times B \dashrightarrow C$

$$
\begin{array}{ccc}
A \times B & \xrightarrow{\;\varphi_0\;} & A \otimes_R B \\
{\scriptstyle\varphi}\big\downarrow & \swarrow{\scriptstyle h} & \\
C & &
\end{array}
\tag{4.7}
$$

there is a unique R-homomorphism $h\colon A \otimes_R B \to C$ such that $\varphi = h\varphi_0$.

The universal property characterises the solution up to a unique isomorphism coherent with the structural bilinear mappings.

In fact, if $\varphi_i\colon A \times B \dashrightarrow C_i$ ($i = 1, 2$) are two solutions of the given problem, there are two (determined) homomorphisms h, k such that

$$h\colon C_1 \to C_2, \qquad k\colon C_2 \to C_1, \qquad \varphi_2 = h\varphi_1, \qquad \varphi_1 = k\varphi_2.$$

Then the homomorphism $kh\colon C_1 \to C_1$ satisfies $(kh)\varphi_1 = \varphi_1 = (\mathrm{id}\,C_1)\varphi_1$ and is the identity; similarly $hk = \mathrm{id}\,C_2$, which means that h and k are isomorphisms, inverse to each other.

To prove that the solution exists, we 'construct' one (by a complicated procedure, rarely used in computations)

$$A \otimes_R B = R(A \times B)/H(A, B), \qquad \varphi_0(a, b) = [(a, b)]. \qquad (4.8)$$

Here $R(A \times B)$ is the free R-module generated by the *set $A \times B$*, and its elements will be written as essentially finite R-linear combinations of pairs $(a, b) \in A \times B$. Secondly, $H(A, B)$ is the submodule of $R(A \times B)$ generated by all the elements of the following types (where a, a' belong to A, b, b' to B and $\lambda \in R$)

(i) $(a + a', b) - (a, b) - (a', b)$,
(ii) $(\lambda a, b) - \lambda(a, b)$,
(iii) $(a, b + b') - (a, b) - (a, b')$,
(iv) $(a, \lambda b) - \lambda(a, b)$.

(Taking in (4.8) the quotient modulo $H(A, B)$ is precisely what is required to make the mapping φ_0 bilinear on R.) We write

$$a \otimes b = \varphi_0(a, b) = [(a, b)] \in A \otimes_R B \qquad \text{(for } a \in A, \ b \in B). \qquad (4.9)$$

The generators (i)–(iv) of $H(A, B)$ give the bilinearity of the tensor product of elements:

$$(a + a') \otimes b = a \otimes b + a' \otimes b, \qquad (\lambda a) \otimes b = \lambda(a \otimes b),$$
$$a \otimes (b + b') = a \otimes b + a \otimes b', \qquad a \otimes (\lambda b) = \lambda(a \otimes b).$$

We can write $A \otimes B$ for $A \otimes_R B$, when no ambiguity can arise.

More generally, we define the tensor product $\otimes A_i$ of a finite family of R-modules, as an R-module equipped with a multilinear mapping

$$\varphi_0 \colon \Pi A_i \dashrightarrow \otimes A_i, \qquad (4.10)$$

universal 'as above', for all n-linear mappings $\Pi A_i \dashrightarrow C$.

4.1.3 Remarks and complements

(a) Every element of $A \otimes B$ can be written as a (finite) sum $\Sigma_i \, a_i \otimes b_i$; obviously, this expression is not unique (generally). The elements of the form $a \otimes b$ form a *canonical system of generators* of $A \otimes_R B$. More generally, if a (resp. b) varies in a system of generators of A (resp. B), the elements $a \otimes b$ still form a system of generators of $A \otimes B$.

There are cases where all these generators vanish, and $A \otimes_R B = 0$ (the trivial R-module).

(b) First, let us note that $a \otimes 0 = 0 \otimes b = 0$ (for $a \in A$ and $b \in B$). Therefore $A \otimes 0 = 0 \otimes B = 0$.

(c) Suppose now that the (commutative) ring R is an integral domain (i.e. has no proper zero-divisors). If A is divisible by a non-zero scalar λ (i.e. the equation $\lambda x = a$ has solutions, for every $a \in A$) and $\lambda B = 0$, then $A \otimes B = 0$. In fact, for $a = \lambda a' \in A$ and $b \in B$: $a \otimes b = (\lambda a') \otimes b = a' \otimes (\lambda b) = 0$.

(d) In the same hypothesis on R, if A is a *divisible* R-module (i.e. the equation $\lambda x = a$ has solutions, for every $a \in A$ and every scalar $\lambda \neq 0$) while B is a *torsion* R-module (every $b \in B$ is annihilated by some non-zero scalar), then $A \otimes B = 0$.

(e) Taking $R = \mathbb{Z}$, for every torsion abelian group B we have $\mathbb{Q} \otimes B = 0 = \mathbb{R} \otimes B$. (Any field of characteristic zero is divisible, as an abelian group.)

4.1.4 Tensor product of homomorphisms

Given two homomorphisms $f\colon A \to A'$, $g\colon B \to B'$ there is a homomorphism $f \otimes g$ determined as below on the canonical generators

$$f \otimes g\colon A \otimes_R B \to A' \otimes_R B', \quad (f \otimes g)(a \otimes b) = f(a) \otimes g(b). \qquad (4.11)$$

In fact, the bilinear mapping

$$A \times B \to A' \times B' \dashrightarrow A' \otimes_R B', \quad (a, b) \mapsto f(a) \otimes g(b),$$

determines such a homomorphism on $A \otimes_R B$.

Plainly, this construction preserves identities and composition:

$$\mathrm{id}\, A \otimes \mathrm{id}\, B = \mathrm{id}\, (A \otimes_R B), \quad (f'f) \otimes (g'g) = (f' \otimes g').(f \otimes g).$$

The tensor product is thus a (covariant) functor in two variables

$$\otimes\colon R\,\mathsf{Mod} \times R\,\mathsf{Mod} \to R\,\mathsf{Mod}. \qquad (4.12)$$

This functor is *bilinear*, i.e. additive and homogeneous in each variable, with respect to the R-linear structure of morphisms defined above (in 4.1.1):

$$\begin{aligned}
(f + f') \otimes g = f \otimes g + f' \otimes g, \quad (\lambda f) \otimes g = \lambda(f \otimes g), \\
f \otimes (g + g') = f \otimes g + f \otimes g', \quad f \otimes (\lambda g) = \lambda(f \otimes g).
\end{aligned} \qquad (4.13)$$

4.1.5 Basic properties of the tensor product

The ring R is fixed, and we write $A \otimes B$ for $A \otimes_R B$.

(a) The tensor product is *commutative*. More precisely, there is a canonical isomorphism determined as follows on the canonical generators:

$$A \otimes B \to B \otimes A, \quad a \otimes b \mapsto b \otimes a. \qquad (4.14)$$

In fact, the obvious bilinear map $A \times B \to B \times A \twoheadrightarrow B \otimes A$ determines a homomorphism $A \otimes B \to B \otimes A$ that acts as above on the canonical generators; symmetrically, we have a homomorphism $B \otimes A \to A \otimes B$ that sends $b \otimes a$ to $a \otimes b$. These homomorphisms are inverse to each other (on generators, whence everywhere).

(b) The tensor product *has a unit*, the R-module R. More precisely, there is a canonical isomorphism:

$$A \otimes R \to A, \qquad a \otimes \lambda \mapsto \lambda a \quad (a \mapsto a \otimes 1_R). \qquad (4.15)$$

In fact the mapping $\varphi_0 \colon A \times R \twoheadrightarrow A$, $(a, \lambda) \mapsto \lambda a$ is bilinear and satisfies the universal property that defines $A \otimes R$.

(c) The tensor product is *associative*, up to a canonical isomorphism:

$$(A \otimes B) \otimes C \to A \otimes (B \otimes C), \qquad (a \otimes b) \otimes c \mapsto a \otimes (b \otimes c). \qquad (4.16)$$

To give a simple proof, we verify that the trilinear mapping

$$\psi_0 \colon A \times B \times C \twoheadrightarrow (A \otimes B) \otimes C, \qquad (a, b, c) \mapsto (a \otimes b) \otimes c,$$

satisfies the universal property of a ternary tensor product (see (4.10)). Therefore $(A \otimes B) \otimes C$ is a realisation of $A \otimes B \otimes C$. Symmetrically, this is also true of $A \otimes (B \otimes C)$, and the two realisations are isomorphic, as above.

In fact, ψ_0 is the composite of the mappings $\varphi_1 \times C$ and φ_2 defined below (where φ_1 and φ_2 are bilinear)

$$
\begin{array}{lll}
A \times B \times C \xrightarrow{\varphi_1 \times C} (A \otimes B) \times C & \qquad \varphi_1(a, b) = a \otimes b, & \\
\hspace{4.5cm} \Big\downarrow{\scriptstyle \varphi_2} & & \\
\psi \Big\downarrow \quad \overset{\varphi'}{\nearrow} & & (4.17) \\
D \xleftarrow{\ \ h\ \ } (A \otimes B) \otimes C & \qquad \varphi_2(z, c) = z \otimes c. &
\end{array}
$$

Now, any trilinear mapping $\psi \colon A \times B \times C \twoheadrightarrow D$ factorises as $\varphi'(\varphi_1 \times C)$ where $\varphi' \colon (A \otimes B) \times C \twoheadrightarrow D$ is bilinear; then we factorise the latter as $\varphi' = h\varphi_2$, by a homomorphism h. The uniqueness of h in the factorisation $\psi = h\psi_0$ is obvious, because the elements $(a \otimes b) \otimes c$ generate $(A \otimes B) \otimes C$ (by 4.1.3(a)).

(d) The tensor product is *distributive* with respect to direct sums, by a canonical isomorphism (an easy consequence of the universal property)

$$(\oplus_{i \in I} A_i) \otimes B \to \oplus_{i \in I}(A_i \otimes B), \qquad (a_i)_{i \in I} \otimes b \mapsto (a_i \otimes b)_{i \in I}. \qquad (4.18)$$

(e) This gives canonical isomorphisms:

$$
\begin{array}{ll}
RI \otimes B \cong \oplus_{i \in I} B, & R^m \otimes B \cong B^m, \\
RI \otimes RJ \cong R(I \times J), & R^m \otimes R^n \cong R^{mn}.
\end{array}
\qquad (4.19)
$$

(f) *Tensor products of free modules (and all vector spaces) are thus determined:* if A, B are free on R with bases $(a_i)_{i \in I}$ and $(b_j)_{j \in J}$, then $A \otimes B$ is free with basis $(a_i \otimes b_j)_{(i,j) \in I \times J}$.

(g) For an abelian group A, the tensor product $A \otimes_{\mathbb{Z}} \mathbb{Q}$ is made into a vector space on $\mathbb{Q}$, letting $\lambda(a \otimes \mu) = a \otimes \lambda\mu$. The *rank* of the abelian group A is defined as the cardinal

$$\mathrm{rk}(A) = \dim_{\mathbb{Q}}(A \otimes_{\mathbb{Z}} \mathbb{Q}). \tag{4.20}$$

If A is finitely generated, this agrees with the previous definition in 1.2.6(c): the Structure Theorem says that A is isomorphic to a direct sum $\mathbb{Z}^n \oplus tA$, where tA is the torsion subgroup of A; therefore $A \otimes_{\mathbb{Z}} \mathbb{Q} \cong \mathbb{Q}^n$ (a cartesian power of abelian groups, and also of vector spaces).

4.1.6 The functor Hom

Another classical issue of linear and homological algebra is the functor Hom (also written as Hom_R).

The set-valued hom-functor $\mathrm{Mor}\colon R\,\mathsf{Mod}^{\mathrm{op}} \times R\,\mathsf{Mod} \to \mathsf{Set}$ (cf. 1.3.8) has an enriched version, recalled in 4.1.1

$$\mathrm{Hom}\colon R\,\mathsf{Mod}^{\mathrm{op}} \times R\,\mathsf{Mod} \to R\,\mathsf{Mod},$$
$$(A, B) \mapsto \mathrm{Hom}(A, B), \qquad (h, k) \mapsto k. - .h. \tag{4.21}$$

$\mathrm{Hom}(A, B)$ is the module of R-homomorphisms $A \to B$, with pointwise operations. For two R-linear mappings $h\colon A' \to A$ and $k\colon B \to B'$, the mapping $k. - .h\colon \mathrm{Hom}(A, B) \to \mathrm{Hom}(A', B')$ is R-linear, as shown in (4.2).

Here we are mostly interested in the covariant functor obtained by fixing the first variable

$$\mathrm{Hom}(A, -)\colon R\,\mathsf{Mod} \to R\,\mathsf{Mod},$$
$$(k\colon B \to B') \mapsto k_{\sharp} = k.- \colon \mathrm{Hom}(A, B) \to \mathrm{Hom}(A, B'), \tag{4.22}$$

but in Section 4.4 we shall use the contravariant functor obtained by fixing the second variable

$$\mathrm{Hom}(-, B)\colon R\,\mathsf{Mod} \dashrightarrow R\,\mathsf{Mod},$$
$$(h\colon A \to A') \mapsto h^{\sharp} = -.h \colon \mathrm{Hom}(B, A') \to \mathrm{Hom}(B, A). \tag{4.23}$$

Exercises and complements. (a) The functor $\mathrm{Hom}(A, -)$ preserves cartesian products, up to a canonical isomorphism

$$\mathrm{Hom}(A, \Pi_{i \in I} B_i) \to \Pi_{i \in I} \mathrm{Hom}(A, B_i), \quad f \mapsto (p_i f)_{i \in I}, \tag{4.24}$$

which takes a homomorphism $f\colon A \to \Pi_{i \in I} B_i$ to the family of its components $p_i f\colon A \to B_i$.

(b) The functor $\mathrm{Hom}(-, B)$ takes direct sums to cartesian products, up to a canonical isomorphism

$$\mathrm{Hom}(\oplus_{i \in I} A_i, B) \to \Pi_{i \in I} \mathrm{Hom}(A_i, B), \quad f \mapsto (fu_i)_{i \in I}, \qquad (4.25)$$

which sends a homomorphism $f \colon \oplus_{i \in I} A_i \to B$ to the family of its co-components $fu_i \colon A_i \to \oplus_{i \in I} A_i \to B$.

(c) The R-module R gives a functorial isomorphism

$$\mathrm{Hom}(R, -) \cong \mathrm{id} \colon R\,\mathsf{Mod} \to R\,\mathsf{Mod}, \quad (f \colon R \to B) \mapsto f(1_R), \qquad (4.26)$$

which is an 'enriched version' of the representability of the forgetful functor $R\,\mathsf{Mod} \to \mathsf{Set}$ by the module R.

4.1.7 The exponential law for sets and modules

In order to describe the relationship between tensor product and Hom, we begin with a similar, more elementary occurrence in Set.

(a) (*Exponential law for sets*) The hom-functor of Set

$$\mathrm{Mor} \colon \mathsf{Set}^{\mathrm{op}} \times \mathsf{Set} \to \mathsf{Set}, \quad (A, Y) \mapsto \mathsf{Set}(A, Y), \quad (h, k) \mapsto k. - .h,$$

is related to the binary cartesian product by a natural isomorphism in the variables X, Y

$$\varphi^A_{XY} \colon \mathsf{Set}(X \times A, Y) \to \mathsf{Set}(X, \mathsf{Set}(A, Y)),$$
$$(f \colon X \times A \to Y) \mapsto (g \colon X \to \mathsf{Set}(A, Y)), \qquad (4.27)$$
$$g(x) = f(x, -) \colon A \to Y.$$

This is a family of adjunctions, indexed by the 'parameter' A: the functor $- \times A \colon \mathsf{Set} \to \mathsf{Set}$ is left adjoint to the functor $\mathsf{Set}(A, -) \colon \mathsf{Set} \to \mathsf{Set}$ (see 7.2.1).

The hom-sets of the category Set are also written in 'exponential notation' $Y^A = \mathsf{Set}(A, Y)$, because an indexed family $(y_a)_{a \in A}$ of Y^A is the same as a mapping $y \colon A \to Y$. With this notation, the previous family of bijections is rewritten as

$$Y^{X \times A} \cong (Y^A)^X, \qquad (4.28)$$

explaining the name of *exponential law*, for sets.

(b) (*Exponential law for modules*) In the same way, we have a natural family of bijections

$$\varphi^A_{XY} \colon R\,\mathsf{Mod}(X \otimes A, Y) \to R\,\mathsf{Mod}(X, \mathrm{Hom}(A, Y)),$$
$$(f \colon X \otimes A \to Y) \mapsto (g \colon X \to \mathrm{Hom}(A, Y)), \qquad (4.29)$$
$$g(x) = f(x \otimes -) \colon A \to Y.$$

Concretely, a homomorphism $g\colon X \to \mathrm{Hom}(A, Y)$ amounts to a mapping $X \times A \to Y$ which is linear in each variable, and therefore to a homomorphism $f\colon X \otimes A \to Y$.

*For every R-module A, the endofunctor $- \otimes A$ is left adjoint to the endofunctor $\mathrm{Hom}(A, -)$ (of $R\,\mathsf{Mod}$). But (4.29) can be viewed as a natural family of isomorphisms of R-modules

$$\varphi^A_{XY}\colon \mathrm{Hom}(X \otimes A, Y) \to \mathrm{Hom}(X, \mathrm{Hom}(A, Y)),$$

giving an adjunction enriched over $R\,\mathsf{Mod}$.*

*4.1.8 *Exercises and complements* (Tensor product of vector spaces)

The ring of scalars is now a (commutative) field K.

Modules on K are called vector spaces and have specific properties, essentially derived from the well-known theory of linear dependence in vector spaces. The category $K\,\mathsf{Mod}$ is usually written as $K\,\mathsf{Vct}$.

(a) In $K\,\mathsf{Vct}$ every monomorphism splits, i.e. has a retraction. In other words, every subspace of a vector space is a retract.

(b) In $K\,\mathsf{Vct}$ every epimorphism splits, i.e. has a section. Every quotient of a vector space is a retract.

(c) All short exact sequences in $K\,\mathsf{Vct}$ split.

(d) For vector spaces A, B, there is a canonical homomorphism

$$u_{AB}\colon A \otimes_K B \to \mathrm{Hom}_K(A^*, B), \quad u_{AB}(a \otimes b)(\alpha) = \alpha(a).b, \qquad (4.30)$$

where $A^* = \mathrm{Hom}_K(A, K)$ is the dual vector space of A and $\alpha\colon A \to K$.

This family defines a natural transformation

$$u\colon (- \otimes_K -) \to \mathrm{Hom}_K((-)^*, -)\colon K\,\mathsf{Vct} \times K\,\mathsf{Vct} \to K\,\mathsf{Vct}. \qquad (4.31)$$

Prove that each component u_{AB} is injective. Moreover, if A and B are *finitely generated*, then u_{AB} is an isomorphism. *Hints:* use a basis of A.

(e) It follows that the tensor product of *finitely generated* vector spaces *can be* (and often *is*) defined as: $A \otimes_K B = \mathrm{Hom}_K(A^*, B)$. This form is also used for vector bundles.

4.2 Additive and exact functors

We deal now with additivity and exactness properties of functors between categories of modules, in particular tensor products and Hom-functors.

R and S are always commutative rings.

We recall that in a category of modules, kernels, cokernels, images, coimages, and exact sequences are dealt with as in Ab, in Section 2.1.

4.2.1 Additive functors

A functor $F\colon R\,\mathsf{Mod} \to S\,\mathsf{Mod}$ between categories of modules is *additive* if it preserves the addition of parallel morphisms, which means that, for every pair A, B of R-modules, the mapping

$$\mathrm{Hom}_R(A, B) \to \mathrm{Hom}_S(FA, FB), \qquad f \mapsto F(f), \qquad (4.32)$$

is a homomorphism of abelian groups.

The natural framework of this issue is additive categories (see 1.6.5).

Exercises and complements. (a) An additive functor $F\colon R\,\mathsf{Mod} \to S\,\mathsf{Mod}$ always preserves the zero object and the zero morphisms (see 1.6.5(a)).

(b) More generally, an additive functor preserves finite direct sums.

(c) An additive functor $F\colon R\,\mathsf{Mod} \to S\,\mathsf{Mod}$ always preserves split exact sequences (see 2.7.4(iii)). In particular, an additive functor $F\colon \mathsf{Ab} \to \mathsf{Ab}$ takes a short exact sequence $A \rightarrowtail B \twoheadrightarrow C$ where C is a free abelian group to a short exact sequence.

4.2.2 Exercises and complements (Exactness properties)

For a functor $F\colon R\,\mathsf{Mod} \to S\,\mathsf{Mod}$ there are diverse forms of preserving exactness, from the strongest (an exact functor) to the weakest (a half exact functor). The solutions are below.

(a) The functor F is said to be *exact* if it satisfies the following equivalent conditions:

(i) F preserves exact sequences,

(ii) F preserves short exact sequences,

(iii) F preserves kernels and cokernels.

The equivalence is proved in Solution (a).

An exact functor also preserves images and coimages, canonical factorisations of homomorphisms, monos and epis (see (2.3)).

Note. Condition (iii) means that F takes a (co)kernel of a morphism f to a (co)kernel of $F(f)$ (see 2.1.2), without having to preserve distinguished choices: we do not expect F to preserve submodules or quotient modules, 'on the nose': this would rarely happen.

(b) The functor F is *left exact* if it satisfies the following equivalent conditions:

(iv) F preserves kernels,

(v) F takes any short exact sequence $A' \rightarrowtail A \twoheadrightarrow A''$ to an exact sequence $0 \to FA' \to FA \to FA''$.

(b*) The functor F is *right exact* if it satisfies the following equivalent conditions:

(iv*) F preserves cokernels,

(v*) F takes any short exact sequence $A' \rightarrowtail A \twoheadrightarrow A''$ to an exact sequence $FA' \to FA \to FA'' \to 0$.

(c) The functor F is *half exact* if any short exact sequence $A' \rightarrowtail A \twoheadrightarrow A''$ is taken to an exact sequence $FA' \to FA \to FA''$.

Every half exact functor preserves split exact sequences and finite direct sums; moreover, *it is an additive functor*. (Henceforth, this also holds for left exact, right exact and exact functors.)

(d) Exact functors are plainly closed under composition. The same holds for left exact and right exact functors, but generally fails for the half exact ones.

(e) A contravariant functor $F\colon R\,\mathsf{Mod} \dashrightarrow S\,\mathsf{Mod}$ is the same as a covariant functor $R\,\mathsf{Mod}^{\mathrm{op}} \to S\,\mathsf{Mod}$. It is said to be *left exact* if it satisfies the equivalent properties:

(vi)　F takes cokernels to kernels,

(vii) F takes any short exact sequence $A' \rightarrowtail A \twoheadrightarrow A''$ to an exact sequence $0 \to FA'' \to FA \to FA'$.

(f) All these definitions can be extended to categories of chain complexes, which are additive categories where kernels, cokernels and exact sequences have been defined. Thus, the chain homology functor $H_n\colon \mathsf{Ch}_+\mathsf{Ab} \to \mathsf{Ab}$ is half exact, by Theorem 2.3.3.

The natural framework of these definitions is abelian categories.

Solutions. (a) We have seen in 2.1.3 that exact sequences detect zero objects, kernels and cokernels, and (obviously) short exact sequences. We have also seen that kernels and cokernels detect short exact sequences, which detect the general ones.

(b) The equivalence is readily proved starting from an exact sequence $A' \rightarrowtail A \to A''$ and factorising epi-mono the last morphism.

(c) Let F be half exact. If $(m,p)\colon A' \rightarrowtail A \twoheadrightarrow A''$ is a split exact sequence, m is a section and p is a retraction, which are preserved by any functor. This shows that (Fm, Fp) is still a split exact sequence.

The other preservation properties are a consequence:

- the zero object Z is characterised by $(\mathrm{id}\,Z, \mathrm{id}\,Z)$ being a split exact sequence,

- a biproduct $A \oplus B$ is characterised, in 2.7.5, by a commutative butterfly diagram with split exact sequences,
- an addition $f + g$ of parallel homomorphism is characterised by binary biproducts, in (2.180).

4.2.3 *Lemma* (Extending additive functors)

(a) An additive functor $F\colon R\,\mathsf{Mod} \to S\,\mathsf{Mod}$ has an obvious extension to an additive functor $F_+\colon \mathrm{Ch}_+R\,\mathsf{Mod} \to \mathrm{Ch}_+S\,\mathsf{Mod}$ that preserves homotopies. (One can say that the extension is homotopy equivariant.*) The extension will often be written as F.*

(b) For an exact functor $F\colon R\,\mathsf{Mod} \to S\,\mathsf{Mod}$, the extension F_+ is exact and commutes with homology. More precisely, on every chain complex of R-modules A we have a natural isomorphism (for every $n \geqslant 0$)

$$\varphi_n\colon H_n(F_+A) \to F(H_n(A)), \tag{4.33}$$

induced by the natural isomorphism $Z_n(FA) \to F(Z_n(A))$ (by exactness).

Proof Point (a) is obvious. As to (b), the homology of A is characterised by the short exact sequence

$$B_n(A) \rightarrowtail Z_n(A) \xrightarrow{\;p\;} H_n(A) \tag{4.34}$$

preserved by the exact functor F.

Similarly, the homology of the complex FA is characterised by the short exact sequence in the upper row of the following diagram

$$\begin{array}{ccccc}
B_n(FA) & \rightarrowtail & Z_n(FA) & \xrightarrow{\;p\;} & H_n(FA) \\
\Big\downarrow{\chi_n} & & \Big\downarrow{\psi_n} & & \Big\downarrow{\varphi_n} \\
F(B_nA) & \rightarrowtail & F(Z_nA) & \xrightarrow[Fp]{} & F(H_n(A))
\end{array} \tag{4.35}$$

Since F preserves images and kernels, we have natural isomorphisms ψ_n and χ_n, which induce a natural isomorphism $\varphi_n\colon H_n(FA) \to F(H_nA)$, by the Induction Lemma 2.1.4. $\qquad\square$

4.2.4 *Theorem* (Exactness properties of the tensor product)

For every R-module A, the tensor functor $F = - \otimes_R A\colon R\,\mathsf{Mod} \to R\,\mathsf{Mod}$ is right exact.

Proof *Category Theory would give this for free: every left adjoint functor preserves categorical colimits, in particular cokernels (and sums, as we have already verified for F, in 4.1.5(d)).*

We verify directly our claim, following nevertheless the outline suggested by the adjunction: the exponential law in 4.1.7(b). We already know that F preserves zero objects and zero morphisms.

We have an R-homomorphism $f\colon B \to C$, with cokernel $p\colon C \to C'$ (in the left diagram below), and we want to prove that $F(p) = p \otimes A\colon C \otimes A \to C' \otimes A$ is a cokernel of $F(f) = f \otimes A\colon B \otimes A \to C \otimes A$ (in the right diagram below), verifying the universal property of cokernels (in 2.1.2(i*))

$$
\begin{array}{ccc}
B \xrightarrow{\;f\;} C \xrightarrow{\;p\;} C' & \quad B \otimes A \xrightarrow{\;Ff\;} C \otimes A \xrightarrow{\;Fp\;} C' \otimes A & \\
\quad{}^{g}\big\downarrow \;{}_{u}\nearrow & \quad {}^{h}\big\downarrow \;{}_{v}\nearrow & (4.36) \\
\operatorname{Hom}(A,D) & \qquad D &
\end{array}
$$

First, $F(p)F(f) = F(pf) = 0$. Secondly, we suppose that $h\colon C \otimes A \to D$ also gives $h.F(f) = 0$.

By the exponential law, $h\colon C \otimes A \to D$ corresponds to a homomorphism in the left diagram

$$g\colon C \to \operatorname{Hom}(A,D), \qquad h(c \otimes a) = g(c)(a) \qquad \text{(for } c \in C,\ a \in A),$$

and $gf = 0$, because $gf(b)(a) = h(f(b) \otimes a) = h.Ff(b \otimes a) = 0$ on every $b \in B$ and $a \in A$.

Similarly, a homomorphism $u\colon C' \to \operatorname{Hom}(A,D)$ in the left diagram corresponds to a homomorphism v in the right one

$$v\colon C' \otimes A \to D, \qquad v(c' \otimes a) = u(c')(a) \qquad \text{(for } c' \in C',\ a \in A).$$

Finally, the condition $g = up$ is equivalent to $h = v.Fp$. By hypothesis, there is a unique $u\colon C' \to \operatorname{Hom}(A,D)$ such that $g = up$; therefore there is also a unique $v\colon C' \otimes A \to D$ such that $h = v.Fp$. $\qquad\square$

4.2.5 Theorem (Exactness properties of Hom)

For every R-module X

(i) the functor $G = \operatorname{Hom}_R(X,-)\colon R\,\mathsf{Mod} \to R\,\mathsf{Mod}$ is left exact,

(ii) the functor $H = \operatorname{Hom}_R(-,X)\colon R\,\mathsf{Mod}^{\mathrm{op}} \to R\,\mathsf{Mod}$ is left exact.

Proof Both points can easily be verified directly. *In Category Theory one proves that a representable functor preserves categorical limits, in particular kernels (and products).*

(i) If $k\colon K \to A$ is a kernel of $f\colon A \to B$, we prove that $G(k) = k_\sharp$ is a kernel of $G(f) = f_\sharp$. In the right diagram below we take a morphism u such that $G(f).u = 0$

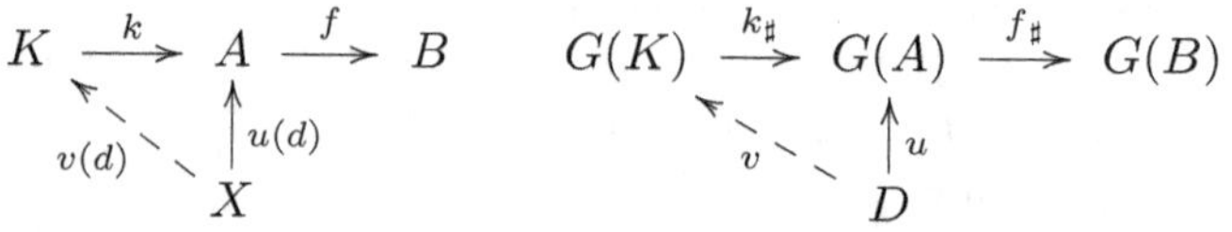

$$K \xrightarrow{\ k\ } A \xrightarrow{\ f\ } B \qquad G(K) \xrightarrow{\ k_\sharp\ } G(A) \xrightarrow{\ f_\sharp\ } G(B) \tag{4.37}$$

Therefore, for every $d \in D$, the homomorphism $u(d)\colon X \to A$ in the left diagram is annihilated by f and factorises as $k.v(d)$, by a unique homomorphism $v(d)\colon X \to K$. Coming back to the right diagram, we have defined the unique mapping $v\colon D \to G(K) = \mathrm{Hom}(X, K)$ such that $G(k).v = u$, and it is a homomorphism because $k\colon K \to A$ is a monomorphism.

(ii) According to 4.2.2(e) we prove that, if $c\colon B \to C$ is a cokernel of $f\colon A \to B$, then $H(c) = c^\sharp$ is a kernel of $H(f) = f^\sharp$. In the right diagram below we take a morphism u such that $H(f).u = 0$

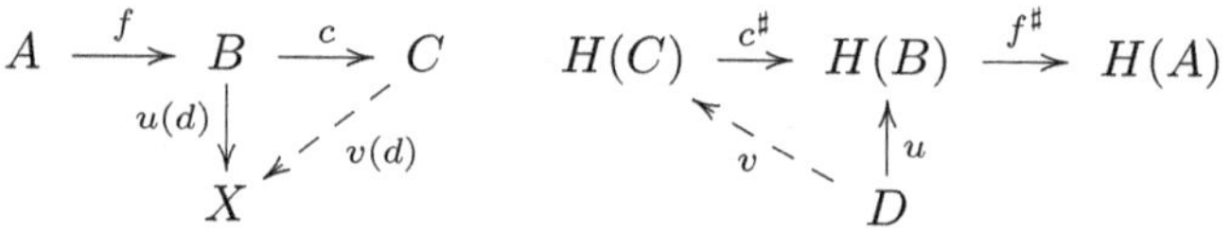

$$A \xrightarrow{\ f\ } B \xrightarrow{\ c\ } C \qquad H(C) \xrightarrow{\ c^\sharp\ } H(B) \xrightarrow{\ f^\sharp\ } H(A) \tag{4.38}$$

For every $d \in D$, the homomorphism $u(d)\colon B \to X$ gives $u(d).f = H(f)(u(d)) = 0$ and factorises in the left diagram as $v(d)c$, by a unique homomorphism $v(d)\colon C \to X$. In the right diagram we have defined the unique mapping $v\colon D \to \mathrm{Hom}(C, X)$ such that $H(c).v = u$, and it is a homomorphism because $c\colon B \to C$ is an epimorphism. $\qquad\square$

4.2.6 Flat, projective and injective modules

(a) An R-module A is said to be *flat* if the functor $F = - \otimes_R A$ is exact; this is equivalent to saying that F preserves monomorphisms.

(b) An R-module P is said to be *projective* if the functor $G = \mathrm{Hom}_R(P, -)\colon R\,\mathrm{Mod} \to R\,\mathrm{Mod}$ is exact. This is equivalent to saying that G preserves epimorphisms, and can be expressed by the following property:

(i) if $p\colon A \to B$ is an epimorphism, every homomorphism $h\colon P \to B$ can be lifted to A, i.e. there is some homomorphism $k\colon P \to A$ such that $pk = h$

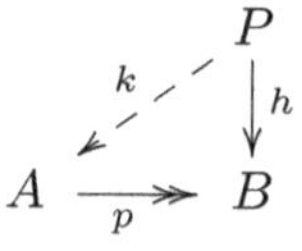

$$A \xrightarrow{\ p\ } B \tag{4.39}$$

(c) An R-module Q is said to be *injective* if the contravariant Hom-functor $H = \operatorname{Hom}_R(-, B)\colon R\operatorname{Mod} \dashrightarrow R\operatorname{Mod}$ preserves exact sequences. This is equivalent to saying that H takes monomorphisms to epimorphisms, and can be expressed by the following property:

(i*) if $m\colon B \to A$ is a monomorphism, every homomorphism $h\colon B \to Q$ can be extended to A, i.e. there exists $k\colon A \to Q$ such that $km = h$

$$\begin{array}{ccc}
 & Q & \\
{\scriptstyle k}\nearrow & \uparrow{\scriptstyle h} & \\
A & \xleftarrow{\ m\ } & B
\end{array} \tag{4.40}$$

4.2.7 Exercises and complements

(a) A free R-module A is flat and projective: the functors $- \otimes_R A$ and $\operatorname{Hom}_R(A, -)$ are exact. (Of course, $A \otimes_R -$ is also.)

(b) Projective modules are closed under direct sums and summands. Injective modules are closed under products and summands.

(c) An R-module is projective if and only if it is a summand of a free module.

*(d) Every projective module is flat. This will be proved in Exercise 5.1.6(b).

(e) The properties (i) and (i) of 4.2.6 are related by reversing all arrows and interchanging epi with mono: projective and injective objects are dual issues, if we state them in the selfdual framework of abelian categories.

4.2.8 Exercises and complements on abelian groups, I

The rest of this section is about abelian groups.

(a) Prove that $\mathbb{Z}_m \otimes \mathbb{Z}_n \cong \mathbb{Z}_d$, where $d = \gcd(m, n)$ is the greatest common divisor of $m, n \geqslant 2$. In particular, if m and n are coprime integers, we get $\mathbb{Z}_m \otimes \mathbb{Z}_n = 0$. *Hints:* use the exactness properties of the tensor product.

(b) Compute $\operatorname{Hom}(\mathbb{Z}_m, A)$, for $n \geqslant 2$. Prove that $\operatorname{Hom}(\mathbb{Z}_m, \mathbb{Z}_n) \cong \mathbb{Z}_d$, where $d = \gcd(m, n)$.

(c) A flat abelian group A is always torsion free (see 1.2.6). *The converse is also true, but less easy to prove: see 5.2.8(a).*

(d) In particular, a finite (non-trivial) abelian group is not flat. For instance, $- \otimes \mathbb{Z}_2\colon \mathsf{Ab} \to \mathsf{Ab}$ takes the monomorphism $2\colon \mathbb{Z} \to \mathbb{Z}$ (multiplication by 2) to $0\colon \mathbb{Z}_2 \to \mathbb{Z}_2$.

4.2.9 Exercises and complements on abelian groups, II

(a) We shall use, without proof, a well known theorem about abelian groups: *every subgroup of a free abelian group is free.*

The proof is not easy, and depends on the axiom of choice. It can be found, for instance, in a classical textbook on abelian groups [Fu], in Theorem II.12.1.

(b) An abelian group is projective if and only if it is free.

(c) An additive functor $F\colon \mathsf{Ab} \to \mathsf{Ab}$ preserves all *upper unbounded* exact sequences of *free abelian groups*. (Note that upper unbounded sequences include all sequences ending with a trivial group in the last position — which can be repeated indefinitely.)

As to considering upper unbounded sequences, we already remarked that the additive functor $- \otimes \mathbb{Z}_2 \colon \mathsf{Ab} \to \mathsf{Ab}$ takes the monomorphism $2 \colon \mathbb{Z} \to \mathbb{Z}$ to $0 \colon \mathbb{Z}_2 \to \mathbb{Z}_2$. In other words, it does not preserve the exactness of a sequence $0 \to \mathbb{Z} \to \mathbb{Z}$ of free abelian groups.

*(d) An abelian group is injective if and only if it is divisible (see 4.1.3(d)). We shall not prove this result, which again depends on the axiom of choice. One can see [CE], Theorem I.3.2, or [M1], Corollary III.7.3.

4.3 Relative singular homology with coefficients

We have defined relative singular homology as a homology theory with coefficients in the group $\mathbb{Z}$ of integers. Working with the tensor functor $- \otimes G$, we can modify it into a homology theory with coefficients in an arbitrary abelian group G. The new theory may be able to detect facts that the original one cannot see.

For instance, homology with coefficients in $\mathbb{Z}_2$ is able to prove that the canonical projection $\mathbb{P}^2 \to \mathbb{S}^2$ is not homotopically trivial (see Exercise 4.3.8(c)). One cannot get this result working with integral coefficients: for $k > 0$, one at least of $H_k(\mathbb{P}^2)$ or $H_k(\mathbb{S}^2)$ is trivial.

G is a fixed abelian group.

4.3.1 Extending tensor products, I

The additive functor $- \otimes G \colon \mathsf{Ab} \to \mathsf{Ab}$ has an obvious componentwise extension to chain complexes (as in 4.2.3(a))

$$- \otimes G \colon \mathrm{Ch}_+\mathsf{Ab} \to \mathrm{Ch}_+\mathsf{Ab}, \quad A \otimes G = ((A_n \otimes G),(\partial_n \otimes G)),$$

$$(f \otimes G)_n = f_n \otimes G \colon A_n \otimes G \to B_n \otimes G, \tag{4.41}$$

where $f \colon A \to B$ is a morphism of chain complexes.

This extension preserves homotopies: a homotopy $\varphi\colon f \simeq g\colon A \to B$ in $\mathrm{Ch}_+\mathsf{Ab}$ gives a homotopy

$$\varphi \otimes G\colon f \otimes G \simeq g \otimes G\colon A \otimes G \to B \otimes G,$$
$$(\varphi \otimes G)_n = \varphi_n \otimes G\colon A_n \otimes G \to B_{n+1} \otimes G. \tag{4.42}$$

Remarks. Letting G vary in Ab, the extension (4.41) becomes a functor $\mathrm{Ch}_+\mathsf{Ab} \times \mathsf{Ab} \to \mathrm{Ch}_+\mathsf{Ab}$, that will be used in Section 4.5. In Section 5.4 we shall introduce a larger extension, the tensor product of chain complexes, defined on $\mathrm{Ch}_+\mathsf{Ab} \times \mathrm{Ch}_+\mathsf{Ab}$.

4.3.2 Chains with coefficients

We now apply this functor $- \otimes G\colon \mathsf{Ab} \to \mathsf{Ab}$ to the singular chains of a space.

We have thus the *singular chain complex* of a space X, *with coefficients in* G, by a functor

$$C_+(-;G)\colon \mathsf{Top} \to \mathrm{Ch}_+\mathsf{Ab}, \qquad C_+(X;G) = C_+(X) \otimes G,$$
$$C_n(X;G) = C_n(X) \otimes G, \tag{4.43}$$
$$f_{\sharp n}\big((\textstyle\sum \lambda_i a_i) \otimes g\big) = (\textstyle\sum \lambda_i f a_i) \otimes g,$$

where $f\colon X \to Y$ and $a_i\colon \mathbb{I}^n \to X$ are in Top, while $g \in G$.

Similarly, we have the *singular chain complex* of relative pairs of spaces, *with coefficients in* G

$$C_+(-;G)\colon \mathsf{Top}_2 \to \mathrm{Ch}_+\mathsf{Ab}, \qquad C_+(X,A;G) = C_+(X,A) \otimes G. \tag{4.44}$$

It is important to note that, for a relative pair (X, A), the short exact sequence

$$C_+(A) \xrightarrow{\ u_\sharp\ } C_+(X) \xrightarrow{\ v_\sharp\ } C_+(X,A) \tag{4.45}$$

splits in every degree (because all $C_n(X, A)$ are free, see Exercise 3.1.2(b)), and is preserved by the additive functor $- \otimes G$ (by 4.2.1(c)): we have a componentwise split short exact sequence of chain complexes

$$C_+(A;G) \xrightarrow{\ u_\sharp\ } C_+(X;G) \xrightarrow{\ v_\sharp\ } C_+(X,A;G). \tag{4.46}$$

Note. We already remarked, in (3.50), that $C_n(X;G) \cong \oplus_a G$, where a varies in $\mathrm{Cub}_n X \setminus \mathrm{Deg}_n X$. Defining $C_n(X;G)$ in this way, without using tensor products, is a shortcut which would soon prove ineffective.

4.3.3 Singular homology with coefficients

Applying chain homology to the previous complexes, we have now the absolute and relative *singular homology groups with coefficients in G*

$$H_n(-;G)\colon \mathsf{Top} \to \mathsf{Ab}, \qquad H_n(X;G) = H_n(C_+(X;G)), \qquad (4.47)$$

$$H_n(-;G)\colon \mathsf{Top}_2 \to \mathsf{Ab}, \qquad H_n(X,A;G) = H_n(C_+(X,A;G)). \qquad (4.48)$$

With $G = \mathbb{Z}$ we find the previous chain complexes and homology with integral coefficients, up to canonical isomorphism

$$\begin{aligned}
C_+(X;\mathbb{Z}) &\cong C_+(X), & C_+(X,A;\mathbb{Z}) &\cong C_+(X,A), \\
H_n(X;\mathbb{Z}) &\cong H_n(X), & H_n(X,A;\mathbb{Z}) &\cong H_n(X,A).
\end{aligned} \qquad (4.49)$$

Exercises and complements. The elementary computations of 2.2.5 can be extended to arbitrary coefficients, still using the partition $X = \bigcup_{i\in I} X_i$ of a space in path components.

(a) Compute $H_n(\{*\};G)$.

(b) Compute $H_0(X;G)$, where X is 0-connected (i.e. path connected and non-empty).

(c) There are canonical isomorphisms

$$\bigoplus_{i\in I} C_+(X_i;G) \to C_+(X;G), \quad (c_i)_{i\in I} \mapsto \Sigma_{i\in I}\, c_i, \qquad (4.50)$$

$$\bigoplus_{i\in I} H_n(X_i;G) \to H_n(X;G), \quad ([z_i])_{i\in I} \mapsto \Sigma_{i\in I}\, [z_i]. \qquad (4.51)$$

(d) If the map $f\colon X \to Y$ is defined on a path connected space, the homomorphism $f_{*0}\colon H_0(X;G) \to H_0(Y;G)$ is injective.

4.3.4 Subdivision Theorem

For a space X and a generalised open cover $\mathcal{U} = (U_i)$, the canonical morphism $j \otimes G\colon C_+(X;\mathcal{U}) \otimes G \to C_+(X;G)$ is a monomorphism, and induces an isomorphism in homology, in every degree

$$(j \otimes G)_{*n}\colon H_n(X;\mathcal{U};G) \to H_n(X;G). \qquad (4.52)$$

Proof We deduce this from the original Subdivision Theorem 2.3.5, with integral coefficients. We start from the short exact sequence of chain complexes

$$C_+(X;\mathcal{U}) \overset{j}{\rightarrowtail} C_+(X) \overset{p}{\twoheadrightarrow} E_+ \qquad (4.53)$$

produced by the inclusion j.

This sequence is componentwise split, because the canonical basis of $C_n(X;\mathcal{U})$ is a subset of the canonical basis of $C_n(X)$. Applying $-\otimes G$ we get a short exact sequence of chain complexes

$$C_+(X;\mathcal{U})\otimes G \xrightarrow{\ j\otimes G\ } C_+(X;G) \xrightarrow{\ p\otimes G\ } E_+\otimes G. \qquad (4.54)$$

In the long exact homology sequence of (4.53) all j_{*n} are isomorphisms (by Theorem 2.3.5), and therefore $H_n(E_+) = 0$, for all n.

The complex E_+ is thus an exact sequence of free abelian groups, and an unbounded one (when extended with zeros). Applying Exercise 4.2.9(c), also $E_+\otimes G$ is an exact sequence and $H_n(E_+\otimes G) = 0$, for all n.

Finally, by the exactness of the homology sequence of (4.54), all the homomorphisms $(j\otimes G)_{*n}$ are isomorphisms. $\qquad\square$

4.3.5 Theorem

Relative singular homology with coefficients in G satisfies the Eilenberg–Steenrod axioms for a homology theory with coefficient group G.

Proof The Functoriality axiom was proved from the start, in 4.3.3. The others are proved as follows.

(a) *Exactness and Naturality.* For a relative pair (X, A), the natural short exact sequence of chain complexes (4.46)

$$C_+(A;G) \xrightarrow{\ u_\sharp\ } C_+(X;G) \xrightarrow{\ v_\sharp\ } C_+(X, A;G) \qquad (4.55)$$

gives a natural, exact homology sequence

$$\begin{aligned}
&\ldots H_n(A;G) \xrightarrow{\ u_{*n}\ } H_n(X;G) \xrightarrow{\ v_{*n}\ } H_n(X, A;G) \xrightarrow{\ D_n\ }\\
&H_{n-1}(A;G) \to \qquad \ldots \qquad \to H_0(X, A;G) \to 0
\end{aligned} \qquad (4.56)$$

where $D_n[z] = [\partial_n z]$, for $z \in Z_n(X, A;G)$.

(b) *Homotopy invariance.* Let $\varphi\colon f \simeq g\colon (X, Y) \to (Y, B)$ be a relative homotopy. In the proof of Theorem 3.1.4f we built a homotopy of chain complexes $\Phi''\colon f_\sharp \simeq g_\sharp\colon C_+(X, A) \to C_+(Y, B)$.

Applying the additive functor $-\otimes G$ we get a homotopy

$$\Phi''\otimes G\colon f_\sharp \simeq g_\sharp\colon C_+(X, A;G) \to C_+(Y, B;G).$$

(c) *Excision.* As in Theorem 3.2.1, using now the Subdivision Theorem 4.3.4, with coefficients in G.

(d) *Dimension.* By Exercise 4.3.3(a), which also determines the coefficient group. $\qquad\square$

4.3.6 Theorem (The Mayer–Vietoris sequence)

There are given a topological space X, two subspaces U and V such that $X = \operatorname{int} U \cup \operatorname{int} V$, and an abelian group G. We let $A = U \cap V$.

There is an exact sequence of singular homology groups

$$\ldots \; H_n(A;G) \; \xrightarrow{\,h_n\,} \; H_n(U;G) \oplus H_n(V;G) \; \xrightarrow{\,k_n\,} \; H_n(X;G) \; \xrightarrow{\,D_n\,} \tag{4.57}$$
$$H_{n-1}(A;G) \; \to \; \ldots$$

called the exact sequence of Mayer–Vietoris, *or MV-sequence, of the space X with coefficients in G (with respect to the generalised open cover (U, V)).*

These homomorphisms are computed as in (2.85).

The sequence is natural for continuous mappings $f \colon X \to X'$, where

$$X' = \operatorname{int} U' \cup \operatorname{int} V', \qquad f(U) \subset U', \qquad f(V) \subset V'. \tag{4.58}$$

Proof We apply the same argument as in 2.3.7, using now the Subdivision Theorem 4.3.4. $\qquad\qquad\square$

4.3.7 Acyclic spaces and maps

(a) A space X is said to be *acyclic* (in singular homology) if $H_n(X) = 0$ for all $n > 0$ (as certainly true if its path components are contractible).

Then the same holds with coefficients in any group G. In fact, the augmented chain complex

$$\ldots \; C_2(X) \; \xrightarrow{\,\partial_2\,} \; C_1(X) \; \xrightarrow{\,\partial_1\,} \; C_0(X) \; \xrightarrow{\,\operatorname{cok}\partial_1\,} \; H_0(X) \; \to \; 0 \tag{4.59}$$

is always exact at $C_0(X)$ and $H_0(X)$. Therefore, the space X is acyclic if and only if the previous sequence is exact. In this case, as the sequence is unbounded and all objects are free abelian groups, tensoring by G gives an exact sequence, and $H_n(X;G) = 0$ for all $n > 0$.

(b) We say that a map $f \colon X \to Y$ is *acyclic* (in singular homology) if $H_n(f;G) = 0$, for any group G and for all $n > 0$.

Otherwise, we say that f is *homologically essential* (in singular homology): this means that $H_n(f;G) \neq 0$, for some group G and some degree $n > 0$; as a consequence, f cannot be homotopic to a constant map, and cannot factorise through a contractible space.

Integral coefficients are not sufficient to settle this point, as shown by Exercise 4.3.8(c).

4.3.8 Exercises and complements

The following exercises show part of the interest of introducing coefficient groups.

(a) We already know, from the Exercises of 4.3.3, that $H_0(\mathbb{S}^0; G) \cong G^2$, with trivial higher groups. For $n > 0$:

$$H_0(\mathbb{S}^n; G) \cong H_n(\mathbb{S}^n; G) \cong G, \qquad H_k(\mathbb{S}^n; G) = 0 \quad (0 \neq k \neq n). \qquad (4.60)$$

(b) For $n \geq 2$, the space $P_n = \mathbb{D}^2/R_n$ defined in 2.5.8 gives:

$$H_1(P_n; G) \cong G/nG, \qquad H_2(P_n; G) \cong \{g \in G \mid ng = 0\}, \qquad (4.61)$$

and the higher groups are trivial.

(c) The projection $p\colon P_n \to \mathbb{S}^2$ (viewing both spaces as quotients of the disc $\mathbb{D}^2$) is homologically essential. In particular, this holds for the projection $\mathbb{P}^2 \to \mathbb{S}^2$ of the projective plane.

> *Hints.* Use homology with coefficients in $\mathbb{Z}_n$, proving that p induces an isomorphism $H_2(P_n; \mathbb{Z}_n) \to H_2(\mathbb{S}^2; \mathbb{Z}_n)$ of cyclic groups of order n. (Integral coefficients cannot see this fact, because $H_1(\mathbb{S}^2) = 0$ and $H_2(P_n) = 0$: in each positive degree, one of the homology functors is 'blind' on our spaces.)

*(d) The Universal Coefficient Theorem, in 5.2.5, will show that singular homology with integral coefficients determines the *groups* of singular homology with any coefficients. (More precisely, we need both groups $H_n(X)$ and $H_{n-1}(X)$ to determine $H_n(X; G)$.)

Exercise (c) shows that *this fails for the morphisms.*

4.4 Relative singular cohomology with coefficients

Working with the contravariant functor $\operatorname{Hom}(-, G)$, singular homology with integral coefficients is transformed into a cohomology theory with coefficients in G.

G is again a fixed abelian group.

4.4.1 Extending the Hom functor, I

Extending the contravariant (additive) functor $\operatorname{Hom}(-, G)\colon \mathsf{Ab} \dashrightarrow \mathsf{Ab}$ to chain complexes, we get a contravariant functor with values in cochain complexes (where $f\colon A \to B$ is a morphism of chain complexes)

$$\operatorname{Hom}(-, G)\colon \mathrm{Ch}_+\mathsf{Ab} \dashrightarrow \mathrm{Ch}^+\mathsf{Ab}, \quad (\operatorname{Hom}(A, G))^n = \operatorname{Hom}(A_n, G),$$

$$d^n = \operatorname{Hom}(\partial_{n+1}, G)\colon \operatorname{Hom}(A_n, G) \to \operatorname{Hom}(A_{n+1}, G), \qquad (4.62)$$

$$f^n = \operatorname{Hom}(f_n, G)\colon \operatorname{Hom}(B_n, G) \to \operatorname{Hom}(A_n, G).$$

This functor preserves homotopies: a homotopy in $\mathsf{Ch}_+\mathsf{Ab}$

$$\varphi\colon f \simeq g\colon A \to B, \quad \varphi_n\colon A_n \to B_{n+1}, \quad \partial_{n+1}\varphi_n + \varphi_{n-1}\partial_n = g_n - f_n,$$

gives a homotopy of cochain complexes (see (3.68))

$$\psi = \mathrm{Hom}(\varphi, G)\colon \mathrm{Hom}(f, G) \simeq \mathrm{Hom}(g, G),$$

$$\psi^n = \mathrm{Hom}(\varphi_{n-1}, G)\colon \mathrm{Hom}(B_n, G) \to \mathrm{Hom}(A_{n-1}, G), \tag{4.63}$$

$$d^{n-1}\psi^n + \psi^{n+1}d^n = \mathrm{Hom}(\varphi_{n-1}\partial_n + \partial_{n+1}\varphi_n, G) = g^n - f^n.$$

4.4.2 Singular cochains

We now apply the functor $\mathrm{Hom}(-, G)$ to the singular chains of a space, with integral coefficients.

We have thus the *singular cochain complex* of a space X, *with coefficients in G*, by a contravariant functor, the composite of $C_+\colon \mathsf{Top} \to \mathsf{Ch}_+\mathsf{Ab}$ and $\mathrm{Hom}(-, G)$

$$C^+(-; G)\colon \mathsf{Top} \dashrightarrow \mathsf{Ch}^+\mathsf{Ab}, \qquad C^+(X; G) = \mathrm{Hom}(C_+(X), G),$$

$$C^n(X; G) = \mathrm{Hom}(C_n(X), G), \qquad d^n = \mathrm{Hom}(\partial_{n+1}, G),$$

$$f^{\sharp n}\colon \mathrm{Hom}(C_n(Y), G) \to \mathrm{Hom}(C_n(X), G), \tag{4.64}$$

$$f^{\sharp n}(\mu\colon C_n(Y) \to G) = \mu f_{\sharp n}\colon C_n(X) \to G,$$

where $f\colon X \to Y$ is a map of topological spaces.

More generally, we have the *singular cochain complex* of relative pairs, *with coefficients in G*

$$C^+(-; G)\colon \mathsf{Top}_2 \dashrightarrow \mathsf{Ch}^+\mathsf{Ab},$$

$$C^+(X, A; G) = \mathrm{Hom}(C_+(X, A), G). \tag{4.65}$$

The (componentwise split) short exact sequence of the relative pair (X, A) (in (4.45)) gives here a (componentwise split) short exact sequence of cochain complexes

$$C^+(X, A; G) \overset{v^\sharp}{\rightarrowtail} C_+(X; G) \overset{u^\sharp}{\twoheadrightarrow} C_+(A; G). \tag{4.66}$$

Comments and complements. (a) A cochain $\lambda\colon C_n(X) \to G$ can be viewed as a mapping on n-cubes which vanishes on the degenerate ones (and is linearly extended on cubical chains)

$$\lambda\colon \mathrm{Cub}_n X \to |G|, \qquad \lambda(\mathrm{Deg}_n X) = \{0\},$$

$$f^{\sharp n}(\mu)(a) = \mu(fa) \qquad (\text{for } \mu\colon \mathrm{Cub}_n Y \to G, \ a\colon \mathbb{I}^n \to X). \tag{4.67}$$

(b) A relative cochain in $C^n(X, A; G)$ is thus the same as an absolute cochain $\lambda\colon C_n(X) \to G$ which vanishes on degenerate cubes and all the cubes of A

$$\lambda\colon \mathrm{Cub}_n X \to |G|, \qquad \lambda(\mathrm{Deg}_n X \cup \mathrm{Cub}_n A) = \{0\}. \qquad (4.68)$$

(c) For a relative map $f\colon (X, A) \to (Y, B)$, the associated homomorphism of relative cochains is described as in (4.67)

$$f^{\sharp n}\colon C^n(Y, B; G) \to C^n(X, A; G), \qquad f^{\sharp n}(\mu)(a) = \mu(fa), \qquad (4.69)$$

where $\mu\colon \mathrm{Cub}_n Y \to G$ vanishes on $\mathrm{Deg}_n Y \cup \mathrm{Cub}_n B$, and $a\colon \mathbb{I}^n \to X$.

(d) (*Restrictions and extensions*) The embedding $u\colon A \to X$ gives a *restriction*

$$u^{\sharp n}\colon C^n(X; G) \twoheadrightarrow C^n(A; G), \qquad u^{\sharp n}(\lambda) = \lambda u_{\sharp n} = \lambda|_A, \qquad (4.70)$$

whose value on the cochain λ will also be written as $\lambda|_A$. This epimorphism has a canonical section

$$C^n(A; G) \rightarrowtail C^n(X; G), \qquad \xi \mapsto \xi_X\colon C_n(X) \to G, \qquad (4.71)$$

which will be called *extension* (in degree n): the cochain $\xi\colon C_n(A) \to G$ is extended to X letting $\xi_X(a) = 0$ for every cube $a\colon \mathbb{I}^n \to X$ whose image is not contained in A.

In this way we have constructed a *componentwise* splitting of the short exact sequence (4.66): the extensions do not commute with the differentials.

4.4.3 Singular cohomology

Applying the (covariant) functor of cochain cohomology to the previous complexes, we have now the absolute and relative *singular cohomology groups with coefficients in G*

$$H^n(-; G)\colon \mathsf{Top} \dashrightarrow \mathsf{Ab}, \qquad H^n(X; G) = H^n(C^+(X; G)), \qquad (4.72)$$

$$H^n(-; G)\colon \mathsf{Top}_2 \dashrightarrow \mathsf{Ab}, \quad H^n(X, A; G) = H^n(C^+(X, A; G)). \qquad (4.73)$$

A cocycle $\lambda \in C^n(X, A; G)$ is a relative cochain (as in (4.68)) which vanishes on all n-boundaries

$$\lambda\colon \mathrm{Cub}_n X \to G, \quad \lambda(\mathrm{Deg}_n X \cup \mathrm{Cub}_n A \cup \partial_{n+1}(\mathrm{Cub}_{n+1} X)) = \{0\}. \quad (4.74)$$

With $G = \mathbb{Z}$ we get the singular cohomology *with integral coefficients*.

Exercises and complements. The elementary computations of 2.2.5 can be adapted to cohomology with coefficients in G.

(a) Compute $H^n(\{*\}; G)$.

(b) Compute $H^0(X; G)$, where X is 0-connected (i.e. path connected and non-empty).

(c) For a space X with path components X_i and inclusions $u_i \colon X_i \to X$ ($i \in I$), there is a canonical isomorphism

$$H^n(X; G) \to \textstyle\prod_{i \in I} H^n(X_i; G), \qquad [\lambda] \mapsto ([\lambda_i])_{i \in I}, \qquad (4.75)$$

where $\lambda_i = u_i^{\sharp n}(\lambda)$: the value $\lambda_i(a)$ on an n-cube $a \colon \mathbb{I}^n \to X_i$ is $\lambda(u_i a)$; or simply $\lambda(a)$ when we view a as a map $\mathbb{I}^n \to X$ with image in X_i.

More formally, the homomorphisms $u_i^{*n} \colon H^n(X; G) \to H^n(X_i; G)$ are the projections of a cartesian product.

(d) In the same situation, $H^0(X; G) \cong G^I$.

4.4.4 Subdivision Theorem

For a space X and a generalised open cover $\mathcal{U} = (U_i)$, the canonical morphism $\mathrm{Hom}(j, G) \colon C^+(X; G) \to \mathrm{Hom}(C_+(X; \mathcal{U}), G)$ is an epimorphism and induces an isomorphism in cohomology, in every degree

$$\mathrm{Hom}(j, G)^{*n} \colon H^n(X; G) \to H^n(X; \mathcal{U}; G). \qquad (4.76)$$

Proof As in Theorem 4.3.4, we start from the short exact sequence (4.53), of chain complexes

$$C_+(X; \mathcal{U}) \overset{j}{\rightarrowtail} C_+(X) \overset{p}{\twoheadrightarrow} E_+ \qquad (4.77)$$

which we know to be componentwise split.

Applying the additive functor $\mathrm{Hom}(-, G)$ we get a short exact sequence of cochain complexes

$$\mathrm{Hom}(E_+, G) \overset{p^{\sharp}}{\rightarrowtail} C^+(X; G) \overset{j^{\sharp}}{\twoheadrightarrow} \mathrm{Hom}(C_+(X; \mathcal{U}), G). \qquad (4.78)$$

Applying the original Subdivision Theorem 2.3.5 we have seen that the complex E_+ is an exact sequence of free abelian groups, and an unbounded one (extended with zeros). Applying Exercise 4.2.9(c) (more precisely, its contravariant form), the cochain complex $\mathrm{Hom}(E_+, G)$ is also an exact sequence and $H^n(\mathrm{Hom}(E_+, G)) = 0$ for all n.

Finally, by the exactness of the cohomology sequence of (4.78), all the homomorphisms $\mathrm{Hom}(j, G)^{*n}$ are isomorphisms. $\qquad \square$

4.4.5 Theorem

Relative singular cohomology with coefficients in G satisfies the Eilenberg–Steenrod axioms for a cohomology theory with coefficients in G.

Proof The Functoriality axiom is proved in 4.4.3.

(a) *Exactness and Naturality.* A relative pair (X, A) has a natural short exact sequence of cochain complexes (4.66). Therefore its cohomology sequence (3.55) is exact and natural.

(b) *Homotopy invariance.* A relative homotopy $\varphi\colon f \simeq g\colon (X, Y) \to (Y, B)$ gives a homotopy of chain complexes $\Phi''\colon f_\sharp \simeq g_\sharp\colon C_+(X, A) \to C_+(Y, B)$ (cf. 3.1.4), turned by the additive contravariant functor $\mathrm{Hom}(-, G)$ into a homotopy

$$\mathrm{Hom}(\Phi'', G)\colon f^\sharp \simeq g^\sharp\colon C^+(Y, B; G) \to C^+(X, A; G).$$

(c) *Excision.* As in Theorem 3.2.1, using the Subdivision Theorem 4.4.4.

(d) *Dimension* and *coefficient group.* By Exercise 4.4.3(a). $\square$

4.4.6 Theorem (The Mayer–Vietoris sequence)

Let X be a topological space, U and V subspaces of X such that $X = \mathrm{int}\, U \cup \mathrm{int}\, V$, and $A = U \cap V$. Let G be an abelian group.

There is a long exact sequence of singular cohomology groups

$$\ldots\, H^n(X; G) \xrightarrow{k^n} H^n(U; G) \oplus H^n(V; G) \xrightarrow{h^n} H^n(A; G) \xrightarrow{D_n}$$
$$H^{n+1}(X; G) \to \ldots \tag{4.79}$$

called the cohomology sequence of Mayer–Vietoris, *or* MV-sequence, *of the space X with coefficients in G (for the generalised open cover (U, V)).*

With the notation of (2.82) for the inclusions $i\colon A \to U$, $u\colon U \to X$, …

$$k^n([\lambda]) = (u^{*n}[\lambda], -v^{*n}[\mu]) = ([\lambda|_U], -[\lambda|_V]),$$
$$h^n([\mu], [\nu]) = i^{*n}[\mu] + j^{*n}[\nu] = [\mu|_A] + [\nu|_A], \tag{4.80}$$
$$D^n([\xi]) = [(d^n \xi_U)_X],$$

where $\lambda \in Z^n(X; G)$, $\mu \in Z^n(U; G)$, $\nu \in Z^n(V; G)$, and $\xi \in Z^n(A; G)$. The restrictions $\lambda|_U, \lambda|_V$, … are defined in (4.70); the extensions ξ_U and $(d^n \xi_U)_X$ in (4.71).

The sequence is natural for continuous mappings $f\colon X \to X'$, where

$$X' = \mathrm{int}\, U' \cup \mathrm{int}\, V', \qquad f(U) \subset U', \qquad f(V) \subset V'.$$

Proof We start from the short exact sequence of chain complexes (2.87)

$$C_+(A) \xrightarrow{\ h'\ } C_+(U) \oplus C_+(V) \xrightarrow{\ k'\ } C_+(X; \mathcal{U}) \tag{4.81}$$

$$h'_n(c) = (i_{\sharp n}(c), j_{\sharp n}(c)) = (c, c),$$

$$k'_n(c, c') = u_{\sharp n}(c) - v_{\sharp n}(c) = c - c'.$$

The sequence splits componentwise, and is taken by $\mathrm{Hom}(-, G)$ to a short exact sequence of cochain complexes

$$C^+(X; G; \mathcal{U}) \xrightarrow{\ k''\ } C^+(U; G) \oplus C^+(V; G) \xrightarrow{\ h''\ } C^+(A; G) \tag{4.82}$$

$$k''_n(\lambda) = (u^{\sharp n}(\lambda), -v^{\sharp n}(\lambda)) = (\lambda|_U, -\lambda|_V),$$

$$h''_n(\mu, \nu) = i^{\sharp n}(\mu) + j^{\sharp n}(\nu) = \mu|_A + \nu|_A.$$

The cohomology of $C^+(X; G; \mathcal{U})$ is identified to that of $C^+(X; G)$, by the (current) Subdivision Theorem 4.4.4. This gives the cohomology sequence (4.79).

The induced homomorphisms h^n and k^n are computed as stated in (4.80). The differential acts indeed as $D^n([\xi]) = [(d^n \xi_U)_X]$

$$
\begin{array}{ccccc}
(\xi_U, 0) & \xrightarrow{\ h''\ } & \xi & & (n) \\[4pt]
\downarrow & & \downarrow & & \\[4pt]
(d^n \xi_U)_X & \xrightarrow{\ k''\ } (d\xi_U, 0) & \xrightarrow{\ h''\ } 0 & & (n+1)
\end{array}
\tag{4.83}
$$

- the cocycle $\xi \in Z^n(A; G)$ is extended to a cochain $\xi_U \in C^n(U; G)$, letting $\xi_U(a) = 0$ for every n-cube $a \colon \mathbb{I}^n \to U$ whose image is not contained in A,

- the cochain $d^n \xi_U$ is similarly extended to the cochain $(d^n \xi_U)_X$, which is a cocycle of X.

The restriction of $\lambda = (d^n \xi_U)_X$ to V is indeed zero: for any cube $a \colon \mathbb{I}^{n+1} \to X$ with image contained in V:

- if $\mathrm{Im}\,(a) \subset A$, then

$$\lambda(a) = (d^n \xi_U)(a) = \xi_U(\partial^{n+1} a) = \xi(\partial^{n+1} a) = d^n \xi(a) = 0,$$

- otherwise, $\mathrm{Im}\,(a)$ is not contained in U, and $\lambda(a) = 0$. $\qquad \square$

4.5 *Changing the coefficient group

This brief section offers a glimpse on the variation of the coefficient group. It will not be used elsewhere.

4.5.1 Extending tensor products, II

As already remarked in 4.3.1, the tensor product of chain complexes and abelian groups is a functor in two variables

$$- \otimes -\colon \mathsf{Ch}_+\mathsf{Ab} \times \mathsf{Ab} \to \mathsf{Ch}_+\mathsf{Ab},$$

$$A \otimes G = ((A_n \otimes G), (\partial_n \otimes G)), \tag{4.84}$$

$$(f \otimes h)_n = f_n \otimes h\colon A_n \otimes G \to B_n \otimes G',$$

where $f\colon A \to B$ is a morphism of chain complexes and $h\colon G \to G'$ a homomorphism of abelian groups. In fact

$$(\partial_n \otimes G')(f_n \otimes h) = \partial_n f_n \otimes h = f_n \partial_{n-1} \otimes h = (f_{n-1} \otimes h)(\partial_n \otimes G).$$

Here we are interested in keeping the chain complex A fixed, and letting the coefficient group vary, giving a functor

$$A \otimes -\colon \mathsf{Ab} \to \mathsf{Ch}_+\mathsf{Ab}, \qquad G \mapsto A \otimes G, \qquad h \mapsto h_\sharp = A \otimes h,$$

$$H_n(A \otimes -)\colon \mathsf{Ab} \to \mathsf{Ab}, \qquad G \mapsto H_n(A \otimes G), \tag{4.85}$$

$$(h\colon G \to G') \mapsto h_{*n} = (A \otimes h)_{*n}\colon H_n(A \otimes G) \to H_n(A \otimes G').$$

4.5.2 The Bockstein operator in homology

Suppose that A is a chain complex of *free* abelian groups. A short exact sequence of abelian groups

$$(m, p)\colon G' \rightarrowtail G \twoheadrightarrow G'' \tag{4.86}$$

gives raise to a sequence of chain complexes

$$A \otimes G' \xrightarrow{\ A \otimes m\ } A \otimes G \xrightarrow{\ A \otimes p\ } A \otimes G'' \tag{4.87}$$

that is still short exact, since each functor $A_n \otimes -\colon \mathsf{Ab} \to \mathsf{Ab}$ is exact (by Exercise 4.2.7(a)).

We have thus a long exact homology sequence

$$\ldots H_n(A \otimes G') \xrightarrow{\ m_{*n}\ } H_n(A \otimes G) \xrightarrow{\ p_{*n}\ } H_n(A \otimes G'') \xrightarrow{\ \beta_n\ }$$
$$H_{n-1}(A \otimes G') \to \ldots \tag{4.88}$$

natural for morphisms $f\colon A \to B$ of chain complexes.

The connecting homomorphism $\beta_n\colon H_n(A \otimes G'') \to H_{n-1}(A \otimes G')$ is called the *Bockstein operator*, for the short exact sequence (4.86) and the chain complex A (with free components).

In particular we can apply this to the chain complex $C_+(X, A)$ of relative chains of a pair of topological spaces, getting an exact homology sequence

$$\ldots H_n(X, A; G') \xrightarrow{m_{*n}} H_n(X, A; G) \xrightarrow{p_{*n}} H_n(X, A; G'') \xrightarrow{\beta_n}$$
$$H_{n-1}(X, A; G') \to \ldots \tag{4.89}$$

natural for relative maps $f\colon (X, A) \to (Y, B)$ in Top_2. The connecting homomorphism β_n is called the (homology) *Bockstein operator*, for the short exact sequence (4.86) and the relative pair (X, A).

4.5.3 *Extending the Hom functor, II*

Similarly, the Hom functor of chain complexes and abelian groups (in 4.4.1) can be viewed as a functor in two variables, contravariant in the first

$$\mathrm{Hom}(-, -)\colon \mathsf{Ch}_+\mathsf{Ab}^{\mathrm{op}} \times \mathsf{Ab} \to \mathsf{Ch}^+\mathsf{Ab},$$
$$(\mathrm{Hom}(A, G))^n = \mathrm{Hom}(A_n, G),$$
$$d^n = \mathrm{Hom}(\partial_{n+1}, G)\colon \mathrm{Hom}(A_n, G) \to \mathrm{Hom}(A_{n+1}, G), \tag{4.90}$$
$$(\mathrm{Hom}(f, h))^n = \mathrm{Hom}(f_n, h)\colon \mathrm{Hom}(B_n, G) \to \mathrm{Hom}(A_n, G'),$$

where $f\colon A \to B$ is a morphism of chain complexes and $h\colon G \to G'$ a homomorphism of abelian groups.

Keeping the chain complex A fixed, and letting the coefficient group vary, we have covariant functors

$$\mathrm{Hom}(A, -)\colon \mathsf{Ab} \to \mathsf{Ch}_+\mathsf{Ab}, \qquad G \mapsto \mathrm{Hom}(A, G),$$
$$H^n(\mathrm{Hom}(A, -))\colon \mathsf{Ab} \to \mathsf{Ab}, \qquad G \mapsto H^n(\mathrm{Hom}(A, G)), \tag{4.91}$$
$$(h\colon G \to G') \mapsto h^{*n}\colon H^n(\mathrm{Hom}(A, G)) \to H^n(\mathrm{Hom}(A, G')).$$

4.5.4 *The Bockstein operator in cohomology*

Working as in 4.5.2, let A be a chain complex of free abelian groups and $(m, p)\colon G' \rightarrowtail G \twoheadrightarrow G''$ a short exact sequence in Ab.

The exact functor $\mathrm{Hom}(A, -)$ gives a short exact sequence in $\mathsf{Ch}^+\mathsf{Ab}$

$$\mathrm{Hom}(A, G') \overset{m^\sharp}{\rightarrowtail} \mathrm{Hom}(A, G) \overset{p^\sharp}{\twoheadrightarrow} \mathrm{Hom}(A, G'') \tag{4.92}$$

and a long exact homology sequence

$$\ldots H^n\mathrm{Hom}(A, G') \xrightarrow{m^{*n}} H^n\mathrm{Hom}(A, G) \xrightarrow{p^{*n}} H^n\mathrm{Hom}(A, G'') \xrightarrow{\beta^n}$$
$$H^{n+1}\mathrm{Hom}(A, G') \to \ldots$$

contravariantly natural for morphisms $f\colon A \to B$ of chain complexes.

Applying this to the chain complex $C_+(X, A)$ of a relative pair, we get a long exact cohomology sequence

$$\ldots H^n(X, A; G') \xrightarrow{\;m^{*n}\;} H^n(X, A; G) \xrightarrow{\;p^{*n}\;} H^n(X, A; G'') \xrightarrow{\;\beta^n\;}$$
$$H^{n+1}(X, A; G') \to \ldots \tag{4.93}$$

contravariantly natural for relative maps $f \colon (X, A) \to (Y, B)$ in Top_2; β^n is the (cohomology) *Bockstein operator* for the short exact sequence (m, p) of coefficient groups and the relative pair (X, A).

4.5.5 Remarks

The usual short exact sequence $\mathbb{Z} \rightarrowtail \mathbb{Z} \twoheadrightarrow \mathbb{Z}_n$ can be used to link homology and cohomology groups with coefficients in $\mathbb{Z}_n$ and the ordinary homology groups with integral coefficients — although this is more effectively done with the Universal Coefficient Theorems of Sections 5.2 and 5.3.

An exercise. (a) An interested reader can find again, in this way, the results of Exercises 4.3.8(c), (d).

5

Derived functors, universal coefficients and products

We show now that the singular theory with integral coefficients determines the singular homology *groups* and singular cohomology *groups* for any coefficient group G. (We already now that this fails on maps, by 4.3.8(c).)

To prove this we have to introduce new tools of Homological Algebra. We begin with a brief exposition of derived functors in categories of modules; this gives the Tor and Ext functors, derived — respectively — of the tensor product and Hom-functor. We are then ready for the Universal Coefficient Theorem, for singular homology and cohomology.

We end by sketching higher results, on the multiplicative structure of singular cohomology and the homology of a product of spaces. The latter is studied by the method of 'acyclic models', which also yields the equivalence of the cubical and simplicial forms of singular homology.

This chapter is less elementary than the previous ones. A reader arrived at this point should be able to follow these developments.

5.1 Derived functors

Derived functors are a powerful topic of Homological Algebra, with diverse applications in Algebraic Topology, Algebraic Geometry, Commutative Algebra, and other fields.

We work in categories of modules. In this book we shall only use two particular cases, the torsion product (a derived functor of the tensor product) and the Ext functor (a derived functor of Hom), for abelian groups. Yet, the pattern is better perceived in the general construction, rather than following the shortcuts of particular cases.

Literature. The classical text on derived functors (and satellites) in categories of modules is Cartan–Eilenberg [CE]. Its extension to abelian categories is exposed in Grothendieck's article [Gt] and Mac Lane's book [M1], Chapter XII.

207

5.1.1 Introduction

We have a right exact functor (which is additive, by 4.2.2(c))

$$F\colon R\,\mathrm{Mod} \to S\,\mathrm{Mod}, \tag{5.1}$$

and we want to construct the sequence of its left derived functors

$$L_n F = F_n\colon R\,\mathrm{Mod} \to S\,\mathrm{Mod} \qquad (n \geqslant 0), \tag{5.2}$$

starting with $F_0 = F$. This sequence is meant to correct the failure of exactness of F, in the sense that every short exact sequence of R-modules

$$(m,p)\colon A \rightarrowtail B \twoheadrightarrow C \tag{5.3}$$

will have an associated exact sequence of left derived functors

$$\tag{5.4}$$

The *connecting morphisms* $\partial_n\colon F_n(C) \to F_{n-1}(A)$ form a natural transformation

$$\partial_n\colon F_n P''' \to F_{n-1} P'\colon \mathrm{Sh}R\,\mathrm{Mod} \to S\,\mathrm{Mod}, \tag{5.5}$$

where P''' and P' are projections $\mathrm{Sh}R\,\mathrm{Mod} \to R\,\mathrm{Mod}$ of the category of short exact sequences of R-modules, as in (2.71).

All this forms an *exact connected sequence* of functors $((F_n),(\partial_n))$, like the sequence $((H_n),(D_n))$ of homology functors of chain complexes, in Theorem 2.3.3.

5.1.2 Projective resolutions

We start from a *projective resolution* of an R-module A. This is a chain complex $A_+ = ((A_n),(\partial_n))$ equipped with an augmentation $\varepsilon\colon A_0 \to A$, such that:

(i) all components A_n are projective R-modules (see 4.2.6(b)),

(ii) the following sequence is exact

$$\ldots A_2 \xrightarrow{\partial_2} A_1 \xrightarrow{\partial_1} A_0 \xrightarrow{\varepsilon} A \to 0. \tag{5.6}$$

The resolution will be written in the form $\varepsilon\colon A_+ \to A$, as a morphism of chain complexes with values in the chain complex $A = J(A)$, where A has degree zero and the rest is trivial (see (2.18)).

We prove below that every module has projective resolutions, determined up to homotopy equivalence of chain complexes.

Exercises and complements. These exercises are the basis of the theory and a significant piece of Homological Algebra. Solutions are given below, but the reader is invited to solve them as independently as possible.

(a) A module A has a *standard free resolution*, with free components. It is constructed using the fact that A is a quotient of the free module $P(A) = F(|A|)$ generated by its underlying set, via the canonical epimorphism $\varepsilon\colon P(A) \twoheadrightarrow A$ that sends any generator to the same element of A.

(b) There are given an R-homomorphism $f\colon A \to B$ and two projective resolutions $\varepsilon\colon A_+ \to A$ and $\eta\colon B_+ \to B$. Then there is a chain morphism $f_+\colon A_+ \to B_+$ that *lifts* f, forming a commutative diagram with exact rows

$$
\begin{array}{ccccccccc}
\ldots\, A_2 & \xrightarrow{\partial_2} & A_1 & \xrightarrow{\partial_1} & A_0 & \xrightarrow{\varepsilon} & A & \longrightarrow & 0 \\
\downarrow{\scriptstyle f_2} & & \downarrow{\scriptstyle f_1} & & \downarrow{\scriptstyle f_0} & & \downarrow{\scriptstyle f} & & \\
\ldots\, B_2 & \xrightarrow[\partial_2]{} & B_1 & \xrightarrow[\partial_1]{} & B_0 & \xrightarrow[\eta]{} & B & \longrightarrow & 0
\end{array}
\tag{5.7}
$$

Hints. Prove the existence of $f_0\colon A_0 \to B_0$ and proceed upwards, using the fact that $\varepsilon\colon A_+ \to A$ satisfies (i) and $\eta\colon B_+ \to B$ satisfies (ii).

(c) If the chain morphisms $f_+, g_+\colon A_+ \to B_+$ lift the same homomorphism $f\colon A \to B$ to projective resolutions, there exists a chain homotopy $\varphi\colon f_+ \simeq g_+$. *Hints:* by additivity, it is sufficient to work on a chain morphism $h_+\colon A_+ \to B_+$ that lifts the zero morphism $0\colon A \to B$.

(d) If $\varepsilon\colon A_+ \to A$ and $\eta\colon B_+ \to A$ are projective resolutions of the same module, the chain complexes A_+ and B_+ are homotopy equivalent. *Hints:* use two liftings $f_+\colon A_+ \to B_+$ and $g_+\colon B_+ \to A_+$ of $\operatorname{id} A$.

(e) For a composite $h = gf\colon A \to B \to C$, any three liftings $f_+\colon A_+ \to B_+$, $g_+\colon B_+ \to C_+$ and $h_+\colon A_+ \to C_+$ to projective resolutions have a chain homotopy $h_+ \simeq g_+ f_+$.

Similarly, any commutative diagram in $R\,\mathsf{Mod}$ can be lifted to a *weakly commutative diagram* in $\mathrm{Ch}_+ R\,\mathsf{Mod}$: i.e. commutative up to chain homotopy — and properly commutative in the homotopy category $\mathrm{ho}(\mathrm{Ch}_+ R\,\mathsf{Mod})$.

Solutions. (a) We form the standard free resolution of A letting:

- $A_0 = P(A)$, with the canonical epimorphism $\varepsilon\colon A_0 \twoheadrightarrow A$,
- $A_1 = P(\operatorname{Ker}\varepsilon)$, with $\partial_1 = (A_1 \twoheadrightarrow \operatorname{Ker}\varepsilon \rightarrowtail A_0)$ and $\operatorname{Im}\partial_1 = \operatorname{Ker}\varepsilon$,
- $A_n = P(\operatorname{Ker}\partial_{n-1})$, with $\partial_n = (A_n \twoheadrightarrow \operatorname{Ker}\partial_{n-1} \rightarrowtail A_{n-1})$ and $\operatorname{Im}\partial_n = \operatorname{Ker}\partial_{n-1}$ (for $n \geqslant 2$).

In a particular case, there are often more economical resolutions.

(b) It is an easy consequence of the projective property of all A_n, combined with the exactness of the lower sequence.

In diagram (5.7), the homomorphism $f\varepsilon\colon A_0 \to B$ starts from a projective module and can be lifted to $f_0\colon A_0 \to B_0$ such that $\eta f_0 = f\varepsilon$

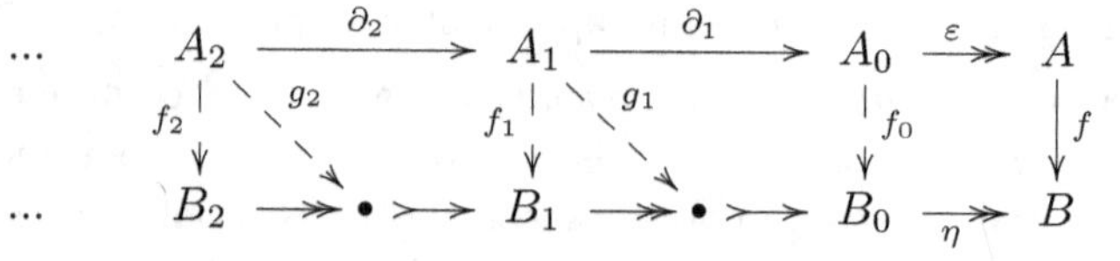

$$
\begin{array}{ccccccccc}
\cdots & A_2 & \xrightarrow{\;\partial_2\;} & A_1 & \xrightarrow{\;\partial_1\;} & A_0 & \xrightarrow{\;\varepsilon\;} & A \\
& \Big\downarrow{\scriptstyle f_2}\;{\scriptstyle g_2} & & \Big\downarrow{\scriptstyle f_1}\;{\scriptstyle g_1} & & \Big\downarrow{\scriptstyle f_0} & & \Big\downarrow{\scriptstyle f} \\
\cdots & B_2 & \longrightarrow\bullet\rightarrowtail & B_1 & \longrightarrow\bullet\rightarrowtail & B_0 & \xrightarrow{\;\eta\;} & B
\end{array}
\tag{5.8}
$$

Now, after factorising $\partial_1\colon B_1 \to B_0$ through $\operatorname{Im}\partial_1 = \operatorname{Ker}\eta$, the composite $f_0\partial_1\colon A_1 \to B_0$ factorises through $\operatorname{Ker}\eta$, because $\eta f_0\partial_1 = f\varepsilon\partial_1 = 0$.

The homomorphism g_1 obtained in this way can be lifted to $f_1\colon A_1 \to B_1$, making a commutative diagram. We go on in the same way.

(c) The difference $h_+ = f_+ - g_+\colon A_+ \to B_+$ lifts $0\colon A \to B$; we prove that it is homotopic to $0\colon A_+ \to B_+$, working as previously

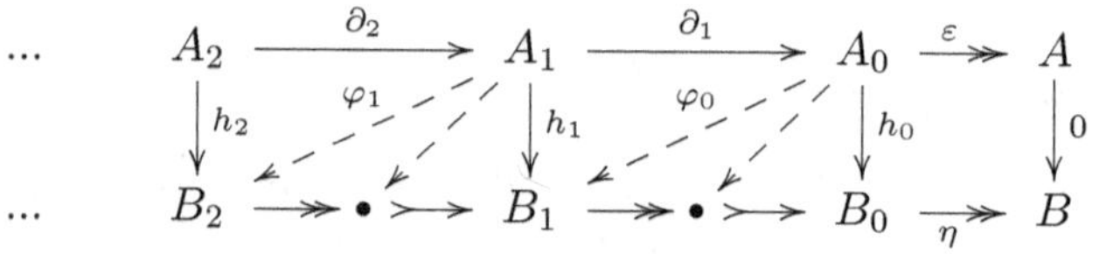

$$
\begin{array}{ccccccccc}
\cdots & A_2 & \xrightarrow{\;\partial_2\;} & A_1 & \xrightarrow{\;\partial_1\;} & A_0 & \xrightarrow{\;\varepsilon\;} & A \\
& \Big\downarrow{\scriptstyle h_2}\;{\scriptstyle \varphi_1} & & \Big\downarrow{\scriptstyle h_1}\;{\scriptstyle \varphi_0} & & \Big\downarrow{\scriptstyle h_0} & & \Big\downarrow{\scriptstyle 0} \\
\cdots & B_2 & \longrightarrow\bullet\rightarrowtail & B_1 & \longrightarrow\bullet\rightarrowtail & B_0 & \xrightarrow{\;\eta\;} & B
\end{array}
\tag{5.9}
$$

The homomorphism $h_0\colon A_0 \to B_0$ gives $\eta h_0 = 0$, and can be lifted to $\operatorname{Im}\partial_1 = \operatorname{Ker}\eta$, and then to $\varphi_0\colon A_0 \to B_1$ (because A_0 is projective), giving $\partial_1\varphi_0 = h_0$. Now $h_1 - \varphi_0\partial_1\colon A_1 \to B_1$ is annihilated by ∂_1 (because $\partial_1 h_1 - \partial_1\varphi_0\partial_1 = \partial_1 h_1 - h_0\partial_1 = 0$) and can similarly be lifted, first to $\operatorname{Im}\partial_2 = \operatorname{Ker}\partial_1$ and then to $\varphi_1\colon A_1 \to B_2$, giving $\partial_2\varphi_1 = h_1 - \varphi_0\partial_1$. And so on.

(d) Take two liftings $f_+\colon A_+ \to B_+$ and $g_+\colon B_+ \to A_+$ of $\operatorname{id}A$. Then $g_+f_+\colon A_+ \to A_+$ and $\operatorname{id}A_+$ lift $\operatorname{id}A$ and are homotopic. Similarly, $f_+g_+ \simeq \operatorname{id}B_+$.

(e) Both morphisms h_+ and $g_+f_+\colon A_+ \to C_+$ lift $h = gf$.

5.1.3 Left derived functors

We are now ready to define the *left derived functors* $L_n F = F_n\colon R\,\mathsf{Mod} \to S\,\mathsf{Mod}$ of the right exact functor F (for $n \geqslant 0$).

For every R-module A we choose a projective resolution $\varepsilon\colon A_+ \to A$. Using the obvious extension $F\colon \mathrm{Ch}_+ R\,\mathsf{Mod} \to \mathrm{Ch}_+ S\,\mathsf{Mod}$ of the additive functor F to chain complexes (see 4.2.3(a)), we define

$$
F_n(A) = H_n(FA_+).
\tag{5.10}
$$

For an R-homomorphism $f\colon A \to B$, there exists a lifting $f_+\colon A_+ \to B_+$ of f with respect to the chosen projective resolutions $\varepsilon\colon A_+ \to A$ and $\eta\colon B_+ \to B$, and all of them give the same homomorphism (by Exercises 5.1.2(b), (c))

$$F_n(f) = H_n(Ff_+)\colon F_n(A) \to F_n(B). \tag{5.11}$$

Exercises and complements. (a) The mappings F_n defined above are additive functors.

(b) The functor $F_0(A) = H_0(FA_+)$ is canonically isomorphic to the original functor F, and we shall identify them.

More generally, one can define as above the left derived functors of any additive functor; then the sequence (5.4) would end with the derived F_0, instead of the original F.

(c) Another choice of projective resolutions gives a sequence of functors F'_n linked to the previous ones by natural isomorphisms $F_n \to F'_n$.

(d) Without using the axiom of choice, one can define the functors F_n using the standard free resolutions of 5.1.2(a).

5.1.4 Theorem (Lifting short exact sequences)

A short exact sequence of R-modules $(m, p)\colon A \rightarrowtail B \twoheadrightarrow C$ has a projective resolution. This means a componentwise split exact sequence of chain complexes $(m_+, p_+)\colon A_+ \rightarrowtail B_+ \twoheadrightarrow C_+$ between projective resolutions of $A, B,$ and C.

One can start with two arbitrary projective resolutions $\varepsilon\colon A_+ \twoheadrightarrow A$ and $\vartheta_+\colon C_+ \twoheadrightarrow C$, and complete the rest

$$
\begin{array}{ccccccc}
\dots\, A_2 & \xrightarrow{\;\partial_2\;} & A_1 & \xrightarrow{\;\partial_1\;} & A_0 & \xrightarrow{\;\varepsilon\;} & A \\
\downarrow{\scriptstyle m_2} & & \downarrow{\scriptstyle m_1} & & \downarrow{\scriptstyle m_0} & & \downarrow{\scriptstyle m} \\
\dots\, B_2 & \dashrightarrow & B_1 & \dashrightarrow & B_0 & \dashrightarrow & B \\
\downarrow{\scriptstyle p_2} & & \downarrow{\scriptstyle p_1} & & \downarrow{\scriptstyle p_0} & & \downarrow{\scriptstyle p} \\
\dots\, C_2 & \xrightarrow[\;\partial_2\;]{} & C_1 & \xrightarrow[\;\partial_1\;]{} & C_0 & \xrightarrow[\;\vartheta\;]{} & C
\end{array}
\tag{5.12}
$$

Proof (a) For every $n \geqslant 0$, we take $B_n = A_n \oplus C_n$ (a projective module, by Exercise 4.2.7(b)) and we let $m_n\colon A_n \to B_n$ be the first injection and $p_n\colon B_n \to C_n$ the second projection.

We have thus a commutative solid diagram

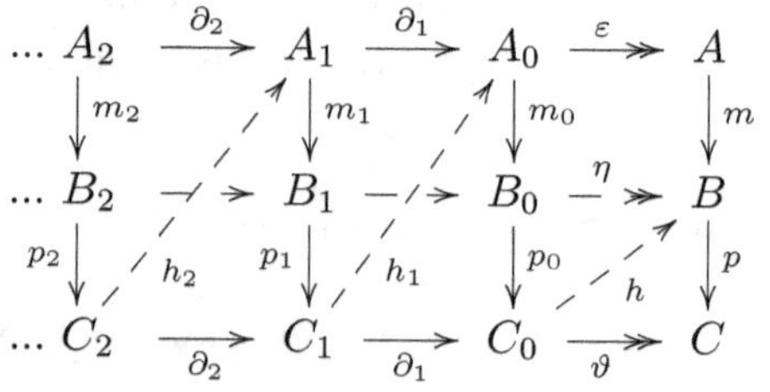

$$
\begin{array}{ccccccc}
\cdots A_2 & \xrightarrow{\partial_2} & A_1 & \xrightarrow{\partial_1} & A_0 & \xrightarrow{\varepsilon} & A \\
\downarrow{m_2} & & \downarrow{m_1} & & \downarrow{m_0} & & \downarrow{m} \\
\cdots B_2 & \rightarrow & B_1 & \rightarrow & B_0 & \xrightarrow{\eta} & B \\
\downarrow{p_2} & {}^{h_2} & \downarrow{p_1} & {}^{h_1} & \downarrow{p_0} & {}^{h} & \downarrow{p} \\
\cdots C_2 & \xrightarrow{\partial_2} & C_1 & \xrightarrow{\partial_1} & C_0 & \xrightarrow{\vartheta} & C
\end{array}
\tag{5.13}
$$

where the upper and lower rows are projective resolutions. All the columns
are split exact sequences, and we still have to define the homomorphisms
of the middle row.

(b) We define the homomorphism

$$\eta \colon A_0 \oplus C_0 \to B, \qquad \eta(a,c) = m\varepsilon(a) + h(c),$$

for some homomorphism $h \colon C_0 \to B$ such that $ph = \vartheta$ (C_0 is projective).
This makes the two rightmost squares commute, in diagram (5.13):

$$\eta m_0(a) = \eta(a,0) = m\varepsilon(a), \quad p\eta(a,c) = pm\varepsilon(a) + ph(c) = \vartheta p_0(a,c).$$

(d) Now we define

$$\partial_1 \colon A_1 \oplus C_1 \to A_0 \oplus C_0, \quad \partial_1(a,c) = (\partial_1(a) + h_1(c),\, \partial_1(c)),$$

for some $h_1 \colon C_1 \to A_0$ such that $m\varepsilon h_1 = -h\partial_1 \colon C_1 \to A$.
   (Indeed $-h\partial_1 \colon C_1 \to A$ can first be lifted to $A = \operatorname{Ker} p$, by exactness of
the rightmost column, and then to A_0, because ε is epi and C_1 is projective.)
   This makes two more commutative squares:

$$\partial_1 m_1(a) = (\partial_1(a),0) = m_0 \partial_1(a), \quad p_0 \partial_1(a,c) = \partial_1(c) = \partial_1 p_1(a,c).$$

Moreover $\operatorname{Im} \partial_1 = \operatorname{Ker} \eta$. First

$$\eta \partial_1(a,c) = \eta(\partial_1(a) + h_1(c),\, \partial_1(c)) = m\varepsilon \partial_1(a) + m\varepsilon h_1(c) + h\partial_1(c) = 0.$$

Secondly, take an element $(a,c) \in \operatorname{Ker} \eta$: thus $c \in \operatorname{Ker} \vartheta \colon C_0 \to C$
and there is some $c' \in C_1$ such that $\partial_1(c') = c$. Now $m\varepsilon(a) = -h(c) =
-h\partial_1(c') = m\varepsilon h_1(c')$, and there is some $a' \in A_1$ such that $\partial_1(a') = a -
h_1(c')$. Finally we have found a pair $(a',c') \in A_1 \oplus C_1$ such that:

$$\partial_1(a',c') = (\partial_1(a') + h_1(c'),\, \partial_1(c')) = (a,c).$$

(e) In the same way, there is some $h_2 \colon C_2 \to A_1$ such that $m_0 \partial_1 h_2 =
-h_1 \partial_2 \colon C_2 \to A_0$; we let

$$\partial_2 \colon A_2 \oplus C_2 \to A_1 \oplus C_1, \quad \partial_2(a,c) = (\partial_2(a) + h_2(c),\, \partial_2(c)),$$

and we go on as above. $\qquad \square$

5.1.5 The sequence of derived functors

Let $(m, p)\colon A \rightarrowtail B \twoheadrightarrow C$ be a short exact sequence of R-modules, with a (componentwise split) projective resolution $(m_+, p_+)\colon A_+ \rightarrowtail B_+ \twoheadrightarrow C_+$.

The right exact functor $F\colon R\,\mathsf{Mod} \to S\,\mathsf{Mod}$ gives a short exact sequence of chain complexes, which splits componentwise

$$(Fm_+, Fp_+)\colon FA_+ \rightarrowtail FB_+ \twoheadrightarrow FC_+. \tag{5.14}$$

We have thus proved our goal: the short exact sequence (m, p) gives raise to the long exact homology sequence (5.4); this is the *exact sequence of left derived functors* of F, with connecting morphisms

$$\partial_n\colon F_n(C) \to F_{n-1}(A) \qquad (n \geqslant 1). \tag{5.15}$$

The whole process is natural: a morphism $(u, v, w)\colon (m, p) \to (m', p')$ of short exact sequences (as in (2.70)) has a lifting

$$(u_+, v_+, w_+)\colon (m_+, p_+) \dashrightarrow (m'_+, p'_+)$$

which is a *weak morphism* of short exact sequences of chain complexes: the two squares below commute up to chain homotopy (by 5.1.2(e))

$$\begin{array}{ccccc}
A_+ & \xrightarrow{\ m_+\ } & B_+ & \xrightarrow{\ p_+\ } & C_+ \\
{\scriptstyle u_+}\downarrow & \simeq & \downarrow{\scriptstyle v_+} & \simeq & \downarrow{\scriptstyle w_+} \\
A'_+ & \xrightarrow[\ m'_+\ ]{} & B'_+ & \xrightarrow[\ p'_+\ ]{} & C'_+
\end{array} \tag{5.16}$$

Applying the functor F and chain homology we get a morphism of exact sequences of S-modules, with components

$$\begin{array}{ccccccccc}
\ldots F_n(A) & \xrightarrow{\ F_n m\ } & F_n(B) & \xrightarrow{\ F_n p\ } & F_n(C) & \xrightarrow{\ \partial_n\ } & F_{n-1}(A) \ldots \\
{\scriptstyle F_n(u)}\downarrow & & {\scriptstyle F_n(v)}\downarrow & & {\scriptstyle F_n(w)}\downarrow & & {\scriptstyle F_{n-1}(u)}\downarrow \\
\ldots F_n(A') & \xrightarrow[\ F_n m'\ ]{} & F_n(B') & \xrightarrow[\ F_n p'\ ]{} & F_n(C') & \xrightarrow[\ \partial_n\ ]{} & F_{n-1}(A') \ldots
\end{array} \tag{5.17}$$

5.1.6 Exercises and complements

(a) A right exact functor $F\colon R\,\mathsf{Mod} \to S\,\mathsf{Mod}$ is exact if and only if all its left derived functors (in positive degree) are trivial ($F_n = 0$, for $n > 0$), if and only if $F_1 = 0$.

(b) For a right exact functor $F\colon R\,\mathsf{Mod} \to S\,\mathsf{Mod}$, the derived functors F_n with $n > 0$ vanish on all projective R-modules. In particular, every projective module is flat.

*(c) The theory of left satellites, developed in [CE], characterises the left

derived functor $L_n F$ as the n-th left satellite $S_n F$. It is important to note that $S_{m+n} F = S_m(S_n F)$.

5.1.7 *The case of abelian groups*

Letting $R = \mathbb{Z}$, an abelian group A has always a *short* free resolution

$$0 \; \to \; A_1 \; \xrightarrow{\partial} \; A_0 \; \xrightarrow{\varepsilon} \; A \; \to \; 0 \tag{5.18}$$

where $\varepsilon \colon A_0 \to A$ is the canonical epimorphism defined on the free abelian group $\mathbb{Z}|A|$ and $\partial \colon A_1 \to A$ is the kernel of ε (recalling that a subgroup of a free abelian group is free).

Therefore, for a right exact functor $F \colon \mathsf{Ab} \to S\,\mathsf{Mod}$, all left derived functors vanish in degree $n \geqslant 2$, and there is a single left derived functor $LF = F_1 \colon \mathsf{Ab} \to S\,\mathsf{Mod}$ of interest:

$$F_1(A) = H_1(A_+) = \mathrm{Ker}\,(F\partial \colon FA_1 \to FA_0). \tag{5.19}$$

A short exact sequence $(m, p) \colon A \rightarrowtail B \twoheadrightarrow C$ gives an exact sequence

$$0 \to F_1 A \xrightarrow{F_1 m} F_1 B \xrightarrow{F_1 p} F_1 C \xrightarrow{\partial} FA \xrightarrow{Fm} FB \xrightarrow{Fp} FC \to 0 \tag{5.20}$$

showing that F_1 is left exact.

On the category Ab one can write the theory of left derived functors using short projective resolutions, which vanish in degree $\geqslant 2$. Yet, this reduced version can mask the nature and power of the process, resulting less clear than the complete picture.

5.1.8 *Right derived functors*

Loosely speaking, we get the dual notion of right derived functors of a left exact functor, reversing all arrows and replacing projective objects with the injective ones.

   This duality has a precise sense if we work in the selfdual context of abelian categories.

Concretely, we are given a right exact additive functor

$$F \colon R\,\mathsf{Mod} \to S\,\mathsf{Mod}, \tag{5.21}$$

and we want to construct the sequence of its right derived functors

$$R^n F = F^n \colon R\,\mathsf{Mod} \to S\,\mathsf{Mod} \qquad (n \geqslant 0), \tag{5.22}$$

starting with $F^0 = F$.

Again, this sequence corrects the failure of exactness of F, in the sense that every short exact sequence of R-modules $(m, p)\colon A \rightarrowtail B \twoheadrightarrow C$ has an associated *exact sequence of right derived functors*

$$
\begin{array}{ccccc}
0 & \longrightarrow & F(A) \xrightarrow{Fm} & F(B) \xrightarrow{Fp} & F(C) \\
& & & \xleftarrow{\ \ d^0\ \ } & \\
& F^1(A) \xrightarrow{F^1 m} & F^1(B) \xrightarrow{F^1 p} & F^1(C) & \\
& & & \xleftarrow{\ \ d^1\ \ } & \\
& F^2(A) \xrightarrow{\ \ \ \ \ \ } & \cdots & &
\end{array}
\tag{5.23}
$$

The *connecting morphisms* $d^n\colon F^n(C) \to F^{n+1}(A)$ are the components of a natural transformation

$$
d^n\colon F^n P''' \to F^{n+1} P'\colon \mathrm{Sh}R\,\mathsf{Mod} \to S\,\mathsf{Mod}, \qquad (n \geqslant 0),
\tag{5.24}
$$

forming an exact connected sequence of functors $((F^n), (d^n))$ (with a connecting morphism of degree 1).

The theory is developed in a similar way to what we have done above, using for every R-module A an *injective resolution*. This is a cochain complex $A^+ = ((A^n), (d^n))$ equipped with an augmentation $\varepsilon\colon A \rightarrowtail A^0$, such that:

(i) all components A^n are injective R-modules (defined in 4.2.6(c)),

(ii) the following sequence is exact

$$
0 \to A \xrightarrow{\varepsilon} A^0 \xrightarrow{d^0} A^1 \xrightarrow{d^1} A^2 \to \cdots
\tag{5.25}
$$

The resolution can be written in the form $\varepsilon\colon A \rightarrowtail A^+$, as a monomorphism of cochain complexes.

Showing that any R-module has an injective resolution is less easy than in the projective case. The basic step, every R-module is a submodule of an injective one, is proved in [M1], Theorem III.7.4.

Exercises and complements. (a) A left exact functor $F\colon R\,\mathsf{Mod} \to S\,\mathsf{Mod}$ is exact if and only if all its right derived functors (in positive degree) are trivial ($F^n = 0$, for $n > 0$), if and only if $F^1 = 0$. *Hints:* the solution is similar to that of Exercise 5.1.6(a).

*(b) The theory of right satellites, developed in [CE], characterises the right derived functor $R^n F$ as the n-th right satellite $S^n F$. Again, $S^{m+n} F = S^m(S^n F)$.

5.1.9 *The case of abelian groups*

We have recalled, in 4.2.9(d), that an abelian group is injective if and only if it is divisible. But a quotient of a divisible abelian group is plainly divisible. It follows that in Ab every object A has a *short* injective resolution $\varepsilon\colon A \rightarrowtail A^+$ with $A^n = 0$ for $n \geqslant 2$

$$0 \to A \xrightarrow{\varepsilon} A^0 \xrightarrow{d} A^1 \to 0. \tag{5.26}$$

For a left exact functor $F\colon \mathsf{Ab} \to S\,\mathsf{Mod}$, all right derived functors vanish in degree $\geqslant 2$, and we only consider the right derived functor $RF = F^1\colon \mathsf{Ab} \to S\,\mathsf{Mod}$.

A short exact sequence $(m,p)\colon A \rightarrowtail B \twoheadrightarrow C$ gives an exact sequence

$$0 \to FA \xrightarrow{Fm} FB \xrightarrow{Fp} FC \xrightarrow{d} F^1A \xrightarrow{F^1m} F^1B \xrightarrow{F^1p} F^1C \to 0 \tag{5.27}$$

showing that F^1 is right exact.

5.2 Torsion product and universal coefficients

It is now convenient to work in Ab. We introduce the torsion product $\mathrm{Tor}(A,B)$ as the left derived functor of $A \otimes B$, working equivalently on the variable A or B. A concrete construction of $\mathrm{Tor}(A,B)$ will be briefly presented in 5.2.8.

G is always an abelian group, used as a coefficient group for homology.

Literature. For R-modules, the torsion product $\mathrm{Tor}_n^R(A,B)$ is the n-th left derived functor of $A \otimes B$, on one of the variables. The theory, much more complex than in Ab, is fully developed in [CE].

5.2.1 The torsion product

We begin by fixing the second variable. The functor $- \otimes B\colon \mathsf{Ab} \to \mathsf{Ab}$ is right exact, and has a left derived functor, *the* Tor *product*

$$\mathrm{Tor}(-,B)\colon \mathsf{Ab} \to \mathsf{Ab}, \tag{5.28}$$

which is left exact. For a free resolution $(\partial,\varepsilon)\colon A_1 \rightarrowtail A_0 \twoheadrightarrow A$, it is computed as

$$\mathrm{Tor}(A,B) = H_1(A_+ \otimes B) = \mathrm{Ker}\,(\partial \otimes B\colon A_1 \otimes B \to A_0 \otimes B). \tag{5.29}$$

A short exact sequence $A' \rightarrowtail A \twoheadrightarrow A''$ of abelian groups gives an exact sequence

$$0 \to \mathrm{Tor}(A',B) \to \mathrm{Tor}(A,B) \to \mathrm{Tor}(A'',B)$$
$$\xrightarrow{\partial} A' \otimes B \to A \otimes B \to A'' \otimes B \to 0 \tag{5.30}$$

which is natural on ShAb. The Tor product is also denoted as $A * B$.

We prove below that, if we work on the left variable, we get the same result; as a consequence, $\mathrm{Tor}(A, B) \cong \mathrm{Tor}(B, A)$.

Exercises and complements. (a) The abelian group B is flat if and only if $\mathrm{Tor}(-, B) = 0$. This is certainly the case if B is a free abelian group.

(More precisely, we shall see in 5.2.8(a) that B is flat if and only if it is torsion free.)

(b) Prove that $\mathrm{Tor}(\mathbb{Z}_m, \mathbb{Z}_n) \cong \mathbb{Z}_d$, where $d = \gcd(m, n)$.

5.2.2 Proposition

Fixing the first variable, we get a functor $\underline{\mathrm{Tor}}(A, -)\colon \mathsf{Ab} \to \mathsf{Ab}$ left derived of $A \otimes -\colon \mathsf{Ab} \to \mathsf{Ab}$.

Then there is a natural isomorphism $\mathrm{Tor}(A, B) \cong \underline{\mathrm{Tor}}(A, B)$, for all pairs (A, B) of abelian groups.

Proof We choose two free resolutions $A_1 \rightarrowtail A_0 \twoheadrightarrow A$ and $B_1 \rightarrowtail B_0 \twoheadrightarrow B$. This gives a commutative diagram, where

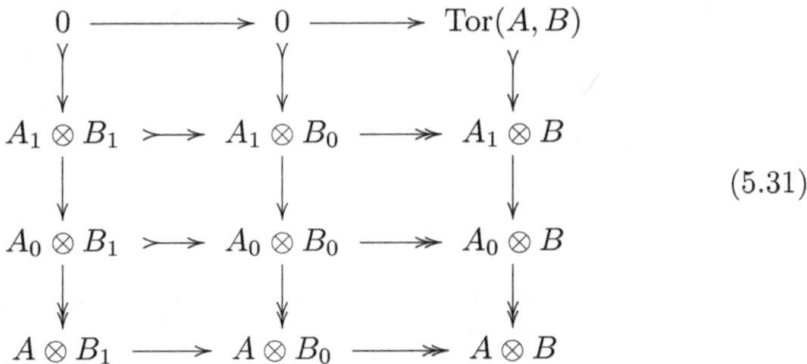

$$\tag{5.31}$$

- the two middle rows are short exact, because A_1 and A_0 are free,

- the right column is exact, by definition of $\mathrm{Tor}(A, B)$,

- the other two columns are exact, because B_1 and B_0 are free.

Applying the Snake Lemma 2.7.1 to the two middle rows we get a natural exact sequence linking the kernels and the cokernels of the three vertical arrows

$$0 \to \mathrm{Tor}(A, B) \to A \otimes B_1 \to A \otimes B_0 \to A \otimes B \to 0.$$

This shows that the connecting homomorphism $\mathrm{Tor}(A, B) \to A \otimes B_1$ is a kernel of $A \otimes B_1 \to A \otimes B_0$; but this 'is' $\underline{\mathrm{Tor}}(A, B)$, computed by a free resolution of B. $\qquad\square$

5.2.3 Corollary

(a) For all pairs (A, B) of abelian groups, there is a natural isomorphism $\mathrm{Tor}(A, B) \cong \mathrm{Tor}(B, A)$.

(b) Every subgroup of a flat abelian group is flat.

Note. Point (b) is proved here without using the characterisation of flat abelian groups as the torsion-free ones.

Proof (a) $\mathrm{Tor}(A, B)$ is the left derived functor of $- \otimes B$ computed in A. By Proposition 5.2.2, it is also the left derived functor of $A \otimes -$ computed in B. By the commutativity of the tensor product, $\mathrm{Tor}(A, B)$ is also the left derived functor of $- \otimes A$ computed in B, that is $\mathrm{Tor}(B, A)$, by definition.

(b) If H is a subgroup of the flat abelian group A, the short exact sequence $H \rightarrowtail A \twoheadrightarrow A/H$ gives, for every abelian group B, an exact sequence $0 \to \mathrm{Tor}(H, B) \to \mathrm{Tor}(A, B) = 0$, showing that $\mathrm{Tor}(H, -) = 0$. By (a), we also have $\mathrm{Tor}(-, H) = 0$. $\qquad\square$

5.2.4 A natural morphism

Let A be a chain complex of abelian groups. We begin to explore the homology group $H_n(A \otimes G)$, introducing a canonical morphism

$$\alpha_n \colon H_n(A) \otimes G \to H_n(A \otimes G), \qquad \alpha_n([z] \otimes g) = [z \otimes g]. \tag{5.32}$$

This is well defined. If z is an n-cycle in A, $\partial_n(z \otimes g) = \partial_n z \otimes g = 0$. If z is an n-boundary, $z \otimes g$ is also. Moreover the expression $z \otimes g$ is linear in each variable.

We have defined a natural transformation

$$\alpha_n \colon H_n(-) \otimes - \ \to \ H_n(- \otimes -) \colon \mathrm{Ch}_+\mathsf{Ab} \times \mathsf{Ab} \to \mathsf{Ab}. \tag{5.33}$$

Remarks. (a) If G is a flat abelian group, then $\alpha_n \colon H_n(A) \otimes G \to H_n(A \otimes G)$ is an isomorphism.

In fact, the functor $- \otimes G \colon \mathsf{Ab} \to \mathsf{Ab}$ is exact, and 'preserves homology', as we have seen in 4.2.3(b).

(b) This becomes obvious for a free abelian group $G \cong \bigoplus_{i \in I} \mathbb{Z}$.

Then $A \otimes G \cong \bigoplus_{i \in I} A$ is a direct sum of chain complexes (see 2.1.8), and we have seen that its n-th homology group is canonically isomorphic to $\bigoplus_{i \in I} H_n(A) \cong H_n(A) \otimes G$.

5.2.5 Universal Coefficient Theorem (for chain homology)

For every chain complex A of flat *abelian groups, and every abelian group G, there is a split exact sequence*

$$H_n(A) \otimes G \xrightarrow{\ \alpha_n\ } H_n(A \otimes G) \xrightarrow{\ \beta_n\ } \mathrm{Tor}(H_{n-1}(A), G) \qquad (5.34)$$

which is natural on morphisms $f \colon A \to B$ of chain complexes, and morphisms $h \colon G \to G'$ of abelian groups.

Note. The ensuing isomorphism

$$H_n(A \otimes G) \cong (H_n(A) \otimes G) \oplus (\mathrm{Tor}(H_{n-1}(A), G)) \qquad (5.35)$$

is *not* natural on chain morphisms, and should be used with prudence (see 5.2.7(b)).

Proof It is a particular case of the Künneth Theorem 5.4.4, replacing the chain complex B with $J(G)$. $\square$

5.2.6 Corollary (universal coefficients for singular homology)

For every relative pair (X, A) of spaces and every abelian group G, there is a split exact sequence

$$H_n(X, A) \otimes G \xrightarrow{\ \alpha_n\ } H_n(X, A; G) \xrightarrow{\ \beta_n\ } \mathrm{Tor}(H_{n-1}(X, A), G) \qquad (5.36)$$

which is natural on relative maps $f \colon (X, A) \to (Y, B)$ and morphisms $h \colon G \to G'$ of abelian groups.

Proof A straightforward consequence. $\square$

5.2.7 Comments and exercises

(a) If the relative pairs (X, A) and (Y, B) have isomorphic homology groups with integral coefficients, the same holds with coefficients in any abelian group.

(b) On the other hand, given two maps $f, g \colon (X, A) \to (Y, B)$ which induce the same homomorphisms $f_{*n} = g_{*n} \colon H_n(X, A) \to H_n(Y, B)$, we cannot deduce that the same holds with any coefficient group: the isomorphism (5.35) is not natural.

In fact, we have already seen, in Exercise 4.3.8(c), that the projection $p \colon \mathbb{P}^2 \to \mathbb{S}^2$ and any constant map $\mathbb{P}^2 \to \mathbb{S}^2$ have the same (trivial) homology homomorphisms with integral coefficients, but different ones with coefficients in $\mathbb{Z}_2$.

(c) If the abelian group G is flat, we have a natural isomorphism

$$\alpha_n \colon H_n(X, A) \otimes G \to H_n(X, A; G), \qquad \alpha_n([z] \otimes g) = [z \otimes g]. \qquad (5.37)$$

(d) Using the Universal Coefficient Theorem 5.2.6, prove that the torus and the Klein bottle have the following homology groups with coefficients in $\mathbb{Z}_2$, and trivial higher groups:

$$H_i(\mathbb{T}; \mathbb{Z}_2) \cong H_i(\mathbb{K}; \mathbb{Z}_2) \cong \mathbb{Z}_2 \qquad \text{for } i = 0, 2,$$
$$H_1(\mathbb{T}; \mathbb{Z}_2) \cong H_1(\mathbb{K}; \mathbb{Z}_2) \cong \mathbb{Z}_2 \oplus \mathbb{Z}_2. \qquad (5.38)$$

(e) The projective space $\mathbb{P}^n$ has the following homology groups with coefficients in $\mathbb{Z}_2$, and trivial higher groups

$$H_i(\mathbb{P}^n; \mathbb{Z}_2) \cong \mathbb{Z}_2 \qquad \text{for } 0 \leqslant i \leqslant n. \qquad (5.39)$$

Other calculations can be worked out, using the homology groups determined in Chapter 2.

5.2.8 *Complements*

We end with a concrete construction of the group $\mathrm{Tor}(A, B)$, based on the 'common torsion' of the abelian groups A and B. The proof can be found in [M1], Section V.6.

The construction is similar to that of the tensor product, in 4.1.2. We let

$$\mathrm{Tor}(A, B) = T(A, B)/H(A, B), \qquad (5.40)$$

where $T(A, B)$ is the free abelian group generated by the following triples

$$(a, m, b) \qquad\qquad (a \in A, \ m \in \mathbb{Z}, \ b \in B; \ ma = 0 = mb), \qquad (5.41)$$

and $H(A, B)$ is the subgroup generated by the elements of the following kinds

$$
\begin{aligned}
&(a + a', m, b) - (a, m, b) - (a', m, b) &&\text{(for } ma = ma' = mb = 0\text{)}, \\
&(a, m, b + b') - (a, m, b) - (a, m, b') &&\text{(for } ma = mb = mb' = 0\text{)}, \\
&(a, mn, b) - (na, m, b) &&\text{(for } mna = mb = 0\text{)}, \\
&(a, mn, b) - (a, m, nb) &&\text{(for } ma = mnb = 0\text{)}.
\end{aligned}
$$

From this construction one can deduce two important facts.

(a) A torsion-free abelian group is flat (and conversely, by 4.2.8(c)).

If B is torsion free, this construction shows that $\mathrm{Tor}(A, B) = 0$ for every A. Indeed, consider a generator $(a, m, b) \in T(A, B)$. If $m \neq 0$, then $b = 0$ and $[a, m, 0] = 0$, by linearity in the last variable. Otherwise $[a, 0, b] = [0a, 0, b] = 0$.

(b) $\mathrm{Tor}(A, B)$ is always a torsion abelian group.

In fact, any generator $[a, m, b]$, with $ma = 0 = mb$, is annihilated by m:

$$m[a, m, b] = [ma, m, b] = [0, m, b] = 0.$$

5.3 Ext functor and universal coefficients

We go on working in Ab. We introduce the functor $\mathrm{Ext}(A, B)$ as the derived functor of $\mathrm{Hom}(A, B)$, working equivalently on the covariant variable B or the contravariant variable A.

G is always an abelian group, used as a coefficient group for cohomology.

Literature. For R-modules, the derived functors $\mathrm{Ext}_R^n(A, B)$ are dealt with in [CE] and most books on Homological Algebra.

5.3.1 The Ext functor

Fixing the first variable, the covariant functor $\mathrm{Hom}(A, -) \colon \mathsf{Ab} \to \mathsf{Ab}$ is left exact. It has a right derived functor

$$\mathrm{Ext}(A, -) \colon \mathsf{Ab} \to \mathsf{Ab}, \tag{5.42}$$

which is right exact. For an injective resolution $(\varepsilon, d) \colon B \rightarrowtail B^0 \twoheadrightarrow B^1$, it is computed as

$$\begin{aligned}
\mathrm{Ext}(A, B) &= H^1(\mathrm{Hom}(A, B^+)) \\
&= \mathrm{Cok}\,(\mathrm{Hom}(A, d) \colon \mathrm{Hom}(A, B^0) \to \mathrm{Hom}(A, B^1)).
\end{aligned} \tag{5.43}$$

For every short exact sequence $(m, p) \colon B' \rightarrowtail B \twoheadrightarrow B''$ of abelian groups, we have an exact sequence

$$\begin{aligned}
0 \;&\to\; \mathrm{Hom}(A, B') \;\to\; \mathrm{Hom}(A, B) \;\to\; \mathrm{Hom}(A, B'') \\
&\xrightarrow{d} \; \mathrm{Ext}(A, B') \;\to\; \mathrm{Ext}(A, B) \;\to\; \mathrm{Ext}(A, B'') \;\to\; 0
\end{aligned} \tag{5.44}$$

which is natural on Ab (contravariantly) and ShAb (covariantly).

Exercises and complements. (a) A is a free abelian group if and only if $\mathrm{Ext}(A, -) = 0$.

(b) Prove that $\mathrm{Ext}(\mathbb{Z}_m, \mathbb{Z}_n) \cong \mathbb{Z}_d$, where $d = \gcd(m, n)$.

5.3.2 Proposition

Fixing the second variable B, every short exact sequence $A' \rightarrowtail A \twoheadrightarrow A''$ of abelian groups gives an exact sequence

$$
0 \to \operatorname{Hom}(A'', B) \to \operatorname{Hom}(A, B) \to \operatorname{Hom}(A', B)
$$
$$
\xrightarrow{d} \operatorname{Ext}(A'', B) \to \operatorname{Ext}(A, B) \to \operatorname{Ext}(A', B) \to 0 \tag{5.45}
$$

natural on ShAb *(contravariantly) and* Ab *(covariantly).*

Proof It is again an application of the Snake Lemma 2.7.1. Taking an injective resolution $(\varepsilon, d)\colon B \rightarrowtail B^0 \twoheadrightarrow B^1$ of B, we have a commutative diagram where

$$
\begin{array}{ccccc}
\operatorname{Hom}(A'', B) & \rightarrowtail & \operatorname{Hom}(A, B) & \longrightarrow & \operatorname{Hom}(A', B) \\
\downarrow & & \downarrow & & \downarrow \\
\operatorname{Hom}(A'', B^0) & \rightarrowtail & \operatorname{Hom}(A, B^0) & \twoheadrightarrow & \operatorname{Hom}(A', B^0) \\
\downarrow & & \downarrow & & \downarrow \\
\operatorname{Hom}(A'', B^1) & \rightarrowtail & \operatorname{Hom}(A, B^1) & \twoheadrightarrow & \operatorname{Hom}(A', B^1) \\
\downarrow & & \downarrow & & \downarrow \\
\operatorname{Ext}(A'', B) & \longrightarrow & \operatorname{Ext}(A, B) & \twoheadrightarrow & \operatorname{Ext}(A', B)
\end{array}
\tag{5.46}
$$

- the columns are exact, by definition of Ext,
- the two middle rows are short exact, because B^0 and B^1 are injective.

   This gives the exact sequence (5.45), with its naturality properties. $\square$

5.3.3 Corollary

Defining $\operatorname{Ext}(A, B)$ *as a derived functor of the left exact functor*

$$
\operatorname{Hom}(-, B)\colon \mathsf{Ab}^{\mathrm{op}} \to \mathsf{Ab},
$$

we get the same result, up to natural isomorphism on $\mathsf{Ab}^{\mathrm{op}} \times \mathsf{Ab}$.

Proof We apply Proposition 5.3.2 to a free resolution $A_1 \rightarrowtail A_0 \twoheadrightarrow A$ of the abelian group A. $\operatorname{Ext}(A_0, B)$ and $\operatorname{Ext}(A_1, B)$ vanish, because A_0 and A_1 are projective, and the exact sequence

$$
0 \to \operatorname{Hom}(A, B) \to \operatorname{Hom}(A_0, B) \to \operatorname{Hom}(A_1, B) \to \operatorname{Ext}(A, B) \to 0
$$

shows that $\operatorname{Ext}(A, B)$ is the cokernel of $m^\sharp\colon \operatorname{Hom}(A_0, B) \to \operatorname{Hom}(A_1, B)$. But this is precisely the alternative form $\underline{\operatorname{Ext}}(A, B)$, computed on the projective resolution of A. $\square$

5.3.4 A natural morphism

Let A be a chain complex. We begin to explore the homology group $H^n(\mathrm{Hom}(A, G))$ by introducing a canonical morphism

$$\alpha^n \colon H^n(\mathrm{Hom}(A, G)) \to \mathrm{Hom}(H_n(A), G), \qquad \alpha^n[\lambda][z] = \lambda(z). \tag{5.47}$$

Here $z \in Z_n(A)$ and $\lambda \colon A_n \to G$ is a cocycle of $\mathrm{Hom}(A_n, G)$, which means that it vanishes on any boundary of A_n. Therefore $\lambda(z)$ only depends on the homology class $[z]$.

We have defined a natural transformation

$$\alpha^n \colon H^n(\mathrm{Hom}(-, -)) \to \mathrm{Hom}(H_n(-), -) \colon \mathrm{Ch}_+ \mathsf{Ab}^{\mathrm{op}} \times \mathsf{Ab} \to \mathsf{Ab}. \tag{5.48}$$

Remarks. (a) If G is an injective group (i.e. a divisible one), the homomorphism $\alpha^n \colon H^n(\mathrm{Hom}(A, G)) \to \mathrm{Hom}(H_n(A), G)$ is an isomorphism. In fact, the functor $\mathrm{Hom}(-, G) \colon \mathsf{Ab}^{\mathrm{op}} \to \mathsf{Ab}$ is exact, and 'preserves homology'.

5.3.5 Universal Coefficient Theorem (for chain cohomology)

For every chain complex A of free *abelian groups, and every abelian group G, there is a split exact sequence*

$$\mathrm{Ext}(H_{n-1}A, G) \overset{\beta^n}{\rightarrowtail} H^n(\mathrm{Hom}(A, G)) \overset{\alpha^n}{\twoheadrightarrow} \mathrm{Hom}(H_n A, G) \tag{5.49}$$

which is natural on morphisms $f \colon A \to B$ of chain complexes (contravariantly), and on morphisms $h \colon G \to G'$ of abelian groups (covariantly).

Note. Again, we have an isomorphism which is not natural on $\mathrm{Ch}_+\mathsf{Ab}$

$$H^n(\mathrm{Hom}(A, G)) \cong (\mathrm{Hom}(H_n(A), G)) \oplus (\mathrm{Ext}(H_{n-1}(A), G)). \tag{5.50}$$

Proof It is a particular case of the Künneth Theorem 5.4.5, replacing the cochain complex B with $J(G)$. (A direct proof can be found in [M1], Theorem III.4.1, or [Mas3], Theorem VII.4.3.) $\qquad\square$

5.3.6 Corollary (universal coefficients for singular cohomology)

For every relative pair (X, A) of spaces, and every abelian group G, there is a split exact sequence

$$\mathrm{Ext}(H_{n-1}(X, A), G) \overset{\beta^n}{\rightarrowtail} H^n(X, A; G) \overset{\alpha^n}{\twoheadrightarrow} \mathrm{Hom}(H_n(X, A), G) \tag{5.51}$$

which is natural on relative maps $f \colon (X, A) \to (Y, B)$ (contravariantly), and on morphisms $h \colon G \to G'$ of abelian groups (covariantly). $\qquad\square$

5.3.7 Betti numbers in cohomology

The abelian group $\mathbb{Q}$ is both torsion free and divisible. We have thus isomorphisms of vector spaces on $\mathbb{Q}$

$$\alpha_n \colon H_n(X) \otimes \mathbb{Q} \to H_n(X;\mathbb{Q}) \qquad \text{(by 5.2.4(a))}, \tag{5.52}$$

$$\alpha^n \colon H^n(X;\mathbb{Q}) \to \operatorname{Hom}(H_n(X),\mathbb{Q}) \qquad \text{(by 5.3.4(a))}. \tag{5.53}$$

If $H_n(X)$ is finitely generated, $H_n(X) \cong \mathbb{Z}^{\beta_n(X)} \oplus \operatorname{t}(H_n(X))$ and

$$H_n(X) \otimes \mathbb{Q} \cong \mathbb{Q}^{\beta_n(X)} \cong \operatorname{Hom}(H_n(X),\mathbb{Q}),$$
$$\beta_n(X) = \dim_{\mathbb{Q}} H_n(X;\mathbb{Q}) = \dim_{\mathbb{Q}} H^n(X;\mathbb{Q}). \tag{5.54}$$

The Betti numbers of X are thus equivalently determined by homology and cohomology.

5.3.8 *Manifolds and Poincaré duality

The homology and cohomology groups of orientable topological manifolds are closely related, by 'Poincaré duality'. Here we only hint at these celebrated results; an interested reader is referred to [Vi], Chapter 6, or [Mas3], Chapter IX, or [Ha], Section 3.3.

(a) The basic theorem is concerned with a compact orientable n-manifold M. In this case, all the homology and cohomology groups of M are finitely generated, and there is a canonical isomorphism

$$H^k(M) \to H_{n-k}(M), \qquad \text{for } 0 \leqslant k \leqslant n, \tag{5.55}$$

all the higher homology and cohomology groups being trivial.

One can see the proof in [Ha], Theorem 3.30.

(b) Recalling that $\beta_k(X) = \dim_{\mathbb{Q}} H_k(X;\mathbb{Q}) = \dim_{\mathbb{Q}} H^k(X;\mathbb{Q})$, a compact orientable n-manifold M has

$$\beta_k(M) = \beta_{n-k}(M), \qquad \text{for } 0 \leqslant k \leqslant n, \tag{5.56}$$

as we have already seen to happen for the n-sphere and the projective space $\mathbb{P}^n$ (in the orientable case, that is for n odd), and will see for the n-torus in Exercise 5.6.7(e).

(c) More generally, an orientable n-manifold M has a canonical isomorphism (for every abelian group G)

$$H_c^k(M;G) \to H_{n-k}(M;G), \tag{5.57}$$

where H_c^k denotes *singular cohomology with compact supports*. This can be found in [Mas3], Sections IX.3–4, or in [Ha], Section 3.3. Various cohomology and homology theories with compact supports are studied in [Mas2].

5.4 Tensor product of chain complexes

The Universal Coefficient Theorem, in 5.2.5, can be extended to the homology group $H_n(A \otimes B)$ of a tensor product of chain complexes, proving that it is determined by the homology groups $H_p(A)$ and $H_p(B)$, for $0 \leqslant p \leqslant n$.

(Again, the morphisms induced in homology are not determined in this way.)

This is made precise in the Künneth Theorem 5.4.4.

Literature. There are various versions of the Künneth theorem, under different hypotheses. Theorem V.10.4 in [M1] is the same as the present statement (up to allowing unbounded chain complexes). Theorem VI.3 in [CE] and Theorem VI.9.13 in [Do] are much more general.

5.4.1 Extending the tensor product, III

There is a tensor product $A \otimes B$ of chain complexes

$$(A \otimes B)_n = \oplus_{p+q=n} (A_p \otimes B_q),$$

$$\partial_n(a \otimes b) = \partial_p a \otimes b + (-1)^p a \otimes \partial_q b \qquad (a \in A_p,\ b \in B_q), \tag{5.58}$$

which extends the tensor product of abelian groups via the canonical embedding $J\colon \mathrm{Ab} \to \mathrm{Ch}_+\mathrm{Ab}$, in (2.18). (The Koszul sign rule applies, as in (3.94).)

The soundness of this definition is verified in the easy exercises below. The homology of $A \otimes B$ is determined by the Künneth Theorem 5.4.4.

Exercises and complements. (a) The differential ∂_n of $A \otimes B$ is well defined on canonical generators, by the formula above. Moreover, $\partial_{n-1}\partial_n = 0$.

(b) Up to canonical isomorphisms, the tensor product of chain complexes is commutative, associative and admits a unit: the chain complex $J(\mathbb{Z})$ associated to the additive group $\mathbb{Z}$.

5.4.2 The tensor functor

The tensor product $f \otimes g$ of two morphisms $f\colon A \to A'$ and $g\colon B \to B'$ of chain complexes has the following components

$$(f \otimes g)_n = \oplus_{p+q=n} (f_p \otimes g_q)\colon (A \otimes B)_n \to (A' \otimes B')_n,$$

$$(f_p \otimes g_q)(a \otimes b) = f_p(a) \otimes g_q(b), \tag{5.59}$$

for $a \in A_p,\ b \in B_q$. Commutativity with differentials is obvious.

This defines the tensor functor of chain complexes

$$- \otimes -\colon \mathrm{Ch}_+\mathrm{Ab} \times \mathrm{Ch}_+\mathrm{Ab} \to \mathrm{Ch}_+\mathrm{Ab}. \tag{5.60}$$

Exercises and complements. (a) A sequence of chain complexes

$$A \otimes B \xrightarrow{f \otimes g} A' \otimes B' \xrightarrow{f' \otimes g'} A'' \otimes B''$$

is exact if and only if every sequence $(f_p \otimes g_q, f'_p \otimes g'_q)$ is exact in Ab. In particular, the functor $- \otimes B \colon \mathsf{Ch}_+\mathsf{Ab} \to \mathsf{Ch}_+\mathsf{Ab}$ is exact if and only if all the components of B are flat.

(b) Let $(f, g) \colon A' \rightarrowtail A \twoheadrightarrow A''$ be a short exact sequence of chain complexes, and B a chain complex. If A'' has flat components, tensoring by B gives a short exact sequence $(f \otimes B, g \otimes B) \colon A' \otimes B \rightarrowtail A \otimes B \twoheadrightarrow A'' \otimes B$.

(c) *Remarks.* In this section we often use chain complexes of flat abelian groups. We know that an abelian group is flat if (and only if) it is torsion free, as stated in 5.2.8(a) (the proof is only referred to).

Most of the time one could use the stronger hypothesis of *free* components (obviously flat): we only need the more general case in a few marginal points.

5.4.3 A natural morphism

Extending 5.2.4, let A and B be chain complexes. We begin to explore the homology group $H_n(A \otimes B)$ introducing a canonical homomorphism

$$\alpha_{pq} \colon H_p(A) \otimes H_q(B) \to H_{p+q}(A \otimes B), \quad \alpha_{pq}([a] \otimes [b]) = [a \otimes b], \quad (5.61)$$

whose result is written as $[a] \times [b]$, and called the *homology external product.*

The homomorphism is well defined. If a and b are cycles in A and B, $a \otimes b$ is a cycle of $A \otimes B$, by (5.58). If $a = \partial a'$ and b is a cycle, then $\partial(a' \otimes b) = a \otimes b$; and symmetrically.

This defines a natural transformation

$$\alpha_n \colon \bigoplus_{p+q=n} H_p \otimes H_q \to H_n(- \otimes -) \colon \mathsf{Ch}_+\mathsf{Ab} \times \mathsf{Ch}_+\mathsf{Ab} \to \mathsf{Ab},$$
$$\alpha_n \colon \bigoplus_{p+q=n} H_p(A) \otimes H_q(B) \to H_n(A \otimes B), \tag{5.62}$$

where α_n has co-components α_{pq}.

Exercises and complements. (a) If G is a flat abelian group (viewed as a complex), we have seen in 5.2.4 that α gives a natural isomorphism

$$\alpha_n = \alpha_{n0} \colon H_n(A) \otimes G \to H_n(A \otimes G), \quad \alpha_n([a] \otimes g) = [a \otimes g]. \tag{5.63}$$

(b) More generally, if B is a complex with zero differentials and flat components, the exact functor $- \otimes B \colon \mathsf{Ch}_+\mathsf{Ab} \to \mathsf{Ch}_+\mathsf{Ab}$ gives a natural isomorphism

$$\alpha_n \colon \bigoplus_{p+q=n} H_p(A) \otimes B_q \to H_n(A \otimes B), \quad \alpha_n([a] \otimes b) = [a \otimes b]. \tag{5.64}$$

5.4.4 Künneth Theorem (for chain complexes)

For a pair A, B of chain complexes of abelian groups, where A (or B) has flat components, there is a short exact sequence

$$\bigoplus_{p+q=n} H_p A \otimes H_q B \overset{\alpha_n}{\rightarrowtail} H_n(A \otimes B) \overset{\beta_n}{\twoheadrightarrow} \bigoplus_{p+q=n-1} \mathrm{Tor}(H_p A, H_q B) \qquad (5.65)$$

which is natural on $\mathrm{Ch}_+\mathsf{Ab} \times \mathrm{Ch}_+\mathsf{Ab}$, and splits (not in a natural way).
In particular, in degree 0, we have an isomorphism

$$\alpha_0 \colon H_0(A) \otimes H_0(B) \ \to \ H_0(A \otimes B), \quad \alpha_0([a] \otimes [b]) = [a \otimes b]. \qquad (5.66)$$

Proof The proof is long and complex: see 5.4.6. $\qquad\square$

5.4.5 Künneth Theorem (for cochain complexes)

For a pair A, B of cochain complexes of abelian groups, where A has free components, there is a short exact sequence

$$\bigoplus_{p+q=n-1} \mathrm{Ext}(H^p A, H^q B) \overset{\beta^n}{\rightarrowtail} H^n(A \otimes B) \overset{\alpha^n}{\twoheadrightarrow} \bigoplus_{p+q=n} H^p A \otimes H^q B \qquad (5.67)$$

which is natural on $\mathrm{Ch}_+\mathsf{Ab}^{\mathrm{op}} \times \mathrm{Ch}_+\mathsf{Ab}$, and splits (not in a natural way).

Proof The proof is similar to that of Theorem 5.4.4. A more general, more difficult result is given in [CE], Theorem VI.3a, under somewhat symmetric hypotheses on A and B. $\qquad\square$

5.4.6 Proof of Theorem 5.4.4

We assume that A has flat components. The Tor product $\mathrm{Tor}(X, Y)$ is written as $X * Y$.

(a) We form two chain complexes with zero differentials

$$Z = (Z_n A)_{n \geqslant 0}, \qquad Z' = (Z'_n A)_{n \geqslant 0},$$

where $Z'_n(A) = A_n / Z_n(A) \cong B_{n-1}(A) \subset A_{n-1}$. They have flat components, by 5.2.3(b).

Applying $- \otimes B$ to the short exact sequence $Z \rightarrowtail A \twoheadrightarrow Z'$ of chain complexes, we get a short exact sequence $Z \otimes B \rightarrowtail A \otimes B \twoheadrightarrow Z' \otimes B$, by Exercise 5.4.3(b).

(b) We have thus an exact homology sequence around the group $H_n = H_n(A \otimes B)$

$$H_{n+1}(Z' \otimes B) \xrightarrow{D_{n+1}} H_n(Z \otimes B) \xrightarrow{\sigma_n} H_n \xrightarrow{\tau_n} H_n(Z' \otimes B) \xrightarrow{D_n} H_{n-1}(Z \otimes B).$$

We rewrite the sequence in the upper part of the following diagram, using the natural isomorphism (5.64)

$$\textstyle\bigoplus_{p+q=n} C_p \otimes H_qB \to H_n(C \otimes B), \qquad c \otimes [b] \mapsto [c \otimes b],$$

for the complexes $C = Z'$ or $C = Z$ (with zero differentials and flat components). The direct sums range over all pairs (p, q) of integers with $p+q = n$

$$\begin{array}{ccccccc}
\bigoplus Z'_{p+1} \otimes H_qB & & & & & & \bigoplus Z_{p-1} \otimes H_qB \\
 & \searrow^{D_{n+1}} & & & & \nearrow^{D_n} & \\
 & \bigoplus Z_p \otimes H_qB & \xrightarrow{\sigma_n} & H_n(A \otimes B) & \xrightarrow{\tau_n} & \bigoplus Z'_p \otimes H_qB & \\
 & \searrow^{v_n} & \nearrow^{\alpha_n} & & \searrow^{\beta_n} & \nearrow^{u_n} & \\
 & \bigoplus H_pA \otimes H_qB & & & & \bigoplus H_{p-1}A * H_qB &
\end{array} \qquad (5.68)$$

(c) We form now the lower part of the previous diagram. The homology group H_pA is characterised by the short exact sequence

$$Z'_{p+1} \xrightarrow{h_p} Z_p \xrightarrow{v_p} H_pA \qquad h_p(\overline{a}) = \partial_{p+1}(a) \quad (a \in A_{p+1}),$$

where we use the canonical isomorphism $Z'_{p+1} \to B_p(A)$ induced by the differential $\partial_{p+1}\colon A_{p+1} \to A_p$.

Tensoring by H_qB, and recalling that the abelian group Z_p is flat, we have a natural exact sequence

$$H_pA * H_qB \xrightarrow{u_{pq}} Z'_{p+1} \otimes H_qB \xrightarrow{h_{pq}} Z_p \otimes H_qB \xrightarrow{v_{pq}} H_pA \otimes H_qB$$

from which we deduce two consequences:

- the relation $\mathrm{Im}\, h_{pq} = \mathrm{Ker}\, v_{pq}$ (added for $p + q = n$), gives in (5.68): $v_n = \mathrm{cok}\, D_{n+1} = \mathrm{coim}\, \sigma_n$,

- the relation $\mathrm{Im}\, u_{p-1,q} = \mathrm{Ker}\, h_{p-1,q}$ (added again), gives in (5.68): $u_n = \ker D_n = \mathrm{im}\, \tau_n$.

(d) The epi-mono factorisations $\sigma_n = \alpha_n v_n$ and $\tau_n = u_n \beta_n$ give the short exact sequence (α_n, β_n).

The monomorphism α_n we have obtained is indeed computed as in 5.4.3, because σ_n is induced by the inclusion $Z \rightarrowtail A$

$$\sigma_n(a \otimes [b]) = [a \otimes b], \qquad \alpha_n([a] \otimes [b]) = [a \otimes b] \quad (a \in Z_pA,\ b \in Z_qB).$$

(e) Finally, we have to show that the sequence (5.65) splits. (Here we follow Dold's proof in [Do], Theorem VI.9.13.)

First, we suppose that A and B have free components. Then the short exact sequence $Z_pA \rightarrowtail A_p \twoheadrightarrow Z'_pA$ splits (because Z'_pA is free), and Z_pA is a direct summand of A_p; the projection $v_p \colon Z_p A \to H_p A$ can thus be extended to a homomorphism $\varphi_p \colon A_p \to H_p A$, which vanishes on boundaries; similarly, by our additional hypothesis on B, the projection $Z_q B \to H_q B$ is extended to a homomorphism $\psi_q \colon B_q \to H_q B$.

Their tensor products give a homomorphism

$$(\varphi \otimes \psi)_n \colon (A \otimes B)_n \to \oplus H_p A \otimes H_q B,$$

which vanishes on boundaries and induces a homomorphism

$$(\varphi \otimes \psi)_{*n} \colon H_n(A \otimes B) \to \oplus H_p A \otimes H_q B. \tag{5.69}$$

This is a retraction of $\alpha_n \colon \oplus H_p A \otimes H_q B \to H_n(A \otimes B)$: for every $a \in Z_p A$ and $b \in Z_q B$:

$$(\varphi \otimes \psi)_{*n} \alpha_n([a] \otimes [b]) = (\varphi \otimes \psi)_{*n}[a \otimes b] = \varphi_p(a) \otimes \psi_q(b) = [a] \otimes [b].$$

(f) Coming back to the general case, we apply Lemma 5.4.7 (written below): there exist two complexes A', B' of free abelian groups and two chain morphisms $f \colon A' \to A$, $g \colon B' \to B$ that induce homology isomorphisms f_{*n} and g_{*n}, in every degree.

The naturality of the sequence (5.65) gives a commutative diagram with short exact rows and $n' = n - 1$

$$
\begin{array}{ccccc}
\oplus_{p+q=n} H_p A' \otimes H_q B' & \overset{u^\sharp}{\rightarrowtail} & H_n(A' \otimes B') & \overset{v^\sharp}{\twoheadrightarrow} & \oplus_{p+q=n'} H_p A' * H_q B' \\
\quad\downarrow{\scriptstyle f_* \otimes g_*} & & \quad\downarrow{\scriptstyle (f \otimes g)_*} & & \quad\downarrow{\scriptstyle f_* * g_*} \\
\oplus_{p+q=n} H_p A \otimes H_q B & \underset{u^\sharp}{\rightarrowtail} & H_n(A \otimes B) & \underset{v^\sharp}{\twoheadrightarrow} & \oplus_{p+q=n'} H_p A * H_q B
\end{array}
$$

The left and the right vertical arrows are isomorphisms, and $(f \otimes g)_{*n}$ is also, by the Five Lemma 2.7.3 (we are actually applying the 'Short Five Lemma', with trivial groups at the left and the right hand).

As we have proved that the upper row splits, the lower also does. $\qquad\square$

5.4.7 Lemma

For every chain complex A of abelian groups there exists a complex A' with free components and a chain morphism $f \colon A' \to A$ that induces isomorphisms $f_{*n} \colon H_n(A') \to H_n(A)$.

Note. This is Lemma V.10.5 in [M1]. ([Do] has similar results in II.4.)

Proof For a fixed $n \geq 0$, we begin by building a complex F with free components, the correct homology in degree n and trivial homology elsewhere.

F is the upper row of the following commutative diagram, where

$$
\begin{array}{ccccccccc}
\cdots & \longrightarrow & 0 & \longrightarrow & F_{n+1} & \xrightarrow{\ \partial_{n+1}\ } & F_n & \longrightarrow & 0 & \longrightarrow & \cdots \\
& & \downarrow & & {\scriptstyle u_{n+1}}\downarrow & \searrow & \downarrow{\scriptstyle u_n} & & \downarrow & & \\
\cdots & \longrightarrow & A_{n+2} & \longrightarrow & A_{n+1} & \twoheadrightarrow \bullet \twoheadrightarrow & A_n & \xrightarrow[\ \partial_n\ ]{} & A_{n-1} & \longrightarrow & \cdots
\end{array}
\tag{5.70}
$$

- F_n is a free abelian group with an epimorphism $F_n \twoheadrightarrow Z_n A$,

- u_n is the composite $F_n \twoheadrightarrow Z_n A \rightarrowtail A_n$ (annihilated by $\partial_n \colon A_n \to A_{n-1}$),

- $F_{n+1} = u_n^{-1}(B_n A)$ is a (free) subgroup of F_n, and $\partial_{n+1} \colon F_{n+1} \to F_n$ is the inclusion,

- for u_{n+1} we first lift $u_n \partial_{n+1} \colon F_{n+1} \to A_n$ to $B_n A$, and then to A_{n+1} (as F_{n+1} is free).

The chain complex F has trivial homology groups in degree $\neq n$, while $u_n \colon F_n \to A_n$ induces an isomorphism

$$
H_n F = F_n / F_{n+1} \cong Z_n A / B_n A = H_n A.
$$

Now we rename the chain complex F as $A^{(n)}$ (it depends on $n \geqslant 0$) and form the required complex A' as the direct sum $A' = \bigoplus_n A^{(n)}$ of chain complexes. $\qquad\qquad\square$

5.5 Products in singular cohomology

We show now that singular cohomology with coefficients in a ring R forms a graded ring, as we have already seen for the cohomology theories of Alexander–Spanier and de Rham. The multiplicative structure is provided by the 'cup product'.

We use here the simplicial form of singular cohomology, where the cup product has a simpler definition. In 5.5.9 we try to investigate 'why' a multiplicative structure should arise in cohomology.

R is a fixed coefficient ring, which is commutative; the rings commonly used are $\mathbb{Z}$, $\mathbb{Z}_n$, $\mathbb{Q}$ and $\mathbb{R}$.

Literature. Hatcher's book [Ha] has many computations of graded algebras of singular cohomology, and a rich information on the difficult problem of realising a graded algebra in such a way: cf. Section 3.2.

The cup product can also be defined in cubical form, but this has to go through heavier computations, by a more complex diagonal approximation $C_+(X) \to C_+(X) \otimes C_+(X)$, as in [HiW], Section 9.3, or by the chain homotopy equivalence $C_+(X \times X) \to C_+(X) \otimes C_+(X)$ (dealt with in Section 5.6), as in [Mas3], Section VIII.3.

5.5.1 Singular cochains with coefficients in a ring

The chain complex $S_+(X)$ of simplicial singular chains of the space X was defined in 2.8.5

$$S_+(X) = \mathrm{Ch}_+\mathrm{Smp}(X) = ((F(\mathrm{Top}(\Delta^n, X))_n, (\partial_n)),$$

$$\partial_n(a) = \Sigma_{0 \leqslant i \leqslant n}\, (-1)^i\, \partial_{ni}(a) = \Sigma_{0 \leqslant i \leqslant n}\, (-1)^i\, a\delta_{ni}, \tag{5.71}$$

$$(f\colon X \to Y) \mapsto f_\sharp\colon S_+(X) \to S_+(Y), \qquad f_{\sharp n}(a) = fa,$$

where $a\colon \Delta^n \to X$ is a simplex of X.

We recall that the tetrahedron Δ^n is the convex hull of the unit points $e_0, ..., e_n$ of $\mathbb{R}^{n+1}$; in particular, $\Delta^0 = \{1\}$. The differential is based on their faces:

$$\delta_{ni}\colon \Delta^{n-1} \to \Delta^n,$$

$$\delta_{ni}(t_0, ..., t_{n-1}) = (t_0, ..., t_{i-1}, 0, t_i, ..., t_{n-1}) \quad (0 \leqslant i \leqslant n). \tag{5.72}$$

From this complex, we form the complex of simplicial singular cochains with coefficients in the abelian group G:

$$S^+(-; G)\colon \mathsf{Top}^{\mathrm{op}} \to \mathrm{Ch}^+\mathsf{Ab},$$

$$S^n(X; G) = \mathrm{Hom}(S_n(X), G) = \mathsf{Set}(\mathrm{Smp}_n(X), G),$$

$$(d^n\lambda)(a) = \Sigma_{0 \leqslant i \leqslant n+1}\, (-1)^i\, \lambda(a\delta_i), \tag{5.73}$$

$$f^{\sharp n}\colon S^n(X; G) \to S^n(Y; G), \qquad (f^{\sharp n}\lambda)(b) = \lambda(fb),$$

for $a\colon \Delta^{n+1} \to X$, $\lambda\colon \mathrm{Smp}_n(X) \to G$, $f\colon Y \to X$ and $b\colon \Delta^n \to Y$.

The simplicial definition of the singular cohomology groups is thus:

$$H^n(X; G) = H^n(S^+(X; G)). \tag{5.74}$$

From now on, we work with coefficients in the ring R.

$S^+(X; R)$ is thus a cochain complex of R-modules. An n-cochain is simply a mapping $\lambda\colon \mathrm{Smp}_n(X) \to R$ defined on the n-simplices of X, and a 0-cochain is a mapping $\lambda\colon X \to R$. Their R-linear combinations are computed pointwise.

5.5.2 Cup product

These cochains have a natural product, called the cup product $\lambda \smile \mu$.

It is based on the product of the ring R, together with the geometry of tetrahedra, which steps in by means of new higher faces.

Namely, the *back p-face* adds q zeros at the right, while the *front q-face* adds p zeros at the left (for $p + q = n$)

$$\tau_p^- : \Delta^p \to \Delta^n, \qquad \tau_p^-(t_0, ..., t_p) = (t_0, ..., t_p, 0..., 0),$$
$$\tau_q^+ : \Delta^q \to \Delta^n, \qquad \tau_q^+(t_0, ..., t_q) \mapsto (0..., 0, t_0, ..., t_q). \tag{5.75}$$

In particular:

$$\tau_0^-(1) = e_0 \in \Delta^n, \qquad \tau_0^+(1) = e_n \in \Delta^n,$$
$$\tau_{n-1}^- = \delta_n : \Delta^{n-1} \to \Delta^n, \qquad \tau_{n-1}^+ = \delta_0 : \Delta^{n-1} \to \Delta^n,$$
$$\tau_n^- = \tau_n^+ = \operatorname{id} \Delta^n.$$

Now we define the *cup product* of cochains with coefficients in R

$$-\smile- : S^p(X; R) \times S^q(X; R) \to S^n(X; R) \qquad (p + q = n),$$
$$(\lambda \smile \mu)(a) = \lambda(a\tau_p^-)\,\mu(a\tau_q^+) \qquad (a : \Delta^n \to X). \tag{5.76}$$

In this way, the complex $S^+(X; R)$ becomes a differential graded R-algebra (see 3.6.2), as we verify below. The unit

$$1 \in S^0(X; R) = \mathsf{Set}(|X|, R), \qquad 1 : X \to R, \tag{5.77}$$

is the constant function at the unit 1_R of R, defined on the set $|X| = \mathrm{Smp}_0(X)$.

For $\lambda \in S^n(X; R)$ and $\rho \in S^0(X; R)$ (i.e. a function $\rho : X \to R$) we simply have:

$$(\rho \smile \lambda)(a) = \rho(a(e_0))\,\lambda(a), \qquad (\lambda \smile \rho)(a) = \lambda(a)\,\rho(a(e_n)). \tag{5.78}$$

The cup product is related to the Alexander–Whitney diagonal approximation map $\tau_{pq} : S_{p+q}(X) \to S_p(X) \otimes S_q(X)$ (for simplicial singular chains), as we shall see in 5.5.6.

Exercises and complements. We have to check the properties listed below. The verifications are obvious, except the last, which works by commuting ordinary faces and higher faces of tetrahedra (it is written in Chapter 8).

We assume that $\lambda, \lambda' \in S^p(X; R)$, $\mu, \mu' \in S^q(X; R)$, $\nu \in S^r(X; R)$ and $\rho \in R$.

(a) $(\rho\lambda) \smile \mu = \rho(\lambda \smile \mu) = \lambda \smile (\rho\mu),$

(b) $\lambda \smile (\mu \smile \nu) = (\lambda \smile \mu) \smile \nu,$

(c) $(\lambda + \lambda') \smile \mu = \lambda \smile \mu + \lambda' \smile \mu, \qquad \lambda \smile (\mu + \mu') = \lambda \smile \mu + \lambda \smile \mu',$

(d) $d^n(\lambda \smile \mu) = d^p\lambda \smile \mu + (-1)^p\,\lambda \smile d^q\mu \qquad\qquad (n = p + q).$

5.5.3 The cohomology algebra

The cohomology graded R-module of X, with coefficients in the ring R

$$H^*(X;R) = H^*(S^+(X;R)), \tag{5.79}$$

becomes a graded R-algebra, with the induced *cup product* of cohomology classes

$$H^p(X;R) \times H^q(X;R) \to H^{p+q}(X;R), \qquad [\lambda] \smile [\mu] = [\lambda \smile \mu], \tag{5.80}$$

and unit $1 \in H^0(X;R) = Z^0(X;R)$, the cohomology class of the constant function $1\colon X \to R$.

The general case was dealt with in 3.6.3. We recall that, because of formula 5.5.2(d) on the differential of a product, a product of cocycles is a cocycles; the coboundaries form a bilateral graded ideal.

A map $f\colon X \to Y$ gives a cochain morphism $f^\sharp\colon S^+(Y;G) \to S^+(X;G)$ which preserves the graded product and the unit

$$\begin{aligned}
(f^{\sharp n}(\lambda \smile \mu))(a) &= (\lambda \smile \mu)(fa) = (\lambda(fa\tau_p^-)\,\mu(fa\tau_q^+)) \\
&= (f^{\sharp p}\lambda) \smile (f^{\sharp q}\mu)(a) \qquad\qquad (a\colon \Delta^n \to X), \\
(f^{\sharp 0}(1))(x) &= 1(fx) = 1 \qquad\qquad\qquad (x \in X).
\end{aligned} \tag{5.81}$$

We have thus defined two contravariant functors, with values in dg-algebras and graded algebras:

$$S^+(-;R)\colon \mathsf{Top}^{\mathrm{op}} \to R\mathsf{Dga}, \qquad H^*(-;R)\colon \mathsf{Top}^{\mathrm{op}} \to R\mathsf{Gra}. \tag{5.82}$$

Complements. (a) The algebra $H^*(X;R)$ is skew-commutative:

$$[\lambda] \smile [\mu] = (-1)^{pq}\,[\mu] \smile [\lambda], \tag{5.83}$$

where λ and μ have degree p and q. An interested reader can see a proof in [Vi], Theorem 5.11, or [Sp2], Subsection 5.6.11.

(b) For a relative pair (X, A), there are relative forms of the cup product of cochains, and therefore of cohomology classes

$$\begin{aligned}
H^p(X, A; R) \times H^q(X; R) &\to H^{p+q}(X, A; R), \\
H^p(X; R) \times H^q(X, A; R) &\to H^{p+q}(X, A; R), \\
H^p(X, A; R) \times H^q(X, A; R) &\to H^{p+q}(X, A; R),
\end{aligned} \tag{5.84}$$

because, if the cochain λ or the cochain μ vanishes on the subspace A, then the cochain $\lambda \smile \mu$ also does.

5.5.4 Path components and cup product

Let X be a space and $(X_i)_{i \in I}$ its partition in path components, with inclusions $u_i \colon X_i \to X$.

(a) $H^0(X; R) = Z^0(X; R)$ is the R-algebra of the functions $\rho \colon X \to R$ constant on path components; the cohomology class of this function can be equivalently written as ρ or $[\rho]$.

There is thus a natural isomorphism

$$H^0(X; R) \to \textstyle\prod_{i \in I} H^0(X_i; R), \qquad \rho \mapsto (\rho_i)_{i \in I}, \tag{5.85}$$

where $\rho_i = \rho u_i \colon X_i \to R$ is the restriction of ρ. In particular, if X is 0-connected, $H^0(X; R)$ will be identified with R.

This isomorphism will now be extended to the whole graded algebra $H^*(X; R)$.

(b) The homomorphisms of graded R-algebras

$$p_i = u_i^* \colon H^*(X; R) \to H^*(X_i; R), \quad [\lambda] \mapsto [\lambda_i] = [u_i^\sharp \lambda] \ \ (i \in I), \tag{5.86}$$

are computed in the obvious way: in degree n, the value $\lambda_i(a)$ on a simplex $a \colon \Delta^n \to X_i$ is $\lambda(u_i a)$; or simply $\lambda(a)$ if we view a as a map $\Delta^n \to X$ with image in X_i.

The consistency with the cup product is based on the fact that, on any simplex a, the restrictions $a\tau_p^-$ and $a\tau_q^+$ stay in the same path component.

As in (4.75), these homomorphisms are the projections of a cartesian product, and there is a canonical isomorphism of graded R-algebras

$$H^*(X; R) \to \textstyle\prod_{i \in I} H^*(X_i; R), \qquad [\lambda] \mapsto ([\lambda_i])_{i \in I}. \tag{5.87}$$

(c) For a 0-cocycle $\rho = (\rho_i) \in Z^0(X; R)$, the cup product with any cochain $\lambda \in S^n(X; R)$ is computed as in (5.78), on a simplex $a \colon \Delta^n \to X$ with image in X_i

$$(\rho \smile \lambda)(a) = \rho_i \, \lambda(a), \qquad (\lambda \smile \rho)(a) = \lambda(a) \, \rho_i. \tag{5.88}$$

These products are thus determined by the R-linear structure of the algebra $Z^0(X; R) = H^0(X; R)$.

(d) (*Basic products*) In $H^*(X; R)$, the cup products $[\lambda] \smile [\mu]$ where one of the factors has degree zero will be called (in this book) *basic products*: they are easily calculated, by the previous formula, and cannot completely vanish (unless the ring R is trivial, or the space X is empty).

We say that the graded algebra $H^*(X; R)$ has a *basic cup product* if all the other cup products vanish: $[\lambda] \smile [\mu] = 0$ whenever both factors have a positive degree.

We shall see in 5.5.5 that a space can have a basic cup product with coefficients in $\mathbb{Z}$, and an interesting cup product with coefficients in $\mathbb{Z}_2$.

(e) The *characteristic function* of the component X_i in the ring R

$$\eta_i \colon X \to R,$$

$$\eta_i(x) = 1_R \ \text{if} \ x \in X_i, \qquad \eta_i(x) = 0_R \ \text{if} \ x \notin X_i,$$

(5.89)

is a 0-cocycle of X. Now $p_i(\eta_i) = 1_i$ and $p_j(\eta_i) = 0$ for $j \neq i$.

For every function $\lambda \colon \mathrm{Smp}_n(X) \to R$

$$(\eta_i \smile \lambda)(a) = (\lambda \smile \eta_i)(a) = \lambda(a), \quad \text{if} \ \mathrm{Im}\,(a) \subset X_i, \qquad (5.90)$$

and is 0 otherwise.

The projection $p_i \colon H^*(X; R) \to H^*(X_i; R)$ restricts thus to an isomorphism

$$\eta_i(H^*(X; R)) \to H^*(X_i; R), \tag{5.91}$$

defined on the principal ideal of $H^*(X; R)$ generated by the cohomology class $\eta_i \in H^0(X; R)$.

*(f) In any product $R = \prod_{i \in I} R_i$ of unital rings, possibly non commutative, the unit of R_i produces as above an idempotent element η_i in the centre of R, which generates a principal ideal isomorphic to R_i.

5.5.5 Exercises (Cohomology algebras)

These exercises are solved below.

(a) (*The spheres*) All the graded algebras $H^*(\mathbb{S}^n; R)$ have a basic cup product.

(b) The same is true of $H^*(\mathbb{P}^2; \mathbb{Z})$ and $H^*(\mathbb{K}; \mathbb{Z})$. *More generally, it is true of $H^*(X; \mathbb{Z})$, for every non-orientable compact surface X; coefficients in $\mathbb{Z}_2$ give a different result: see (e).*

(c) (*The torus*) Compute the graded algebra $H^*(\mathbb{T}) = H^*(\mathbb{T}; \mathbb{Z})$ of the torus, showing that $H^1(\mathbb{T})$ has two generators whose cup product generates $H^2(\mathbb{T})$.

Hints: use the Universal Coefficient Theorem for cohomology, with suitable generators of $H_1(\mathbb{T})$ and $H_2(\mathbb{T})$, in simplicial form.

*(d) This relationship between H^1 and H^2 fails for the pointed sum $X = \mathbb{S}^1 \vee \mathbb{S}^1 \vee \mathbb{S}^2$, which has the same cohomology groups as the torus, but a basic cup product.

Of course there is no need of the cohomology ring to show that the torus is not homeomorphic to X: the latter is not even a manifold, and can be disconnected taking out the basepoint.

*(e) More advanced results can be found in [Ha], Section 3.2: for instance, the graded algebra $H^*(\mathbb{T}^n)$ of the n-torus, in Example 3.11, and the graded algebra $H^*(\mathbb{P}^n; \mathbb{Z}_2)$ of the projective n-space in Theorem 3.12.

The latter is the quotient $\mathbb{Z}_2[t]/(t^{n+1})$ of a polynomial algebra in one indeterminate. Let us note that, with coefficients in $\mathbb{Z}_2$, the commutative and skew-commutative properties are the same.

Solutions. (a) For $n = 0$, this is obvious. For $n > 0$, the graded R-module $H^*(\mathbb{S}^n; R)$ has non-trivial components in degree 0 and n. In positive degree, $[\lambda] \smile [\mu] = 0$, simply because $H^{2n}(\mathbb{S}^n) = 0$.

(b) It is an obvious consequence of $H^2(\mathbb{P}^2; \mathbb{Z}) = H^2(\mathbb{K}; \mathbb{Z}) = 0$. *Every non-orientable compact surface X also has $H_2(X; \mathbb{Z}) = 0$, as recalled in 3.3.6(b).*

(c) For every $n \geqslant 0$, $H_n(\mathbb{T})$ is a free abelian group of finite rank. The Universal Coefficient Theorem for cohomology gives a canonical isomorphism

$$\alpha^n \colon H^n(\mathbb{T}) \to \mathrm{Hom}(H_n(\mathbb{T}), \mathbb{Z}), \qquad \alpha^n[\lambda][z] = \lambda(z), \tag{5.92}$$

for $\lambda \in Z^n(\mathbb{T})$ and $z \in Z_n(\mathbb{T})$.

The standard generators $[a], [b]$ of $H_1(\mathbb{T})$, realised by two loops around the equator and a meridian (as in 2.6.3(b)), give two (free) generators of the abelian group $H^1(\mathbb{T})$

$$\begin{aligned}
[\lambda] \in H^1(\mathbb{T}), \quad \lambda(a) = 1, \quad \lambda(b) = 0, \\
[\mu] \in H^1(\mathbb{T}), \quad \mu(a) = 0, \quad \mu(b) = 1.
\end{aligned} \tag{5.93}$$

The cup product will be determined proving that, at the level of cochains

$$\lambda \smile \lambda = 0 = \mu \smile \mu, \qquad \lambda \smile \mu = \omega, \tag{5.94}$$

where $[\omega]$ is a generator of $H^2(\mathbb{T})$.

We use as generator $[z]$ of $H_2(\mathbb{T})$ the chain $z = f - g$ produced by two simplices $f, g \colon \Delta^2 \to \mathbb{T}$ meeting at an edge; letting $p \colon \mathbb{I}^2 \to \mathbb{T}$ be the projection, we define $f = pf'$ and $g = pg'$, where $f', g' \colon \Delta^2 \to \mathbb{I}^2$ are *affine* simplices of the square, defined as follows on the vertices

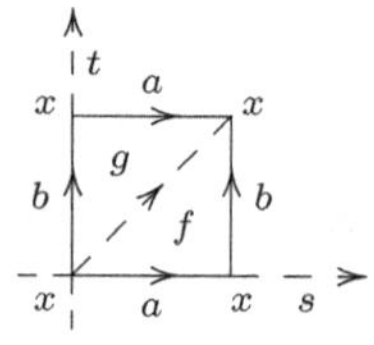

$$\begin{aligned}
f'(e_0)=(0,0), \quad f'(e_1)=(1,0), \quad f'(e_2)=(1,1), \\
g'(e_0)=(0,0), \quad g'(e_1)=(0,1), \quad g'(e_2)=(1,1).
\end{aligned} \tag{5.95}$$

The fact that z is a cycle is obvious. The fact that $[z]$ generates $H_2(\mathbb{T})$ would be proved in the initial study of singular homology in simplicial form, and is left as understood.

We can now compute $\lambda \smile \mu$, $\lambda \smile \lambda$ and $\mu \smile \mu$ on the cycle z

$$(\lambda \smile \mu)(f - g) = \lambda(f\tau_1^-)\,\mu(f\tau_q^+) - \lambda(g\tau_1^-)\,\mu(g\tau_q^+)$$
$$= \lambda(f\delta_2)\,\mu(f\delta_0) - \lambda(g\delta_2)\,\mu(g\delta_0)$$
$$= \lambda(a)\mu(b) - \lambda(b)\,\mu(a) = 1,$$

$$(\lambda \smile \lambda)(f - g) = \lambda(f\tau_1^-)\,\lambda(f\tau_q^+) - \lambda(g\tau_1^-)\,\lambda(g\tau_q^+)$$
$$= \lambda(f\delta_2)\,\lambda(f\delta_0) - \lambda(g\delta_2)\,\lambda(g\delta_0)$$
$$= \lambda(a)\lambda(b) - \lambda(b)\,\lambda(a) = 0,$$

and similarly $\mu \smile \mu = 0$.

5.5.6 The Alexander–Whitney diagonal approximation

There is a natural transformation

$$\tau \colon S_+ \to S_+ \otimes S_+ \colon \mathsf{Top} \to \mathsf{Ch_+Ab},$$
$$\tau_{pq}X \colon S_{p+q}(X) \longrightarrow S_p(X) \otimes S_q(X), \tag{5.96}$$
$$\tau_{pq}(a) = a\tau_p^- \otimes a\tau_q^+ \qquad (a \colon \Delta^{p+q} \to X),$$

called the *Alexander–Whitney diagonal approximation*. It is constructed by means of the higher faces introduced in (5.75), for $p + q = n$

$$\tau_p^- \colon \Delta^p \to \Delta^n, \qquad \tau_p^-(t_0, ..., t_p) = (t_0, ..., t_p, 0..., 0),$$
$$\tau_q^+ \colon \Delta^q \to \Delta^n, \qquad \tau_q^+(t_0, ..., t_q) \mapsto (0..., 0, t_0, ..., t_q). \tag{5.97}$$

Exercises and complements. (a) Prove that the sequence of homomorphisms

$$\tau_n X \colon S_n(X) \to \bigoplus_{p+q=n} S_p(X) \otimes S_q(X)$$

of co-components $\tau_{pq}X$ commutes with the differentials. (Naturality is obvious.)

(b) In degree 0, we simply have the diagonal homomorphism

$$\tau_{00}X \colon S_0(X) \to S_0(X) \otimes S_0(X), \qquad \tau_{00}(x) = x \otimes x \quad (x \in X) \tag{5.98}$$

which induces in homology a natural homomorphism $[x] \mapsto [x \otimes x]$.

(c) The cup product of cohomology arises in a natural way from the natural transformation (5.96).

5.5.7 Cross product

For two spaces X, Y there is also a *cross product*, or *external product*, of singular cochains

$$S^p(X; R) \times S^q(Y; R) \longrightarrow S^{p+q}(X \times Y; R),$$

$$(\lambda, \mu) \mapsto \lambda \times \mu = p^{\sharp 1}(\lambda) \smile p^{\sharp 2}(\mu), \qquad (5.99)$$

$$(\lambda \times \mu)(a) = \lambda(p_1 a \tau_p^-)\,\mu(p_2 a \tau_q^+) \quad (\text{for } a\colon \Delta^n \to X \times Y),$$

where $p_1\colon X \times Y \to X$ and $p_2\colon X \times Y \to Y$ are the cartesian projections.

This product is R-bilinear and consistent with the differential, in the sense that (as proved below):

$$d^n(\lambda \times \mu) = d^p(\lambda) \times \mu + (-1)^p \lambda \times d^q(\mu) \qquad (n = p + q). \qquad (5.100)$$

There is thus a *cross product*, or *external product*, in cohomology:

$$H^p(X; R) \times H^q(Y; R) \to H^{p+q}(X \times Y, A; R),$$

$$[\lambda] \times [\mu] = [\lambda \times \mu]. \qquad (5.101)$$

Exercises and complements. (a) Prove formula (5.100).

(b) We have defined the cross product in terms of the cup product. The converse can also be done, taking $X = Y$.

5.5.8 Complements

The diagonal approximation $\tau\colon S_+ \to S_+ \otimes S_+$ has also an equivalent 'external form', working on two spaces X, Y. This is a natural transformation on $\mathsf{Top} \times \mathsf{Top}$

$$\vartheta\colon S_+(X \times Y) \to S_+(X) \otimes S_+(Y),$$

$$(a, b) \mapsto a\tau_p^- \otimes b\tau_q^+ \qquad (a\colon \Delta^{p+q} \to X,\ b\colon \Delta^{p+q} \to Y), \qquad (5.102)$$

whose components $\vartheta(X, Y)$ are chain homotopy equivalences, as we shall see in Section 5.6.

Given the Alexander–Whitney diagonal approximation, we define ϑ_{pq} as the composite $\vartheta_{pq} = (p_{1\sharp} \otimes p_{2\sharp})\tau_{pq}$

$$S_{p+q}(X \times Y) \to S_p(X \times Y) \otimes S_q(X \times Y) \to S_p(X) \otimes S_q(Y), \qquad (5.103)$$

where $p_1\colon X \times Y \to X$ and $p_2\colon X \times Y \to Y$ are the cartesian projections.

Conversely, given all ϑ_{pq}, we define τ_{pq} by the diagonal map Δ

$$\tau_{pq} = \vartheta_{pq}\Delta_\sharp\colon S_{p+q}(X) \to S_{p+q}(X \times X) \to S_p(X) \otimes S_q(X). \qquad (5.104)$$

5.5.9 *The dual affinity of spaces and rings*

There are geometrical reasons why a multiplication appears in *cohomology* theories: every space has a diagonal map $\Delta\colon X \to X \times X$, which — in the theory of coalgebras — is a *comonoid* structure (satisfying diagrammatic properties of coassociativity and counitarity).

> In fact, we have seen that the diagonal, combined with the chain homotopy equivalence $S_+(X \times X) \to S_+(X) \otimes S_+(X)$ recalled above, gives the Alexander–Whitney diagonal approximation $\tau\colon S_+ \to S_+ \otimes S_+$, which supplies the cup product in a natural way (worked out in Exercise 5.5.6(c)).

More essentially, there seems to be a dual affinity between topological spaces and rings, according to which rings and homomorphisms are able to give a 'mirror image' of topological spaces and maps, rather than a covariant translation.

(a) As a concrete instance of this affinity, Gelfand duality says that the category of compact Hausdorff spaces is equivalent to the opposite of the category of commutative (unital) C^*-algebras, by the contravariant functor taking a compact Hausdorff space X to the C^*-algebra $C(X) = \mathsf{Top}(X, \mathbb{C})$ of complex continuous functions on X.

This leads to the domain of Noncommutative Geometry, introduced by A. Connes [Co]: a general C^*-algebra is viewed as a (possibly) noncommutative space. Directed and weighted algebraic topology have a strong relationship with this domain [G1].

(b) Pointless Topology also simulates spaces by the opposite of a category of an algebraic character.

In fact, a space X determines the ordered set $\mathcal{O}(X)$ of its open subsets; this is a *frame*, that is a complete lattice where arbitrary joins distribute over binary meets. Contravariantly, a map $f\colon X \to Y$ gives the preimage homomorphism $f^*\colon \mathcal{O}(Y) \to \mathcal{O}(X)$, that preserves arbitrary joins and binary meets. Pointless Topology studies the category $\mathsf{Loc} = \mathsf{Frm}^{\mathrm{op}}$, the formal opposite of the category of frames [Jo1, Jo2, Bo3].

(c) The dual affinity we are considering also appears in elementary categorical facts.

The terminal object $\{*\}$ of Top determines the category $\mathsf{Top}_\bullet$ of pointed spaces as the *slice category* $\mathsf{Top}\backslash\{*\}$: an object (X, x_0) is viewed as a map $x_0\colon \{*\} \to X$.

In a dually similar way, the initial object $\mathbb{Z}$ of Rng determines the slice category $\mathsf{Rng}/\mathbb{Z}$, of *copointed rings* $p\colon R \to \mathbb{Z}$. This is equivalent to the category Rng' *of possibly-non-unital rings*, taking the (unital) homomorphism p to the ideal $\operatorname{Ker} p$, and adding a unit to a possibly-non-unital ring. Details can be found in [G5], Exercise 3.3.2(b).

5.6 *Acyclic models and products of spaces

This last section of Chapter 5 establishes some important results, among which:

(i) the equivalence of the simplicial and cubical forms of singular homology,

(ii) the Eilenberg–Zilber Theorem on the singular homology of a product of spaces.

These results are proved as applications of a powerful technique, the Acyclic Models Theorem.

Literature. Acyclic models were introduced by Eilenberg and Mac Lane [EM4] in 1953, formalising the role of the 'simplicial models' Δ^n and the 'cubical models' $\mathbb{I}^n$ in the corresponding constructions of singular homology; their equivalence is a consequence of the Acyclic Model Theorem. Application (ii) was exposed in the subsequent article of the same journal, by Eilenberg and Zilber [EZ2].

The main theorem is also exposed in [Sp2], Section 4.2, Theorem 8, in a simplified version that covers application (ii), but does not cover cubical chains (because of normalistion). [HiW], in Sections 8.4 and 8.7, uses for both applications the method of Acyclic Models, rather than the theorem.

(Here we give the simplified version in Part (i) of Theorem 5.6.3, and the original one in Part (ii).)

5.6.1 *Acyclic models*

We consider a category C equipped with a set $\mathcal{M}$ of objects, called *models*. We begin by exposing the simplified version used in [Sp2, Vi].

First, a functor $F\colon \mathsf{C} \to \mathsf{Ab}$ is said to be *free on the models* of $\mathcal{M}$ if there is a family of *distinguished elements* $\xi_j \in F(M_j)$ (with $j \in J$ and $M_j \in \mathcal{M}$) such that, for every object X in C, $F(X)$ is the free abelian group generated by all the elements $a_j = (Fa)(\xi_j)$, for $j \in J$ and $a \in \mathsf{C}(M_j, X)$; these elements a_j can be called the *F-models of X*.

Now we consider a functor $F\colon \mathsf{C} \to \mathsf{Ch}_+\mathsf{Ab}$ with values in chain complexes. We say that F is *free on the models* if each component $F_n\colon \mathsf{C} \to \mathsf{Ab}$ is free, with distinguished elements $\xi_j \in F_n(M_j)$ (for $j \in J_n$ and $M_j \in \mathcal{M}$). Every abelian group $F_n(X)$ has thus a corresponding basis of F_n-models

$$a_j = (F_n a)(\xi_j), \quad \text{for } j \in J_n \text{ and } a \in \mathsf{C}(M_j, X). \tag{5.105}$$

For the sake of simplicity, we are supposing that all the sets J_n are disjoint, so that we need not specify the degree n in ξ_j and M_j.

The functor $F\colon \mathsf{C} \to \mathrm{Ch}_+\mathsf{Ab}$ is *acyclic on the models* if $H_n(F(M)) = 0$, for all $n > 0$ and $M \in \mathcal{M}$.

Examples. (a) The basic example is the category Top, equipped with the *simplicial models*, i.e. the set of all tetrahedra Δ^n, for $n \geqslant 0$. The simplicial singular complex functor

$$S_+\colon \mathsf{Top} \to \mathrm{Ch}_+\mathsf{Ab}, \qquad S_n(X) = \mathbb{Z}(\mathsf{Top}(\Delta^n, X)), \tag{5.106}$$

is free and acyclic on the models Δ^n $(n \geqslant 0)$.

In fact, for every $n \geqslant 0$, S_n is free on the single model Δ^n, with $\xi_n = \mathrm{id}\,\Delta^n$, because the abelian group $S_n(X)$ is freely generated by the n-simplices $(S_n a)(\xi_n) = a$, for all $a\colon \Delta^n \to X$. Acyclicity on these models was verified in Lemma 2.8.7.

(b) To study cartesian products $X \times Y$ (in 5.6.5), we shall use the category $\mathsf{Top} \times \mathsf{Top}$, with models consisting of all cartesian products $\Delta^p \times \Delta^q$, for $p, q \geqslant 0$.

(c) This approach does not cover the cubical singular complex functor

$$C_+\colon \mathsf{Top} \to \mathrm{Ch}_+\mathsf{Ab}, \qquad C_n(X) = \mathbb{Z}(\mathsf{Top}(\mathbb{I}^n, X))/\mathbb{Z}\mathrm{Deg}_n(X), \tag{5.107}$$

which is acyclic on the models $\mathbb{I}^n$ $(n \geqslant 0)$, but is not free on them, because of normalisation.

Let us note that the non-normalised cubical functor is not even acyclic on the singleton $\mathbb{I}^0$, as shown in the solution of Exercise 2.2.5(a).

5.6.2 *A more complex version*

To deal with the cubical layout we resort to the original setting of [EM4]. We are still examining a functor $F\colon \mathsf{C} \to \mathsf{Ab}$ defined on a category C with a set $\mathcal{M}$ of models, and we want to extend the property of being free on them.

First we build a new functor $\tilde{F}\colon \mathsf{C} \to \mathsf{Ab}$, where

(i) for an object X of C, $\tilde{F}(X)$ is the free abelian group generated by all pairs (a, ξ), where $a\colon M_a \to X$ is a morphism of C defined on a model $M_a \in \mathcal{M}$, and $\xi \in F(M_a)$,

(ii) for a morphism $f\colon X \to Y$ in C, the homomorphism $\tilde{F}f\colon \tilde{F}X \to \tilde{F}Y$ is defined on the previous basis, as $\tilde{F}(f)(a, \xi) = (fa\colon M_a \to Y, \xi)$.

There is a natural transformation

$$\pi\colon \tilde{F} \to F, \qquad (\pi X)(a, \xi) = F(a)(\xi) \in F(X), \tag{5.108}$$

for $a\colon M_a \to X$, $\xi \in F(M)$.

We say that F is *essentially free on the models* if there is also a natural transformation $\rho\colon F \to \tilde{F}$ such that $\pi\rho = \mathrm{id}$ (a section of π).

Such a functor is called 'representable' in the original paper [EM4]; but this term has now a different, well-established meaning in category theory: see 1.3.8.

Exercises and complements. The solutions are below.

(a) Every functor $F\colon \mathsf{C} \to \mathsf{Ab}$ free on the models of $\mathcal{M}$ is also essentially free on them.

(b) The cubical singular complex functor (5.106) is essentially free (and acyclic) on the models $\mathbb{I}^n$ ($n \geqslant 0$).

Solutions. (a) We define $\rho\colon F \to \tilde{F}$ taking the basis element $a_j = (Fa)(\xi_j)$ to the pair (a, ξ_j). Then $\pi\rho(a_j) = (Fa)(\xi_j) = a_j$.

(b) Each component $C_n(X)$ of cubical chains is freely generated by the family of non-degenerate cubes $a\colon \mathbb{I}^n \to X$.

Letting $\mathcal{M}_n = \{\mathbb{I}^n\}$ and $\xi_n = \mathrm{id}\,\mathbb{I}^n$, we can define

$$\rho_n X\colon C_n(X) \to \tilde{C}_n(X), \qquad \rho_n(a) = (a, \xi_n) \tag{5.109}$$

(for $a\colon \mathbb{I}^n \to X$ non-degenerate): this gives $\pi_n \rho_n(a) = (C_n a)(\xi_n) = a$.

5.6.3 Acyclic Models Theorem

Let C be a category with models $\mathcal{M}$ and $F, G\colon \mathsf{C} \to \mathrm{Ch}_+\mathsf{Ab}$ two functors.

(i) Suppose that F is free on these models and G is acyclic on them.

Then every natural transformation $\Phi\colon H_0 F \to H_0 G\colon \mathsf{C} \to \mathsf{Ab}$ can be lifted to a natural transformation $\varphi\colon F \to G$ such that $H_0\varphi = \Phi$, and any two of them φ, ψ are naturally chain homotopic: there exists a family of natural transformations $\tau_n\colon F_n \to G_{n+1}$ forming a homotopy $\tau X = (\tau_n X)_n\colon \varphi X \simeq \psi X\colon FX \to GX$.

(ii) More generally, the same conclusion holds if each F_n is essentially free on a subset $\mathcal{M}_n \subset \mathcal{M}$, and G is acyclic on $\mathcal{M}$.

Proof See 5.6.9. $\qquad\qquad\qquad\qquad\qquad\qquad\qquad\qquad\qquad\qquad\quad$ □

5.6.4 Theorem (Equivalence of the simplicial and cubical form)

The simplicial and the cubical functors of singular chains

$$S_+\colon \mathsf{Top} \to \mathrm{Ch}_+\mathsf{Ab}, \qquad C_+\colon \mathsf{Top} \to \mathrm{Ch}_+\mathsf{Ab}, \tag{5.110}$$

are linked by natural transformations $\varphi\colon S_+ \rightleftarrows C_+ \colon \psi$, whose composites are naturally chain homotopic to identities.

The corresponding homology functors are naturally isomorphic.

Proof (a) First we take in Top the set of simplicial models Δ^n, for $n \geqslant 0$.

We have already seen, in 5.6.1(a), that the simplicial functor S_+ is free on these models. On the other hand, the cubical functor C_+ is acyclic on them (and all contractible spaces).

We have thus a natural transformation $\varphi \colon S_+ \to C_+$ that lifts the obvious isomorphism $\Phi \colon H_0 S_+ \to H_0 C_+$ that takes the simplicial homology class $[x]_\triangle$ of a point to the cubical homology class $[x]_\square$ of the same point (see 2.8.6(e)); φ is determined up to natural chain homotopy.

(b) Secondly, we take in Top the set of cubical models $\mathbb{I}^n$, for $n \geqslant 0$.

On these models, the cubical functor C_+ is essentially free, as we have seen in 5.6.2(b), and the simplicial functor $S_+ \colon \mathsf{Top} \to \mathsf{Ch}_+\mathsf{Ab}$ is acyclic, by Lemma 2.8.7.

We have thus a natural transformation $\psi \colon C_+ \to S_+$ that lifts the inverse isomorphism Φ^{-1}, and is determined up to natural chain homotopy.

(c) The composite $\psi\varphi :: S_+ \to S_+$ lifts $\mathrm{id}\, H_0 S_+$, and is chain homotopic to the identity. The composite $\varphi\psi$ is also. $\square$

5.6.5 Eilenberg–Zilber Theorem

The functors

$$F^\triangle \colon \mathsf{Top} \times \mathsf{Top} \to \mathsf{Ch}_+\mathsf{Ab}, \qquad F^\triangle(X,Y) = S_+(X) \otimes S_+(Y),$$

$$G^\triangle \colon \mathsf{Top} \times \mathsf{Top} \to \mathsf{Ch}_+\mathsf{Ab}, \qquad G^\triangle(X,Y) = S_+(X \times Y),$$

$$F^\square \colon \mathsf{Top} \times \mathsf{Top} \to \mathsf{Ch}_+\mathsf{Ab}, \qquad F^\square(X,Y) = C_+(X) \otimes C_+(Y),$$

$$G^\square \colon \mathsf{Top} \times \mathsf{Top} \to \mathsf{Ch}_+\mathsf{Ab}, \qquad G^\square(X,Y) = C_+(X \times Y),$$

$$(5.111)$$

are linked by natural chain homotopy equivalences, forming the left square below, commutative up to chain homotopy

$$
\begin{array}{ccc}
F^\triangle(X,Y) \xrightarrow{\ f\ } G^\triangle(X,Y) & \qquad & H_n F^\triangle(X,Y) \xrightarrow{\ f_{*n}\ } H_n G^\triangle(X,Y) \\
h \downarrow \quad \simeq \quad \downarrow k & & h_{*n} \downarrow \quad = \quad \downarrow k_{*n} \\
F^\square(X,Y) \xrightarrow[\ g\ ]{} G^\square(X,Y) & & H_n F^\square(X,Y) \xrightarrow[\ g_{*n}\ ]{} H_n G^\square(X,Y)
\end{array}
\qquad (5.112)
$$

These natural transformations induce in homology a commutative diagram of natural isomorphisms, in the right square; they are determined up to chain homotopy by the latter in degree $n = 0$.

Proof It is a consequence of the Acyclic Model Theorem, as proved in the article [EZ2].

(a) First, we take as models of the category $\mathsf{Top} \times \mathsf{Top}$ the pairs (Δ^p, Δ^q), for $p, q \geqslant 0$; in dimension n there are $n + 1$ models.

The functors $F^\square$ and $G^\square$ are acyclic on these models, while:

(i) $F^\triangle$ is acyclic and free on these models, with distinguished elements $\eta_{pq} = \mathrm{id}\,(\Delta^p) \otimes \mathrm{id}\,(\Delta^q)$,

(ii) $G^\triangle$ is acyclic and free on these models, with distinguished elements $\xi_{pq} = \mathrm{id}\,(\Delta^p \times \Delta^q)$.

(b) Secondly, we take as models of $\mathsf{Top} \times \mathsf{Top}$ the pairs $(\mathbb{I}^p, \mathbb{I}^q)$. Now, the functors $F^\triangle$ and $G^\triangle$ are acyclic on these models, while:

(iii) $F^\square$ is acyclic and essentially free on these models, with

$$\rho_n X \colon F_n^\square(X, Y) \to \tilde{F}_n^\square(X, Y), \quad \rho_{pq}(a \otimes b) = (a \otimes b, \eta_{pq}),$$

where $a \colon \mathbb{I}^p \to X$ and $b \colon \mathbb{I}^q \to Y$ are non-degenerate cubes, and $\eta_{pq} = \mathrm{id}\,(\mathbb{I}^p) \otimes \mathrm{id}\,(\mathbb{I}^q)$,

(iv) $G^\square$ is acyclic and essentially free on these models, with

$$\rho_n X \colon G_n^\square(X, Y) \to \tilde{G}_n^\square(X, Y), \quad \rho_{pq}(a \times b) = (a \times b, \xi_{pq}),$$

where $a \colon \mathbb{I}^p \to X$ and $b \colon \mathbb{I}^q \to Y$ are non-degenerate cubes, and $\xi_{pq} = \mathrm{id}\,(\mathbb{I}^p \times \mathbb{I}^q)$.

(c) The bijective correspondence $\Pi_0(X) \times \Pi_0(Y) \to \Pi_0(X \times Y)$ (between the bases of H_0) gives the right commutative square in (5.112), for $n = 0$ (natural on Top^2).

Applying the Acyclic Model Theorem we have the thesis. $\qquad\square$

5.6.6 *Corollary* (Eilenberg–Zilber)

For topological spaces X, Y there is a short exact sequence

$$\bigoplus_{p+q=n} H_p X \otimes H_q Y \overset{\alpha_n}{\rightarrowtail} H_n(X \times Y) \overset{\beta_n}{\twoheadrightarrow} \bigoplus_{p+q=n-1} \mathrm{Tor}(H_p X, H_q Y)$$

which is natural on $\mathsf{Top} \times \mathsf{Top}$, and splits in a non-natural way.

Proof It is a straightforward consequence of Theorem 5.6.5 (using either the cubical or the simplicial chains) and the Künneth Theorem 5.4.4. $\qquad\square$

5.6.7 Exercises and complements

There are given two spaces X, Y.

(a) If $H_n(X) = 0$ for $n > n'$ and $H_n(Y) = 0$ for $n > n''$, then $H_n(X \times Y) = 0$ for $n > n' + n''$.

(b) If the homology groups of the space X are flat, there is a canonical isomorphism

$$\alpha_n \colon \bigoplus_{p+q=n} H_p(X) \otimes H_q(Y) \longrightarrow H_n(X \times Y), \qquad (5.113)$$

which is natural on $\mathsf{Top} \times \mathsf{Top}$. (We only need that $H_p(X)$ be flat for $p < n$.)

(c) (*Poincaré polynomial*) Suppose that each homology group $H_k(X)$ has a finite rank $\beta_k(X)$, eventually zero (see 2.2.7). It is convenient to write the sequence of Betti numbers as a polynomial, in the polynomial ring $\mathbb{Z}[t]$ (with a formal indeterminate t)

$$\beta X(t) = \Sigma_{k \geqslant 0} \, \beta_k(X) t^k \qquad (5.114)$$

called the *Poincaré polynomial* of X. (It determines the graded group $H_*(X)$, when all $H_k(X)$ are free abelian groups.)

If Y satisfies the same hypotheses, this is also true of $X \times Y$ (by Corollary 5.6.6), and we can express $\beta(X \times Y)$ as a product of polynomials

$$\beta(X \times Y)(t) = \Sigma_{k \geqslant 0} \, (\Sigma_{p+q=k} \, \beta_p(X) \, \beta_q(Y)) t^k = \beta X(t) \, \beta Y(t). \quad (5.115)$$

Without assuming the Betti numbers to be eventually zero, we have a formal series in the ring $\mathbb{Z}[[t]]$, with the same result as above: note that every coefficient $\Sigma_{p+q=n} \, \beta_p(X) \, \beta_q(Y)$ is still given by a finite sum.

(d) (*Euler–Poincaré characteristic*) In the same hypotheses on X and Y their Betti numbers are finite and eventually zero), we have

$$\chi(X \times Y) = \chi(X) \, \chi(Y). \qquad (5.116)$$

Hints: $\chi(X)$ can be expressed by the polynomial $\beta X(t)$.

(e) Compute the homology groups of the n-torus $\mathbb{T}^n = \mathbb{S}^1 \times \ldots \times \mathbb{S}^1$ (with n factors).

5.6.8 Complements

The equivalence 5.6.5 between the functors $F^\square$ and $G^\square$, or $F^\triangle$ and $G^\triangle$, was obtained by applying the Acyclic Models Theorem. Yet, there are specific candidates for some of these chain maps (that are determined up to chain homotopy by the homomorphisms induced on H_0).

In fact, there are *easy* constructions $F^\square \to G^\square$ and $G^\triangle \to F^\triangle$, but not the other way round.

(a) (*Product of spaces by cubical chains*) We can build a natural transformation $\zeta \colon F^\square \to G^\square$, exploiting the fact that a product of standard cubes $\mathbb{I}^p \times \mathbb{I}^q$ 'is' the standard cube $\mathbb{I}^{p+q}$.

This gives a chain map $\zeta \colon C_+(X) \otimes C_+(Y) \to C_+(X \times Y)$, where ζ_n has co-components

$$\zeta_{pq} \colon C_p(X) \otimes C_q(Y) \longrightarrow C_n(X \times Y) \qquad (n = p + q),$$
$$\zeta_{pq}(a \otimes b) = a \times b \colon \mathbb{I}^n \to X \times Y \qquad (a \colon \mathbb{I}^p \to X,\, b \colon \mathbb{I}^q \to Y). \tag{5.117}$$

A (more complex) chain map $\eta \colon C_+(X \times Y) \to C_+(X) \otimes C_+(Y)$ such that $\eta\zeta$ is the identity is constructed in [Mas3], Section VI.5; then, a homotopy equivalence $\zeta\eta \simeq \mathrm{id}$ is proved to exist, by the method of Acyclic Models.

(b) (*Product of spaces by simplicial chains*) We have already built, in (5.102), a natural transformation $\vartheta \colon G^\triangle \to F^\triangle$, related to the Alexander–Whitney diagonal approximation.

(c) Singular homology can equivalently be built with the *normalised* simplicial chain complex, namely the quotient of the ordinary complex modulo degenerate simplicial chains.

This can be seen in [HiW], Section 8.2; the equivalence is proved in Section 8.4, by the method of Acyclic Models (and could be proved here, as in Theorem 5.6.4).

5.6.9 Proof of Theorem 5.6.3

(a) We begin to prove (i).

For a space X, consider the following solid diagram

$$
\begin{array}{ccccccc}
\cdots \xrightarrow{\partial_2} & F_1(X) & \xrightarrow{\partial_1} & F_0(X) & \xrightarrow{\partial_0} & (H_0F)X == & F_{-1}(X) \\
& \downarrow{\varphi_1} & & \downarrow{\varphi_0} & & \downarrow{\Phi} & \downarrow{\varphi_{-1}} \\
\cdots \xrightarrow{\partial_2} & G_1(X) & \xrightarrow{\partial_1} & G_0(X) & \xrightarrow{\partial_0} & (H_0G)X == & G_{-1}(X)
\end{array}
\tag{5.118}
$$

where we have augmented the chain complexes FX and GX so that, in both rows, $\partial_0(z) = [z]$ is the natural projection on H_0, and $\mathrm{Im}\,\partial_1 = \mathrm{Ker}\,\partial_0$. We also rename $\Phi = \varphi_{-1}$ for convenience in the induction step, and remark that the lower row is exact (in every degree) when X is a model.

We want to lift the natural transformation $\Phi \colon H_0F \to H_0G$ to a natural transformation $\varphi \colon F \to G$. The argument is a modified version of the lifting procedure in 5.1.2(b) (on projective resolutions): here we are bound by the naturality requirement; moreover, we only know that the complex FX is free on the given models and the complex GX is acyclic on models.

(b) *Case $n = 0$.* The first lifting, namely $\varphi_0 X \colon F_0(X) \to G_0(X)$, already has to pass through the models. First, for every distinguished element $\xi_j \in F_0(M_j)$, with $j \in J_0$, we choose an element $(\varphi_0 M_j)(\xi_j) \in G_0(M_j)$ such that $\partial_0(\varphi_0 M_j)(\xi_j) = (\Phi M_j)\partial_0(\xi_j)$

$$
\begin{array}{ccccccc}
F_0(X) & \xleftarrow{\ F_0(a)\ } & F_0(M_j) & \xrightarrow{\ \partial_0\ } & (H_0 F)M_j & \longrightarrow & 0 \\
\ \downarrow{\varphi_0} & & \ \downarrow{\varphi_0} & & \downarrow{\Phi} & & \\
G_0(X) & \xleftarrow[G_0(a)]{} & G_0(M_j) & \xrightarrow[\partial_0]{} & (H_0 G)M_j & \longrightarrow & 0
\end{array}
\qquad (5.119)
$$

Note that we have not (yet) defined $\varphi_0 M_j$ on the whole $F_0(M_j)$. Rather, we define $\varphi_0 X$ for an arbitrary space, on the basis of $F_0(X)$ formed by the elements $a_j = (F_0 a)(\xi_j)$ (for $j \in J_0$ and all maps $a \colon M_j \to X$), letting

$$
\varphi_0 X \colon F_0(X) \to G_0(X), \qquad (\varphi_0 X)(a_j) = (G_0 a)(\varphi_0 M_j)(\xi_j), \qquad (5.120)
$$

as required by the left square above.

The family of these components is natural: for a map $f \colon X \to Y$ we have $(G_0 f)(\varphi_0 X) = (\varphi_0 Y)(F_0 f)$, as we verify on the basis $a_j = (F_0 a)(\xi_j)$

$$
(G_0 f)(\varphi_0 X)(a_j) = (G_0 f)(G_0 a)(\varphi_0 M_j)(\xi_j)
$$

$$
= (G_0(fa))(\varphi_0 M_j)(\xi_j) = (\varphi_0 Y)(F_0(fa))(\xi_j) = (\varphi_0 Y)(F_0 f)(a_j).
$$

Moreover $H_0\varphi_0 = \Phi$ because the rightmost square in (5.118) commutes, by the naturality of Φ

$$
(\Phi X)\partial_0(a_j) = (\Phi X)\partial_0(F_0 a)(\xi_j) = \Phi(H_0 F(a))\partial_0(\xi_j)
$$

$$
= (H_0 G(a))(\Phi M_j)\partial_0(\xi_j) = (H_0 G(a))\partial_0(\varphi_0 M_j)(\xi_j)
$$

$$
= \partial_0(G_0 a)(\varphi_0 M_j)(\xi_j) = \partial_0(\varphi_0 X)(a_j).
$$

(c) *Inductive step.* Suppose that, for $n \geqslant 1$, we have defined the natural transformations $\varphi_0, \varphi_1, \ldots \varphi_{n-1} \colon F_{n-1} \to G_{n-1}$, commuting so far with the differentials.

We want to lift $\varphi_n X \colon F_n(X) \to G_n(X)$. For each distinguished element $\xi_j \in F_n(M_j)$ ($j \in J_n$) we choose an element $(\varphi_n M_j)(\xi_j) \in G_n(M_j)$ such that $\partial_n(\varphi_n M_j)(\xi_j) = (\varphi_{n-1} M_j)(\partial_n \xi_j)$

$$
\begin{array}{ccccccc}
F_n(X) & \xleftarrow{\ F_n(a)\ } & F_n(M_j) & \xrightarrow{\ \partial_n\ } & F_{n-1}(M_j) & \xrightarrow{\ \partial_{n-1}\ } & F_{n-2}(M_j) \\
\ \downarrow{\varphi_n} & & \ \downarrow{\varphi_n} & & \downarrow{\varphi_{n-1}} & & \downarrow{\varphi_{n-2}} \\
G_n(X) & \xleftarrow[G_n(a)]{} & G_n(M_j) & \xrightarrow[\partial_n]{} & G_{n-1}(M_j) & \xrightarrow[\partial_{n-1}]{} & G_{n-2}(M_j)
\end{array}
\qquad (5.121)
$$

This is possible, because $\partial_{n-1}(\varphi_{n-1} M_j)(\partial_n \xi_j) = (\varphi_{n-2} M_j)\partial_{n-1}(\xi_j) = 0$, and $(\varphi_{n-1} M_j)(\partial_n \xi_j)$ is a boundary in $G_{n-1}(M_j)$ (by exactness).

Now we define the component $\varphi_n X$, on the basis formed by the elements $a_j = (F_n a)(\xi_j)$ (for $j \in J_n$ and all maps $a\colon M_j \to X$), letting

$$\varphi_n X\colon F_n(X) \to G_n(X), \qquad (\varphi_n X)(a_j) = (G_n a)(\varphi_n M_j)(\xi_j), \qquad (5.122)$$

and we verify as above that this is a natural transformation and preserves commutativity, in diagram (5.118).

(d) *Homotopy.* We suppose that $\varphi X\colon FX \to GX$ lifts the trivial homomorphism $(H_0 F)X \to (H_0 G)X$, and we want to prove that it is homotopic to the zero transformation $0\colon FX \to GX$, by a family of natural transformations $\tau_n\colon F_n \to G_{n+1}$ such that $\partial_{n+1}\tau_n + \tau_{n-1}\partial_n = \varphi_n$

$$
\begin{array}{ccccccc}
\cdots\, F_2(X) & \xrightarrow{\partial_2} & F_1(X) & \xrightarrow{\partial_1} & F_0(X) & \xrightarrow{\partial_0} & (H_0 F)X \\
\downarrow{\varphi_2} \quad {}^{\tau_1} & & \downarrow{\varphi_1} \quad {}^{\tau_0} & & \downarrow{\varphi_0} & & \downarrow{0} \\
\cdots\, G_2(X) & \xrightarrow[\partial_2]{} & G_1(X) & \xrightarrow[\partial_1]{} & G_0(X) & \xrightarrow[\partial_0]{} & (H_0 G)X
\end{array}
\qquad (5.123)
$$

Again, the lifting is an enriched version of the corresponding argument for projective resolutions, in Exercise 5.1.2(c).

(e) *Case $n = 0$.* We have to build a natural transformation $\tau_0 X\colon F_0(X) \to G_1(X)$ such that $\partial_1 \tau_0 = \varphi_0$.

First, for every distinguished element $\xi_j \in F_0(M_j)$ $(j \in J_0)$, we choose an element $(\tau_0 M_j)(\xi_j) \in G_1(M_j)$ such that $\partial_1(\tau_0 M_j)(\xi_j) = (\varphi_0 M_j)(\xi_j)$

$$
\begin{array}{ccccccc}
& F_0(X) & \xleftarrow{F_0(a)} & F_0(M_j) & \xrightarrow{\partial_0} & (H_0 F)M_j \\
{}^{\tau_0} & & {}^{\tau_0} & \downarrow{\varphi_0} & & \downarrow{0} \\
G_1(X) & \xleftarrow[G_0(a)]{} & G_1(M_j) & \xleftarrow[G_0(a)]{} & G_0(M_j) & \xrightarrow[\partial_0]{} & (H_0 G)M_j
\end{array}
\qquad (5.124)
$$

This is legitimate: $(\varphi_0 M_j)(\xi_j)$ is a boundary in $G_0(M_j)$, by the right square above. We define the component $\tau_0 X$ on the usual basis formed by the elements $a_j = (F_0 a)(\xi_j)$ (for $j \in J_0$ and all maps $a\colon M_j \to X$), letting

$$\tau_0 X\colon F_0(X) \to G_1(X), \qquad (\tau_0 X)(a_j) = (G_1 a)(\tau_0 M_j)(\xi_j). \qquad (5.125)$$

We verify as in (b) the naturality of τ_0 and the identity $\partial_1 \tau_0 = \varphi_0$.

(f) The inductive step, defining the natural transformation $\tau_n X\colon F_n(X) \to G_{n+1}(X)$ after the lower ones, is performed as in (d).

(g) We end by showing how the previous part is modified, to prove (ii).

By hypothesis, every functor $F_n\colon \mathsf{C} \to \mathsf{Ab}$ has a natural retraction $\pi\rho\colon F_n \to \tilde{F}_n \to F_n$. We recall that, for an object X of C, $\tilde{F}_n(X)$ is the free abelian group generated by all pairs (a, ξ), where $a\colon M \to X$ is a morphism of C defined on a model $M \in \mathcal{M}_n$ and $\xi \in F_n(M)$.

We have the solid diagram:

$$
\begin{array}{ccccccc}
\dots\ F_2(X) & \xrightarrow{\partial_2} & F_1(X) & \xrightarrow{\partial_1} & F_0(X) & \xrightarrow{\partial_0} & (H_0F)X \\
\ \downarrow{\scriptstyle \rho_2} & & \ \downarrow{\scriptstyle \rho_1} & & \ \downarrow{\scriptstyle \rho_0} & & \\
\dots\ \tilde{F}_2(X) & & \tilde{F}_1(X) & & \tilde{F}_0(X) & & \Big\downarrow{\scriptstyle \Phi} \\
\ \vdots\,{\scriptstyle \tilde{\varphi}_2} & & \vdots\,{\scriptstyle \tilde{\varphi}_1} & & \vdots\,{\scriptstyle \tilde{\varphi}_0} & & \\
\dots\ G_2(X) & \xrightarrow[\partial_2]{} & G_1(X) & \xrightarrow[\partial_1]{} & G_0(X) & \xrightarrow[\partial_0]{} & (H_0G)X
\end{array}
\tag{5.126}
$$

For every $\xi \in F_0(M)$, with $M \in \mathcal{M}_0$, we choose an element $(\varphi_0 M)(\xi) \in G_0(M)$ such that $\partial_0(\varphi_0 M)(\xi) = (\Phi M)\partial_0(\xi)$. Then we define the homomorphism $\tilde{\varphi}_0 X \colon \tilde{F}_0(X) \to G_0(X)$ letting, for $a \colon M \to X$

$$
\tilde{\varphi}_0(a, \xi) = (G_0 a)(\varphi_0 M)(\xi),
$$

and we let $\varphi_0 X = (\tilde{\varphi}_0 X)(\rho_0 X) \colon F_0(X) \to G_0(X)$.

The rest is similarly modified. $\qquad\qquad\Box$

6

An introduction to homotopy groups

Homotopy groups explore n-dimensional properties of a space X, by maps $\mathbb{S}^n \to X$. Their study, quite complex, forms a large domain of Algebraic Topology. Here we only want to present the basic ideas and the relationship between these groups and singular homology.

After the definition and basic properties of the fundamental groupoid and fundamental group we sketch how they are computed and the definition of the higher homotopy groups.

This chapter can be read together with Chapter 2, and is mostly written in an elementary way.

Literature. The fundamental group of a pointed space, also called the *Poincaré group*, was introduced in 1895 by H. Poincaré in a celebrated article called 'Analysis situs' [Po]. It was the first algebraic structure used to study spaces, and the beginning of Algebraic Topology — before topology had this name.

Many books on homology theories deal with the fundamental group. Elementary presentations can be found in [Ha], Chapter 1, and [Mu], Chapter 9. For the fundamental groupoid we refer to R. Brown's article and books [Br1]–[Br3].

Higher homotopy groups can be approached with [Ha], Chapter 4. A classical reference on Homotopy Theory is a book by S.T. Hu [Hu].

6.1 The fundamental groupoid

The fundamental groupoid $\Pi_1 X$ of a space is a global structure, that contains the fundamental group $\pi_1(X, x)$ at each basepoint $x \in X$, all of them nicely organised inside the groupoid — a (small) category where all arrows are invertible.

This structure gives raise to a functor $\Pi_1 \colon \mathsf{Top} \to \mathsf{Gpd}$ that transforms topological spaces into small groupoids, their maps into functors, and their homotopies into natural isomorphisms, linking the two sides of the analogy remarked in 1.5.5.

6.1.1 Reviewing paths

The basic terminology on paths, loops and homotopies was fixed in Section 1.4.

A path $a \colon \mathbb{I} \to X$ from x to x' (its endpoints) is viewed as a homotopy $a \colon x \simeq x' \colon \{*\} \to X$. If $x = x'$, the path is called a loop at x (its basepoint).

We write as $P(X)$ the set of paths of the space X, as $P(X, x, x')$ the set of paths from x to x', as $\Omega(X, x) = P(X, x, x)$ the set of loops at x.

Consecutive paths in $P(X)$ have a concatenation. The *regular concatenation* of a sequence of consecutive paths is written as $a_1 * a_2 * \ldots * a_n$ (cf. (1.74)), the reversed path as $a^\sharp$.

Exercises. There are given a path $a \colon x \simeq y$, a loop $b \colon y \simeq y$, and a path $c \colon y \simeq z$ in a space X.

(a) Write the parametrisation of the paths $a * b * c$ and $a * b * b * c$.

(b) Write the parametrisation of the loop $a * b * a^\sharp \in \Omega(X, x)$.

6.1.2 The fundamental groupoid

We associate to a space X a small groupoid $\Pi_1 X$, called the *fundamental groupoid* of X.

An object (or *vertex*) of $\Pi_1 X$ is any point $x \in X$; an arrow $[a] \colon x \to x'$ is a class of paths in X, from x to x', up to homotopy with fixed endpoints.

In other words (as already said in 1.4.7), two paths $a_0, a_1 \colon \mathbb{I} \to X$ from x to x' are identified if there exists a continuous mapping $h \colon \mathbb{I} \times \mathbb{I} \to X$ which deforms one into the other, without moving the endpoints

$$h(s, i) = a_i(s), \quad h(0, t) = x, \quad h(1, t) = x' \qquad (s, t \in \mathbb{I},\ i = 0, 1), \qquad (6.1)$$

The map h is also called a 2-homotopy from a_0 to a_1 and denoted as $h \colon a_0 \simeq_2 a_1 \colon x \simeq x'$. The 2-homotopy relation $a_0 \simeq_2 a_1$ is an equivalence relation in the set of paths $P(X, x, x')$, as verified in Exercise 1.4.7(b).

In $\Pi_1 X$ consecutive arrows $[a]\colon x \to x'$ and $[b]\colon x' \to x''$ are composed by concatenation of paths (written in the natural order):

$$[a]\,[b] = [a * b]\colon x \to x''. \tag{6.2}$$

As verified in the exercises below this operation is well defined and gives a groupoid, with identities represented by constant loops and inverses represented by reversed paths.

At each point $x \in X$ the endoarrows $[a]\colon x \to x$ of $\Pi_1 X$ form the fundamental group

$$\pi_1(X, x) = (\Pi_1 X)(x, x), \tag{6.3}$$

studied in the next section.

Exercises and complements. These exercises are an easy extension of the corresponding issues for the fundamental group — in case this is already known to the reader.

(a) In $\Pi_1 X$, the partial operation of composition is well defined.

(b) It is associative. This is a particular case of 1.4.7(d), but — owing to its importance — it can be useful to give a direct proof. *Hints:* the concatenations $a * (b * c)$ and $(a * b) * c$ of three consecutive paths in X are interchanged by an invertible reparametrisation of $\mathbb{I}$.

(c) Every vertex $x \in X$ has an identity, namely the class $[e_x]$ of the constant loop at x.

(d) The inverse of $[a]\colon x \to y$ is the class $[a]^{-1} = [a^\sharp]$ of the reversed path.

(e) One can give an equivalent construction by 'Moore paths', parametrised on intervals $[0, r]$, so that concatenation is associative from the start; [HiW] follows this way for the fundamental group.

6.1.3 The fundamental-groupoid functor

The previous construction gives a functor with values in the category of small groupoids

$$\Pi_1 \colon \mathsf{Top} \to \mathsf{Gpd}, \qquad X \mapsto \Pi_1 X,$$
$$f_* \colon \Pi_1 X \to \Pi_1 Y, \qquad f_*(x) = f(x), \quad f_*([a]) = [fa]. \tag{6.4}$$

A map $f\colon X \to Y$ is sent to the functor $\Pi_1 f = f_*$ which takes the class of a path $a\colon x \simeq x'$ in X to the class of the path $fa\colon fx \simeq fx'$ in Y.

This requires some straightforward verifications, listed in the exercises below.

Exercises and complements. (a) The mapping

$$f_* \colon \Pi_1 X(x, x') \to \Pi_1 Y(fx, fx')$$

is well defined.

(b) The transformation $f_* \colon \Pi_1 X \to \Pi_1 Y$ is a functor between groupoids.

(c) (*Global functorial properties*) $\Pi_1 \colon \mathsf{Top} \to \mathsf{Gpd}$ is a functor:

- given two consecutive maps $f \colon X \to Y$ and $g \colon Y \to Z$, $(gf)_* = g_* f_*$,
- given a space X, the functor $(\mathrm{id}\, X)_*$ is the identity of the groupoid $\Pi_1 X$.

(d) In degree zero, we have already seen in (1.69) the homotopy invariant functor $\Pi_0 \colon \mathsf{Top} \to \mathsf{Set}$, that takes every space to the set of its path components

$$\Pi_0 \colon \mathsf{Top} \to \mathsf{Set}, \qquad \Pi_0(X) = |X|/ \simeq,$$
$$f_* = \Pi_0(f) \colon \Pi_0(X) \to \Pi_0(Y), \qquad f_*([x]_P) = [f(x)]_P, \tag{6.5}$$

where $[x]_P$ is the path component of x in X, and $f \colon X \to Y$ is a map.

This functor can be seen as a composite $F\Pi_1 \colon \mathsf{Top} \to \mathsf{Gpd} \to \mathsf{Set}$.

6.1.4 Theorem (Homotopy equivariance)

(a) The functor Π_1 has a canonical enrichment: a homotopy $\varphi \colon f \simeq g \colon X \to Y$ has an associated homotopy of groupoids, i.e. a natural isomorphism

$$\varphi_* \colon f_* \to g_* \colon \Pi_1 X \to \Pi_1 Y,$$
$$\varphi_*(x) = [\varphi(x, -)] \in \Pi_1 Y(f(x), g(x)) \qquad (x \in X), \tag{6.6}$$

where $\varphi(x, -) \colon \mathbb{I} \to Y$ is a path from $f(x)$ to $g(x)$ in Y.

The naturality condition on an arrow $[a] \colon x \to x'$ in $\Pi_1 X$ means the commutativity of the following square, in the groupoid $\Pi_1 Y$

$$
\begin{array}{ccc}
f(x) & \xrightarrow{\ \varphi_*(x)\ } & g(x) \\
{\scriptstyle [fa]}\big\downarrow & & \big\downarrow{\scriptstyle [ga]} \\
f'(x) & \xrightarrow[\ \varphi_*(x')\ ]{} & g'(x)
\end{array}
\tag{6.7}
$$

(b) A homotopy equivalence of spaces is taken by Π_1 to a categorical equivalence of groupoids.

Proof We define $\varphi_*(x) = [\varphi(x, -)]$, as in (6.6): this makes sense, because $\varphi_x = \varphi(x, -) \colon \mathbb{I} \to Y$ is a path from $f(x)$ to $g(x)$, in Y, and its class $[\varphi_x]$ is an arrow in $\Pi_1 Y$.

The commutativity of the square (6.7) follows from the continuous mapping $\Phi = \varphi(a \times \mathrm{id}\,\mathbb{I}) \colon \mathbb{I}^2 \to Y$, whose boundary is represented in the diagram below

$$
\begin{array}{ccc}
 & ga & \\
\varphi_x & \Phi & \varphi_{x'} \\
 & fa & s
\end{array}
\tag{6.8}
$$

Applying Lemma 6.1.8, we get a 2-homotopy $fa * \varphi_{x'} \simeq_2 \varphi_x * ga$.

(b) Given a homotopy equivalence $f \colon X \rightleftarrows Y \colon g$, with homotopies $gf \simeq 1_X$ and $fg \simeq 1_Y$, we get two functors $f_* \colon \Pi_1 X \rightleftarrows \Pi_1 Y \colon g_*$ whose composites are isomorphic to identity functors. $\qquad\square$

6.1.5 *Exercises and complements* (Path connectedness)

(a) As in 1.4.6, we say that a space X is 0-connected if it is non-empty and path connected.

Equivalently, it has precisely one path component: $\Pi_0(X)$ is a singleton. Equivalently again, $\Pi_1 X$ is a *connected* groupoid: it is not empty and for any two points $x, x' \in X$ there is some arrow $x \to x'$ in $\Pi_1 X$.

(b) The following conditions on a space X are equivalent:

(i) any two paths $a, b \colon x \simeq y$ in X are 2-homotopic,
(ii) any loop $a \colon x \simeq x$ in X is 2-homotopic to the constant loop e_x,
(iii) for any $x, x' \in X$ there is at most one arrow $x \to x'$ in $\Pi_1 X$,
(iv) for any $x \in X$ there is a unique arrow $x \to x$ in $\Pi_1 X$, its identity.

(c) The following conditions are equivalent, and define a *1-connected* space (as anticipated in 1.4.7)

(i′) X is non-empty, path connected, and satisfies the equivalent properties above,
(ii′) the groupoid $\Pi_1 X$ is non-empty and *indiscrete*: for any $x, x' \in X$ there is a unique arrow $x \to x'$ in $\Pi_1 X$,
(iii′) $\Pi_1 X$ is equivalent to the category **1**.

*A further equivalent condition can be added, using a result that will be proved in 6.3.1
(iv′) X is non-empty, path connected, and $\pi_1(X, x) = 0$ for some $x \in X$.*

(d) The property of being 0-connected, or simply connected, or 1-connected, is invariant up to homotopy equivalence. Every contractible space is 1-connected. $\mathbb{S}^0$ and $\mathbb{S}^1$ are not, but we shall see in 6.4.4(a) that all higher spheres $\mathbb{S}^n$ are (for $n \geqslant 2$).

(e) A space X is said to be *totally path disconnected* if all its path components are singletons. This is equivalent to saying that the groupoid $\Pi_1 X$ is discrete, i.e. its only arrows are the identities. This property is also invariant up to homotopy equivalence.

Every discrete space and the rational line $\mathbb{Q}$ (with euclidean topology) are of this kind. (We may require that there are at least two points!)

6.1.6 Proposition (Cartesian products)

The functor $\Pi_1 \colon \mathsf{Top} \to \mathsf{Gpd}$ preserves cartesian products, as made precise in the following points.

(a) For a product $X = \Pi_{i \in I} X_i$ of topological spaces (with projections $p_i \colon X \to X_i$), there is a canonical isomorphism of groupoids

$$\varphi \colon \Pi_1 X \to \Pi_{i \in I} \Pi_1 X_i, \qquad \varphi([a]) = ([p_i a])_{i \in I}. \tag{6.9}$$

(b) (Naturality) Every family of maps $f_i \colon X_i \to Y_i$ gives a commutative square in Gpd (with $f = \Pi_i f_i$)

$$
\begin{array}{ccc}
\Pi_1(\Pi_i X_i) & \xrightarrow{\ \varphi\ } & \Pi_i(\Pi_1 X_i) \\[2pt]
{\scriptstyle f_*}\downarrow & & \downarrow{\scriptstyle \Pi_i f_{i*}} \\[2pt]
\Pi_1(\Pi_i Y_i) & \xrightarrow[\ \varphi\ ]{} & \Pi_i(\Pi_1 Y_i)
\end{array}
\tag{6.10}
$$

Note. The commutative square (6.10) says that (6.9) is the component on the family $(X_i)_{i \in I}$ of an isomorphism of functors defined on the cartesian power category $\mathsf{Top}^I = \Pi_{i \in I} \mathsf{Top}$

$$\varphi \colon F \to G \colon \mathsf{Top}^I \to \mathsf{Gpd},$$
$$F((X_i)_{i \in I}) = \Pi_1(\Pi_{i \in I} X_i), \qquad G((X_i)_{i \in I}) = \Pi_{i \in I}(\Pi_1 X_i). \tag{6.11}$$

Proof (a) A path $a \colon x \simeq x'$ in $X = \Pi_{i \in I} X_i$ gives a path $p_i a \colon p_i x \simeq p_i x'$ in each X_i, and amounts to their family $(p_i a)_{i \in I}$, by the universal property of the product (in 1.6.1(i)). Similarly a 2-homotopy $h \colon a \simeq_2 b \colon x \simeq x'$ in X amounts to the family of 2-homotopies $p_i h \colon p_i a \simeq_2 p_i b$ $(i \in I)$.

The mapping $a \mapsto (p_i a)_{i \in I}$ is consistent with path-concatenation, and induces an isomorphism of groupoids $\varphi \colon [a] \mapsto ([p_i a])_{i \in I}$.

(b) In the square (6.10), the two composed diagonal functors send an arrow $[a] \in \Pi_1 X$ to the same I-indexed family in $\Pi_i(\Pi_1 Y_i)$

$$\varphi f_*([a]) = \varphi([fa]) = ([p_i fa])_{i \in I},$$
$$(\Pi_i f_{i*})\,\varphi([a]) = (\Pi_i f_{i*})([p_i a])_{i \in I}) = ([f_i p_i a])_{i \in I}.$$

$\square$

6.1.7 Double homotopies and 2-homotopies

A 2-cube $A\colon \mathbb{I}^2 \to X$ will be seen as a *double homotopy* among four paths, its faces

$$\partial_1^\alpha A = a_\alpha, \qquad \partial_2^\alpha A = b_\alpha. \tag{6.12}$$

It is a 2-homotopy $b_0 \simeq_2 b_1$ when both paths $a_\alpha = \partial_1^\alpha A$ are constant.

Double homotopies have a *horizontal concatenation*, in the first variable, provided they are horizontally consecutive $(\partial_1^+ A = \partial_1^- B)$

$$(A *_1 B)(s,t) = \begin{cases} A(2s,t), & \text{for } 0 \leqslant s \leqslant 1/2, \\ B(2s-1,t), & \text{for } 1/2 \leqslant s \leqslant 1. \end{cases} \tag{6.13}$$

Symmetrically, there is a *vertical concatenation $A *_2 C$*, in the second variable.

We shall often use the following double homotopies, produced by a path $a\colon x \simeq y$

$$M(s,t) = a(\max(s,t)), \qquad L(s,t) = a(\min(s,t)), \tag{6.14}$$

These double homotopies come out of the higher structure of the interval mentioned in 1.4.3(d): the connections $g^\alpha\colon \mathbb{I}^2 \to \mathbb{I}$.

6.1.8 Lemma

The double homotopy $A\colon \mathbb{I}^2 \to X$ represented in (6.12) determines a 2-homotopy

$$h\colon b_0 * a_1 \simeq_2 a_0 * b_1 \qquad (\partial_1^\alpha A = a_\alpha,\ \partial_2^\alpha A = b_\alpha). \tag{6.15}$$

Proof We let L and M be the double homotopies (6.14), associated to the paths a_0 and a_1.

We form the (regular) horizontal concatenation $L *_1 A *_1 M$, represented below, with different units on the axes

$$
\begin{array}{c|ccc|c}
 & a_0 & b_1 & & e_y \\
\hline
e_x & L \quad a_0 & A \quad a_1 & M & e_y \\
\hline
 & e_x \quad \tfrac{1}{3} \quad b_0 & \tfrac{2}{3} \quad a_1 & 1 &
\end{array}
\tag{6.16}
$$

This is a 2-homotopy $e_x * b_0 * a_1 \simeq_2 a_0 * b_1 * e_y$. As we already know that

$$
b_0 * a_1 \simeq_2 e_x * b_0 * a_1, \qquad\qquad a_0 * b_1 * e_y \simeq_2 a_0 * b_1,
$$

the proof is done. $\square$

6.1.9 *Representative subsets*

Let A be a subset of the space X. We shall write as $\Pi_1 X | A$ the full subgroupoid of $\Pi_1 X$ with vertices in A. (Brown's notation is $\pi_1 X A$.)

We say that A is a *representative subset* of X if it meets all its path connected components. Then $\Pi_1 X | A$ is categorically equivalent to $\Pi_1 X$ — possibly a simpler representation (see Exercise (b)).

Exercises and complements. (a) If A is a singleton $\{x_0\}$, $\Pi_1 X | A$ is the fundamental group $\pi_1(X, x_0)$.

(b) If A is representative in X, the subgroupoid $\Pi_1 X | A$ is a *deformation retract* of $\Pi_1 X$: there exists a functor $P \colon \Pi_1 X \to \Pi_1 X | A$ such that, letting $J \colon \Pi_1 X | A \to \Pi_1 X$ be the inclusion, $PJ = \mathrm{id}\,(\Pi_1 X | A)$ and $JP \cong \mathrm{id}\,\Pi_1 X$.

Hints: choose, for every $x \in X$, an arrow $\lambda_x \colon x \to P(x)$ in $\Pi_1 X$, with $P(x) \in A$.

  This statement is a particular case of the characterisation 7.1.8(iii) of equivalence of categories: in fact, the subset A is representative in X if and only if the inclusion functor $J \colon \Pi_1 X | A \to \Pi_1 X$ is essentially surjective on objects. The general case is proved as this particular case.

6.2 The fundamental group

At each basepoint $x \in X$, the fundamental group $\pi_1(X, x)$ is the group of automorphisms of the fundamental groupoid at the vertex x. Viewing these groups inside $\Pi_1 X$ makes evident how all of them are related.

We compute here the fundamental group of the circle and the n-torus $\mathbb{T}^n$.

Various exercises, in 6.2.6 and 6.2.7, show how one can get a concrete grasp of the fundamental group, making 'experiments' with elastic bands and strings in our 3-dimensional space, where we place obstructions (like rods or rings) that our material loops cannot trespass.

6.2.1 Definition

For every $x \in X$, the fundamental group $\pi_1(X, x)$ is the group of automorphisms $\Pi_1 X(x) = \Pi_1 X(x, x)$ of the groupoid $\Pi_1 X$, at the vertex x.

In other words, $\pi_1(X, x)$ is the set $\Omega(X, x)/\simeq_2$ of classes $[a]$ of loops of X at the point x, modulo homotopy with fixed basepoint; this set is equipped with the operation induced by loop concatenation:

$$[a][b] = [a * b]. \tag{6.17}$$

By 6.1.2, we have the *fundamental-group functor*, defined on the category of pointed spaces, with values in the category of groups

$$\pi_1 \colon \mathsf{Top}_{\bullet} \to \mathsf{Gp}, \qquad (X, x) \mapsto \pi_1(X, x),$$
$$f_* \colon \pi_1(X, x) \to \pi_1(Y, y), \qquad f_*([a]) = [fa]. \tag{6.18}$$

Here a pointed map $f \colon (X, x) \to (Y, y)$ is sent to a group-homomorphism $\pi_1(f) = f_*$; the latter takes the class of a loop $a \in \Omega(X, x)$ to the class of the loop $fa \in \Omega(Y, y)$.

6.2.2 Comments and complements

(a) Plainly, the group $\pi_1(X, x)$ only depends on the path component of x in X. Basepoints in different path components of a space give unrelated fundamental groups.

More precisely, if X' is the path component of x in X, the inclusion $u \colon (X', x) \to (X, x)$ induces a canonical isomorphism $u_* \colon \pi_1(X', x) \to \pi_1(X, x)$. Identifying a loop $a \colon \mathbb{I} \to X'$ with $ua \colon \mathbb{I} \to X$ (as we generally did in the context of singular cubes), we also identify $\pi_1(X', x)$ with $\pi_1(X, x)$.

(b) Changing the basepoint within a path component of X will give a non-canonical isomorphism, depending on the 2-homotopy class of the path 'along which we choose to move' (see 6.3.1).

(c) The functor π_1 can be identified with the homotopy functor associated to the pointed circle $(\mathbb{S}^1, e_1)$, as defined in (1.71)

$$\pi_1(X, x_0) = [(\mathbb{S}^1, e_1), (X, x_0)]_{\bullet}. \tag{6.19}$$

Indeed, a loop $a \colon \mathbb{I} \to X$ at x_0 amounts to a pointed map $\hat{a} \colon (\mathbb{S}^1, e_1) \to (X, x_0)$, and a homotopy $h \colon a \simeq_2 b$ with fixed basepoint amounts to a pointed homotopy $\hat{h} \colon \hat{a} \simeq \hat{b}$.

(d) The pointed singleton has a unique loop, and a trivial fundamental group.

(e) The pointed set $\pi_0(X, x_0)$, introduced in (1.70), will be dealt with in Section 6.5.

6.2.3 *Exercises and complements* (Homotopy invariance)

It is easy to prove the invariance of the fundamental group up to pointed homotopy. A more important result, about non-pointed homotopies, will be given in Theorem 6.3.2.

(a) For a pointed homotopy $\varphi\colon f \simeq g\colon (X, x_0) \to (Y, y_0)$ (see 1.4.5), we have $f_* = g_*\colon \pi_1(X, x_0) \to \pi_1(Y, y_0)$.

(b) (*Corollary*) A pointed homotopy equivalence $f\colon (X, x_0) \rightleftarrows (Y, y_0) : g$ induces two isomorphisms $\pi_1(X, x_0) \rightleftarrows \pi_1(Y, y_0)$, inverse to each other.

6.2.4 *Proposition* (Cartesian products)

The functor $\pi_1\colon \mathsf{Top_\bullet} \to \mathsf{Gp}$ preserves cartesian products: for a product $(X, x) = \prod_{i \in I} (X_i, x_i)$ of pointed spaces (with projections p_i and basepoint $x = (x_i)_{i \in I}$), there is a natural isomorphism of groups

$$\varphi\colon \pi_1(X, x) \to \textstyle\prod_{i \in I} \pi_1(X_i, x_i), \qquad \varphi([a]) = ([p_i a])_{i \in I}. \tag{6.20}$$

Proof It is a straightforward consequence of Proposition 6.1.6 on fundamental groupoids. $\qquad\square$

6.2.5 *Theorem* (The fundamental group of the circle)

There is an isomorphism

$$\varphi\colon \mathbb{Z} \to \pi_1(\mathbb{S}^1, e_1), \qquad \varphi(k) = [a_k], \tag{6.21}$$

where $a_k(t) = (\cos 2k\pi t,\ \sin 2k\pi t)$ is the usual loop winding k times around the circle, with the usual convention for orientation (see (2.104)).

**Note.* The isomorphism (6.21) agrees with the Hurewicz isomorphism $h\colon \pi_1(\mathbb{S}^1, e_1) \to H(\mathbb{S}^1)$ dealt with in 6.3.4.

Proof The argument is based on the exponential map

$$p\colon \mathbb{R} \to \mathbb{S}^1, \qquad p(t) = (\cos 2\pi t,\ \sin 2\pi t), \tag{6.22}$$

that can be represented as the projection of a helix on the plane: see figure (6.29).

We shall study a path $a\colon \mathbb{I} \to \mathbb{S}^1$ by a *lifting* $f\colon \mathbb{I} \to \mathbb{R}$, namely a path such that $pf = a$. In particular, the loop $a_k\colon e_1 \simeq e_1$ in $\mathbb{S}^1$ can be lifted to the path $f_k\colon 0 \simeq k$ in $\mathbb{R}$, where $f_k(t) = kt$ (for $k \in \mathbb{Z}$).

First, we prove that the mapping φ defined in (6.21) is a homomorphism. For $k = h + h'$ in $\mathbb{Z}$, we lift the loop a_h to the path $f = f_h$, the loop a_k to the path $g = f_k$, and the loop $a_{h'}\colon e_1 \simeq e_1$ to the path $f'\colon h \simeq k$, where $f'(t) = h + h't$ (a path consecutive to f).

Then, in the group $\pi_1(\mathbb{S}^1, e_1)$

$$\varphi(h)\varphi(h') = [a_h][a_{h'}] = [a_h * a_{h'}] = [pf * pf'] = [p(f * f')]$$

$$= [pg] = [a_k] = \varphi(k),$$

because $f * f' \simeq_2 g$: in the line, any two paths with the same endpoints are 2-homotopic, by means of the affine homotopy between them.

To prove that φ is an isomorphism we use two crucial *lifting properties* of the exponential map p:

(i) every path $a\colon \mathbb{I} \to \mathbb{S}^1$ starting at $x_0 = a(0)$ has a unique lifting to a path $f\colon \mathbb{I} \to \mathbb{R}$ starting at a given point $y_0 \in p^{-1}\{x_0\}$,

(ii) every homotopy $h\colon a \simeq_2 a'\colon \mathbb{I} \to \mathbb{S}^1$ of paths starting at x_0 has a unique lifting to a 2-homotopy $k\colon f \simeq_2 f'\colon \mathbb{I} \to \mathbb{R}$ of paths starting at the same point $y_0 \in p^{-1}\{x_0\}$.

Here we take these properties as granted. They will be proved in Theorem 6.2.9, as a consequence of a property of the map $p\colon \mathbb{R} \to \mathbb{S}^1$, namely its being a 'covering map'.

Now, to show that φ is surjective, we take a loop $a\colon e_1 \simeq e_1$ in $\mathbb{S}^1$; by (i) there exists a lifting $f\colon 0 \simeq k$ in $\mathbb{R}$, with $k \in p^{-1}\{e_1\} = \mathbb{Z}$, and the latter is 2-homotopic to $f_k\colon 0 \simeq k$ (by an affine homotopy). Then $k \in p^{-1}\{e_1\} = \mathbb{Z}$, and $\varphi(k) = [a_k] = [pf_k] = [pf] = [a]$.

Finally, to show that φ is injective, suppose that $\varphi(k) = [a_k]$ is the identity of $\pi_1(\mathbb{S}^1, e_1)$: this means that a_k is 2-homotopic to the trivial loop $a_0\colon e_1 \simeq e_1$; by (ii) there is a 2-homotopy between their liftings $f_k\colon 0 \simeq k$ and $f_0\colon 0 \simeq 0$, which means that $k = 0$. $\square$

6.2.6 *Exercises* (Playing with π_1)

The Poincaré group formalises aspects which — in some cases — can be made concrete, viewing loops as closed curves drawn in the plane, or elastic bands in our 3-dimensional space.

(a) We replace the circle with the (homotopy equivalent) pierced plane $X = \mathbb{R}^2 \setminus \{0\}$, where we can more clearly draw paths.

In the figure below the omitted point $(0,0)$ is marked with a cross. The loop a of (X, e_1)

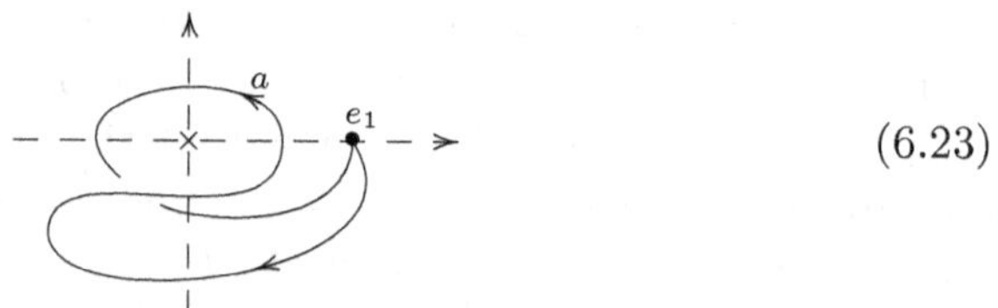

$$\text{(6.23)}$$

is 2-homotopic to the simple loop a' turning once around the origin, counterclockwise. We can realise the loop by a stretched elastic band, and realise the deformation by letting the band come back to its natural form. On the other hand, the band cannot be contracted to the basepoint e_1 (staying in X): this 'shows' that $\pi_1(X, e_1)$ is not the trivial group.

The same 'experiment' can be more concretely realised in our three dimensional space, excluding a rod: in fact the complement Y in $\mathbb{R}^3$ of the third axis contains X and $\mathbb{S}^1$ as deformation retracts

$$Y = X \times \mathbb{R} = \{x \in \mathbb{R}^3 \mid (x_1, x_2) \neq (0,0)\} \supset X \supset \mathbb{S}^1.$$

(b) More interesting aspects appear piercing a plane in more points, or equivalently taking out of $\mathbb{R}^3$ some parallel lines. Consider the space $X = \mathbb{R}^2 \setminus \{e_1, -e_1\}$, pierced in two points and pointed at the origin (homotopy equivalent to the figure-eight space)

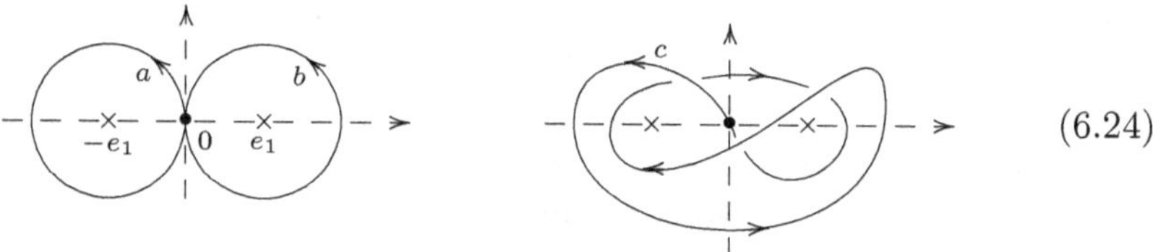

$$\text{(6.24)}$$

The left figure shows two loops a, b at the origin, which will be shown in 6.4.4(b) to generate the fundamental group $\pi_1(X, 0)$. Concretely, we 'see' that they should determine different, non-trivial classes $[a], [b]$ in $\pi_1(X, 0)$. 'Experiment' also shows that these classes do not commute in $\pi_1(X, 0)$.

In fact, the right figure shows a path c which is 2-homotopic to the concatenation $c' = a * b * a^\sharp * b^\sharp$. Realising this loop with a string winding around two parallel rods (in our 3-space), and tightening the loop, we see that the string cannot be taken 'outside' the rods: the loop c is not trivial, and the class $[a][b][a]^{-1}[b]^{-1}$ is not the identity of $\pi_1(X, 0)$.

(c) The reader may note that, in the previous cases, deforming a loop by a 2-homotopy (keeping the basepoint fixed) or by a loop homotopy (allowing the latter to move) would not make a substantial difference.

This is not always the case: in the same space $X = \mathbb{R}^2 \setminus \{e_1, -e_1\}$, the loop c represented below is 2-homotopic to $c' = b^\sharp * a * b$

$$- - \left(\!\!-\!\times\!-\!\bullet\!\leftarrow\!-\!\times\!-\right)\!-\!\to \qquad (6.25)$$

Now c (and c') is not 2-homotopic to a, but c and a are plainly loop homotopic: if we allow the basepoint of c to leave the origin, the elastic band can be deformed to the simple loop a around $-e_1$.

6.2.7 Exercises and complements (The n-torus)

(a) For the n-torus $\mathbb{T}^n = \mathbb{S}^1 \times ... \times \mathbb{S}^1$, based at $x_0 = (e_1, ..., e_1)$, we have

$$\pi_1(\mathbb{T}^n, x_0) \cong \mathbb{Z}^n, \qquad (6.26)$$

a straightforward consequence of Theorems 6.2.4 (on cartesian products) and 6.2.5 (on the fundamental group of the circle).

The fundamental group has a canonical basis $[b_i] = [u_i a_1]$, where a_1 is the standard loop of $\mathbb{S}^1$ based at e_1, and the mapping

$$u_i \colon \mathbb{S}^1 \to \mathbb{T}^n, \qquad u_i(x) = (e_1, ..., e_1, x, e_1, ... e_1), \qquad (6.27)$$

is the i-th embedding (of pointed spaces), that inserts the variable x in the i-th position, and sends $e_1 \in \mathbb{S}^1$ to $x_0 = (e_1, ..., e_1)$.

(b) The surface $\mathbb{T} = \mathbb{S}^1 \times \mathbb{S}^1$ is realised in $\mathbb{R}^3$ as in 2.6.2 by the rotation around the axis z of a circle C, which lies in the plane $y = 0$, with centre $p = (2, 0, 0)$ and radius 1. The circle C is a deformation retract of the pierced open half-plane $P = \{(x, z) \in \mathbb{R}^2 \mid x > 0\} \setminus \{(2, 0)\}$.

The torus $\mathbb{T}$ is thus a deformation retract of the space generated by P, by rotation around the axis z: this is the complement in $\mathbb{R}^3$ of a straight line L (the axis z) and the circle S (in the plane xy, with centre the origin and radius 2), represented below by a dashed and a dotted line

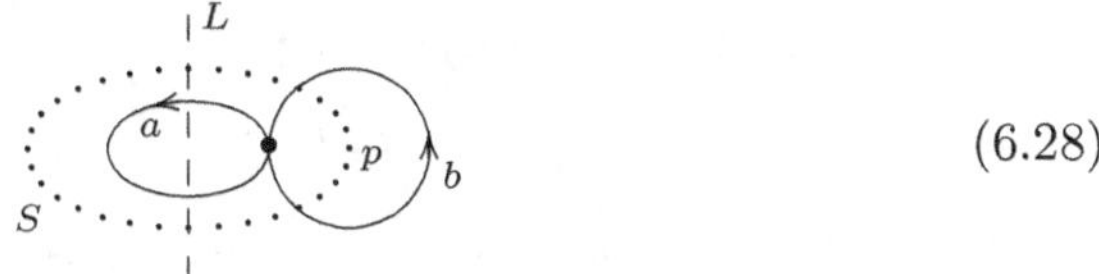

$$(6.28)$$

Making these obstructions concrete, by a rod L and a ring S around the rod, we can realise the generators as the class $[a]$ of a loop around the rod (corresponding to the inner equator of the torus), and the class $[b]$ of a loop around the ring (corresponding to a meridian of the torus).

The loop $c = a * b * a^{\sharp} * b^{\sharp}$ can be materially realised by a string turning around the rod and the ring as prescribed (knotting its ends, of course); moving part of the string around the ring, one will find that the string can be taken 'outside' rod and ring: the loop c is 2-homotopic to the trivial loop, and the classes $[a], [b]$ commute in the fundamental group.

6.2.8 Covering maps

Covering maps, their classification and relationship with the fundamental group are important facts, which can be studied in [Ha], Section 1.3, or [HiW], Chapter 6, or [Hu], Sections III.16 and III.17.

Here we only give the basic definitions, and prove (in 6.2.9) the lifting properties of a covering map, already used in 6.2.5.

(a) A *covering map* $p\colon X \to B$ is a continuous surjective mapping between topological spaces such that every $b \in B$ has an open nbd V whose inverse image $p^{-1}(V)$ is a union of disjoint open sets in X (the *sheets* of p over V), each of which is mapped homeomorphically onto V by p

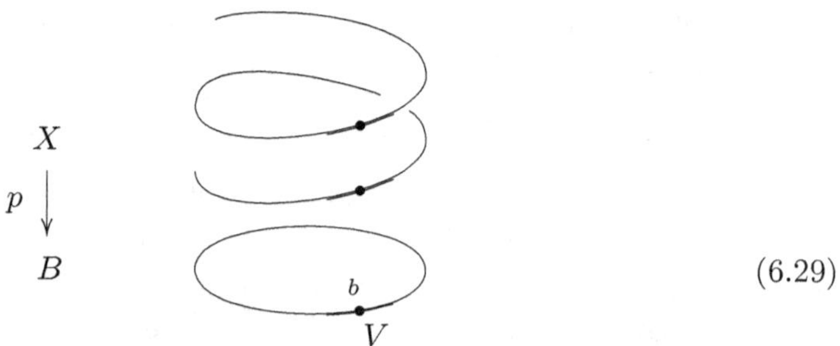

$$(6.29)$$

The nbd V is said to be *evenly covered* by p. The mapping p is a surjective local homeomorphism, and an open map.

(b) The prime examples, based on the standard circle $\mathbb{S}^1$, are the exponential map, represented in the figure above

$$p\colon \mathbb{R} \to \mathbb{S}^1, \qquad p(t) = e^{2\pi i t} = (\cos 2\pi t,\, \sin 2\pi t), \qquad (6.30)$$

and the endomap (for $n > 0$)

$$p\colon \mathbb{S}^1 \to \mathbb{S}^1, \qquad p(z) = z^n. \qquad (6.31)$$

A *trivial covering map* is a cartesian projection $B \times F \to B$, where B and F are topological spaces and F is discrete.

(c) Coming back to the general case, the spaces X and B are called the *total space* and the *base* of the covering. For every $b \in B$, the *fibre* $F_b = p^{-1}\{b\}$ is a discrete subspace of X. Over an evenly covered nbd V of b, the projection

$p_V \colon p^{-1}(V) \to V$ is equivalent to a cartesian projection $p \colon V \times F_b \to V$, by a homeomorphism h that agrees with these projections

$$
\begin{array}{ccc}
p^{-1}(V) & \xrightarrow{\ h\ } & V \times F_b \\
{\scriptstyle p_V}\downarrow & & \downarrow{\scriptstyle p} \\
V & =\!=\!=\!= & V
\end{array}
\tag{6.32}
$$

The cardinal of the fibre F_b is constant on every connected component of B. The exponential map (6.30) has a countable fibre, while the map (6.31) is an *n-fold covering*: the cardinal of 'the' fibre is n.

*(d) A covering map is the same as a fibre bundle with discrete fibres.

A *universal covering* $X \to B$ is a covering where the total space X is simply connected; under suitable hypothesis, a space B has a universal covering, which is a covering space of any other. The exponential map (6.30) is the universal covering of the circle.

6.2.9 Theorem (Lifting properties)

Let $p \colon X \to B$ be a covering map.

(i) Every path $b \colon \mathbb{I} \to B$ in the base, starting at $b_0 = b(0)$, has a unique lifting $a \colon \mathbb{I} \to X$ starting at a given point $x_0 \in F_{b_0} = p^{-1}\{b_0\}$.

(ii) Every 2-homotopy $h \colon b \simeq_2 b' \colon \mathbb{I} \to B$ of paths starting at b_0 has a unique lifting $k \colon a \simeq_2 a' \colon \mathbb{I} \to X$ to a 2-homotopy of paths starting at $x_0 \in F_{b_0}$.

Proof (i) For the given path $b \colon \mathbb{I} \to B$, we can find a decomposition $0 = t_0 < t_1 < ... < t_n = 1$ such that each interval $[t_{i-1}, t_i]$ is mapped by b into some open set of B evenly covered by p.

In fact, these sets form an open cover of B, and their preimages by b form an open cover of $\mathbb{I}$. Choosing a Lebesgue number ε of the latter, and $n \geqslant 1$ such that $1/n < \varepsilon$, we can take $t_i = i/n$.

Now, we lift $b_0 = b(0)$ at the given point $x_0 \in p^{-1}\{b_0\}$. We suppose we have uniquely lifted the restriction $b_i \colon [0, t_i] \to B$ of b to a map $a_i \colon [0, t_i] \to X$ (with $pa_i = b_i$ and $a_i(0) = x_0$). This is certainly possible for $i = 0$; for every $i < n$ we can proceed one step further, because $b[t_i, t_{i+1}]$ is contained in an open set of B evenly covered by p; our goal is thus reached in a finite number of steps.

Property (ii) is proved in a similar way, using a decomposition $0 = t_0 < t_1 < ... < t_n = 1$ such that each rectangle $[t_{i-1}, t_i] \times [t_{j-1}, t_j]$ is mapped by h into some open set of B evenly covered by p.

We begin by lifting $b_0 = h(0, 0)$ at $x_0 = k(0, 0)$ and complete the lifting in n^2 steps, following a lexicographic order on the set $\{1, ..., n\}^2$. $\qquad\square$

6.3 Moving or dropping the basepoint

If we move the basepoint x in a path connected component of the space X, the fundamental group $\pi_1(X, x)$ stays the same, up to an isomorphism that *depends* on the path we are following.

Because of this, if the space X is 0-connected (non-empty and path connected), the symbol $\pi_1(X)$ is currently used for the 'isomorphism type' of the fundamental group of X at any basepoint — a notation that should be used with prudence, because it is not related to canonical isomorphisms, nor denotes a functor: this is discussed in 6.3.1(b).

In the same hypotheses, each group $\pi_1(X, x)$ determines $H_1(X)$ in a canonical way, by the Hurewicz Theorem in degree 1.

6.3.1 The transition isomorphism

Let x_0 and x_1 belong to the same path component of the space X: then $\pi_1(X, x_0)$ and $\pi_1(X, x_1)$ are isomorphic.

More precisely, every path $c \colon \mathbb{I} \to X$ from x_0 to x_1 determines, within the groupoid $\Pi_1 X$, a *transition isomorphism*

$$\hat{c} \colon \pi_1(X, x_0) \to \pi_1(X, x_1), \qquad [a] \mapsto [c]^{-1}[a][c] = [c^\sharp * b * c], \qquad (6.33)$$

whose inverse is associated to the reversed path $c^\sharp$.

It is an *inner isomorphism* of the groupoid, see Exercise (c).

In the previous situation, a map $f \colon X \to Y$ gives a path $d = fc$ in Y, from $y_0 = f(x_0)$ to $y_1 = f(x_1)$, and two pointed maps (in $\mathsf{Top_\bullet}$) whose underlying map (in Top) is f

$$f' \colon (X, x_0) \to (Y, y_0), \qquad f'' \colon (X, x_1) \to (Y, y_1). \qquad (6.34)$$

We have thus a square diagram of groups and homomorphisms

$$
\begin{array}{ccc}
\pi_1(X, x_0) & \xrightarrow{\hat{c}} & \pi_1(X, x_1) \\
{\scriptstyle f'_*}\downarrow & & \downarrow{\scriptstyle f''_*} \\
\pi_1(Y, y_0) & \xrightarrow[\hat{d}]{} & \pi_1(Y, y_1)
\end{array}
\qquad (6.35)
$$

This square is commutative: on a class $[a] \in \pi_1(X, x_0)$ we have:

$$\hat{d} f'_*[a] = \hat{d}[fa] = [fc]^{-1}[fa][fc] = f''_*([c]^{-1}[a][c]) = f''_* \hat{c}[a].$$

Exercises and complements. (a) Let x_0, x_1 belong to the same path component of X. The transition isomorphism $\hat{c} \colon \pi_1(X, x_0) \to \pi_1(X, x_1)$ only depends on the points x_0 and x_1 if and only if one of these groups is commutative (or both, equivalently).

(b) For a 0-connected space X, the notation $\pi_1(X)$ is used to denote the 'isomorphism type' of the fundamental group of X, which — as we have seen — does not depend on the choice of the basepoint. But let us stress once more that the isomorphism (6.33) depends on the choice of a path: the fundamental group does *not* give a functor $\mathsf{Top}' \to \mathsf{Gp}$ defined on the full subcategory of 0-connected spaces.

However, if $f\colon X \to Y$ is a map between *0-connected spaces with commutative fundamental group*, the symbol $f_*\colon \pi_1(X) \to \pi_1(Y)$ can be given a sense. First, choosing a point $x_0 \in X$, we have a precise homomorphism $f'_*\colon \pi_1(X, x_0) \to \pi_1(Y, f(x_0))$. Secondly, if we change the basepoint to $x_1 \in X$, any path $c\colon x_0 \simeq x_1$ in X gives the commutative diagram (6.35), that links f'_* with $f''_*\colon \pi_1(X, x_1) \to \pi_1(Y, f(x_1))$; this diagram is independent of c and d, by (a).

All this will become clearer with the Hurewicz Theorem 6.3.4, which — in the previous hypotheses — gives a canonical isomorphism $\pi_1(X, x_0) \to H_1(X)$: in fact, we are just looking at the functor H_1.

*(c) An arbitrary groupoid G presents facts similar to those described in (a): each arrow $c\colon x_0 \to x_1$ gives a group-isomorphism

$$\hat{c}\colon G(x_0) \to G(x_1), \qquad a \mapsto c^{-1}ac, \tag{6.36}$$

which can be called an *inner automorphism* of the groupoid (with composition written in the 'straightforward' order). If $x_0 = x_1$ we are just considering an inner automorphism of the group $G(x_0)$.

6.3.2 Theorem (Non-pointed homotopies)

(a) For a homotopy $\varphi\colon f \cong g\colon X \to Y$ and a point $x_0 \in X$, there is a commutative diagram

$$\tag{6.37}$$

where $y_1 = f(x_0)$, $y_2 = g(x_0)$, and the vertical arrow

$$\hat{c}\colon \pi_1(Y, y_1) \to \pi_1(Y, y_2), \qquad [b] \mapsto [c]^{-1}[b][c] = [c^{\sharp} * b * c], \tag{6.38}$$

is the transition isomorphism of the path $c = \varphi(x_0, -)$, from y_1 to y_2 in Y.

(b) Suppose that the spaces X, Y are 0-connected and homotopy equivalent. Then, for every $x_0 \in X$ and $y_0 \in Y$

$$\pi_1(X, x_0) \cong \pi_1(Y, y_0). \tag{6.39}$$

(c) If the space X is contractible, every group $\pi_1(X, x_0)$ is trivial. The same holds whenever the path component of x_0 in X is contractible.

Proof (a) It is an application of the naturality square (6.7) associated to the homotopy $\varphi\colon f \cong g$.

At the point $x_0 \in X$, the loop $a\colon x_0 \simeq x_0$ gives a commutative square in the groupoid $\Pi_1 Y$

$$
\begin{array}{ccc}
y_1 & \xrightarrow{\;[c]\;} & y_2 \\
{\scriptstyle [fa]}\Big\downarrow & & \Big\downarrow{\scriptstyle [ga]} \\
y_1 & \xrightarrow[\;[c]\;]{} & y_1
\end{array}
\qquad\qquad [fa][c] = [c][ga] \tag{6.40}
$$

where $c = \varphi(x_0, -)$. Therefore

$$
\hat{c}[fa] = [c]^{-1}[fa][c] = [ga].
$$

(b) By 6.3.1, it is sufficient to prove the existence of an isomorphism $\pi_1(X, x_0) \cong \pi_1(Y, y_0)$, for a given pair of basepoints $x_0 \in X$ and $y_0 \in Y$.

We have a homotopy equivalence $f\colon X \rightleftarrows Y \colon g$. We choose a point x_0 in X; we let $y_0 = f(x_0)$ and $x_1 = g(y_0)$. By (a), the existence of a homotopy $\mathrm{id}\, X \simeq gf$ implies that the composed homomorphism

$$
g_* f_* \colon \pi_1(X, x_0) \to \pi_1(Y, y_0) \to \pi_1(X, x_1) \tag{6.41}
$$

is a (transition) isomorphism (not the identity, in general), so that f_* is mono and g_* is epi.

Symmetrically, a homotopy $\mathrm{id}\, Y \simeq fg$ implies that g_* is mono and f_* is epi. Finally, they are both isomorphisms.

(c) A straightforward consequence of (b). $\square$

6.3.3 The Hurewicz homomorphism

For a pointed space (X, x_0), there is a natural relationship between the fundamental group $\pi_1(X, x_0)$ and the singular homology group $H_1(X)$: any loop $a \in \Omega(X, x_0)$ determines a homotopy class $[a] \in \pi_1(X, x_0)$ and a homology class $[a]_H \in H_1(X)$.

There is in fact a natural homomorphism, called the *Hurewicz homomorphism* (in degree 1)

$$
h\colon \pi_1(X, x_0) \to H_1(X), \qquad h[a] = [a]_H \quad (a \in \Omega(X, x_0)), \tag{6.42}
$$

examined in the following exercises and Theorem 6.3.4.

Exercises and complements. (a) The mapping (6.42) is a well-defined homomorphism of groups.

(b) This homomorphism is consistent with the transition isomorphisms (6.33): for any path $c\colon x_0 \simeq x_1$ in X the following triangle commutes

$$\begin{array}{ccc}
\pi_1(X, x_0) & \xrightarrow{\ h\ } & H_1(X) \\
{\scriptstyle \hat{c}}\big\downarrow & \nearrow_{\!h'} & \\
\pi_1(X, x_1) & &
\end{array} \qquad (6.43)$$

where h' is the Hurewicz homomorphism at x_1. *Hints:* use Lemma 6.3.6.

(c) The Hurewicz homomorphism gives a natural transformation

$$h\colon \pi_1 \to H_1 U\colon \mathsf{Top}_{\bullet} \to \mathsf{Gp}, \qquad (6.44)$$

where $U\colon \mathsf{Top}_{\bullet} \to \mathsf{Top}$ is the forgetful functor which drops the basepoint.

(d) To further investigate h we have to recall some basic issues of group theory. In a group G, two elements x, y commute (i.e. $xy = yx$) if and only if their *commutator*

$$[x, y] = x\, y\, x^{-1}\, y^{-1} \qquad (6.45)$$

is the identity 1_G. By definition, the subgroup $[G, G]$ *of commutators* of G is the subgroup generated by all commutators $[x, y]$, for $x, y \in G$. Plainly, G is commutative if and only if $[G, G]$ is the trivial subgroup.

Prove that the subgroup $[G, G]$ is normal in G.

(e) The quotient $G/[G, G]$ is a commutative group, determined by the following universal property:

(*) for every homomorphism of groups $f\colon G \to A$ with values in an abelian group there is a unique homomorphism $g\colon G/[G, G] \to A$ such that $gp = f$

$$\begin{array}{ccc}
G & \xrightarrow{\ p\ } & G/[G, G] \\
& \searrow_{\!f} & \big\downarrow{\scriptstyle g} \\
& & A
\end{array} \qquad (6.46)$$

The quotient $G^{\mathrm{ab}} = G/[G, G]$ is called the *abelianised group* of G.

(f) The pair $(G^{\mathrm{ab}}, p\colon G \to G^{\mathrm{ab}})$ is thus a universal arrow from the object G to the inclusion functor $U\colon \mathsf{Ab} \to \mathsf{Gp}$.

Their family gives a functor $(-)^{\mathrm{ab}}\colon \mathsf{Gp} \to \mathsf{Ab}$, left adjoint to U (see Section 7.1).

6.3.4 *Hurewicz Theorem* (in dimension one)

Let (X, x_0) be a path connected pointed space.

Then the Hurewicz homomorphism $h\colon \pi_1(X, x_0) \to H_1(X)$ defined in (6.42) is surjective, and its kernel coincides with the subgroup of commutators of $\pi_1(X, x_0)$.

In other words, the Hurewicz homomorphism induces an isomorphism h'

$$\pi_1(X, x_0) \xrightarrow{\ p\ } \pi_1(X, x_0)^{\mathrm{ab}} \qquad\qquad\quad (6.47)$$

with arrows h and h' pointing to $H_1(X)$.

between the abelianised group $\pi_1(X, x_0)^{\mathrm{ab}}$ and $H_1(X)$.

Proof See [HiW], Theorem 8.8.3, or [Hu], Theorem II.6.1. $\square$

6.3.5 *Remarks and complements*

(a) If a path connected pointed space (X, x_0) has a commutative fundamental group $\pi_1(X, x_0)$, the Hurewicz homomorphism is an isomorphism. In this case we understand better the meaning of $\pi_1(X)$ discussed in 6.3.1(b): essentially, it refers to H_1, which for sure is a functor.

This is the case of the circle and of any torus $\mathbb{T}^n$.

(b) The homotopy group $\pi_1(-)$ gives a richer information than the homology group $H_1(-)$: if (X_i) is the partition of X in path components, then

$$H_1(X) \cong \oplus_i H_1(X_i) \cong \oplus_i \pi_1(X_i, x_i)^{\mathrm{ab}}, \qquad (6.48)$$

where x_i is any point in the path component X_i. But we shall see that, in higher dimension, the homotopy groups of a space no longer determine its homology groups.

*(c) Every group G can be realised as the fundamental group of a suitable *Eilenberg–Mac Lane space* $K(G, 1)$: a 0-connected space whose fundamental group is isomorphic to G, while all higher homotopy groups are trivial (see [EM2], [EM3] and [Ha], Section 1.B).

If we take as G a non-commutative *simple* group (i.e. a group whose only normal subgroups are the trivial and the total one), then $[G, G] = G$ and $G^{\mathrm{ab}} = 0$.

This shows that there are spaces X with a quite complicated $\pi_1(X)$, and trivial $H_1(X)$.

6.3.6 Lemma

For a loop $a\colon x \simeq x$ and a path $c\colon x \simeq y$ in the space X, the loop a and the loop $a' = c^\sharp * a * c\colon y \simeq y$ are loop homotopic.

Proof (This extends an issue already considered in (6.25), for the plane pierced in two points.)

Using double homotopies, as in (6.14) and (6.16), the figure below shows how to construct a loop homotopy $H = L' *_1 A *_1 L$, from the loop $e_x * a * e_x$ (at x) to the loop $a' = c^\sharp * a * c$ (at y),

$$\tag{6.49}$$

$$L'(s,t) = c(\min(1 - s, t)), \quad A(s,t) = a(s), \quad L(s,t) = c(\min(s,t)).$$

Therefore the loop $a \simeq_2 e_x * a * e_x$ is loop homotopic to $c^\sharp * a * c$. $\quad\square$

6.3.7 Theorem (Classification by degree)

(a) Let X be a 0-connected space with a commutative fundamental group $\pi_1(X, x_0)$ (at any point). Let $f, g\colon \mathbb{S}^1 \to X$ be two maps with the same induced homomorphism $u\colon H_1(\mathbb{S}^1) \to H_1(X)$. Then $f \simeq g$.

(b) Two maps $f, g\colon \mathbb{S}^1 \to \mathbb{S}^1$ are homotopic if (and only if) $\deg f = \deg g$: the topological degree classifies the homotopy class of these endomaps.

Proof It is sufficient to prove (a), as (b) is a straightforward consequence. For the standard basepoint $e_1 \in \mathbb{S}^1$ we let $x_1 = f(e_1)$ and $x_2 = g(e_1)$.

In the following diagram

$$\tag{6.50}$$

- h, h' and h'' are Hurewicz isomorphisms, by hypothesis,
- $\hat{c}\colon \pi_1(X, x_1) \to \pi_1(X, x_2)$ is the inner isomorphism associated to a chosen path $c\colon x_1 \simeq x_2$ in X,
- the squares commute, by the naturality of the Hurewicz homomorphism (in 6.3.3), and the triangle commutes, by (6.43).

We deduce that $\hat{c}f_* = g_* \colon \pi_1(\mathbb{S}^1, e_1) \to \pi_1(X, x_2)$.

For the exponential loop $p \colon \mathbb{I} \to \mathbb{S}^1$, the loops $a = pf \colon \mathbb{I} \to \mathbb{S}^1$ and $b = pg \colon \mathbb{I} \to \mathbb{S}^1$ correspond to the maps $f = \hat{a}$ and $g = \hat{b}$ (as in 1.4.2(a)).

We have thus

$$[b] = g_*[p] = \hat{c}f_*[p] = [c^{\sharp} * a * c],$$

and $b \simeq_2 c^{\sharp} * a * c$; but the latter is loop homotopic to a, by Lemma 6.3.6. Therefore a and b are loop homotopic, and $f \simeq g$, by 1.4.8(b).　　　　　$\square$

6.4 The van Kampen Theorem

After some preliminaries on groups, we introduce one of the main tools to compute fundamental groups, the van Kampen Theorem. It plays a role similar to the Mayer–Vietoris sequence for homology groups: to determine $\pi_1(X)$ on the basis of a generalised open cover of the space X.

Yet, it is less effective: it needs stronger hypothesis (which do not apply to the circle), and cannot be easily extended to higher degree.

R. Brown's extension to groupoids (exposed in 6.4.5, 6.4.6) solves the first drawback: it is stated for any generalised open cover and can be used to determine the fundamental groupoid of the circle (and then the fundamental group). Its extension to higher dimensional groupoids has been developed in diverse forms, far beyond an elementary exposition.

Computing higher homotopy groups can be quite difficult or still unsolved (even for the spheres $\mathbb{S}^n$); the main tools, as the exact sequence of a fibration or the homotopy spectral sequences, are studied in [Hu].

6.4.1 *Free products of groups*

To state the van Kampen Theorem, we need some preliminary topics of group theory: free products and pushouts.

For a family of groups (G_i), one can form a new group $G = *_i\, G_i$ called the *free product* of the family. It comes with a family of homomorphisms $u_i \colon G_i \to G$ ($i \in I$), called *injections*, which satisfy the universal property we have considered in 1.6.2(i*)

$$
\begin{array}{ccc}
G_i & \xrightarrow{\;u_i\;} & G \\[2pt]
 & \searrow^{f_i} & \big\downarrow^{f} \\[2pt]
 & & H
\end{array}
\qquad (6.51)
$$

(*) for every group H and every family of homomorphisms $f_i \colon G_i \to H$, there is a unique homomorphism $f \colon G \to H$ such that $fu_i = f_i$ (for $i \in I$).

In other words, we are considering the categorical sum in the category Gp (which had a name and a symbol in group theory, before the existence of categories).

The group G is essentially built as follows (a more detailed construction is the subject of Exercise (a)). An element of $*_i G_i$ can be written as a finite word $(x_1, x_2, ..., x_n)$ on the *alphabet* $X = \Sigma_i |G_i|$, the disjoint union of the underlying sets, provided we replace every occurrence x_h, x_{h+1} of two elements in the same group G_i with their product in this group, and we take out any unit-element of a G_i.

Writing $x_1 x_2 ... x_n$ the equivalence class of the word $(x_1, x_2, ..., x_n)$, two classes are multiplied by juxtaposition; the operation is associative, has a unit 1 (the class of the empty word) and inverses exist, with $(x_1 x_2 ... x_n)^{-1} = x_n^{-1} ... x_2^{-1} x_1^{-1}$ (each inverse is computed in the corresponding G_i).

The homomorphism $u_i \colon G_i \to G$ obviously takes any $x \in G_i$ to the class of its unary word. The universal property (*) is plainly satisfied.

Exercises and complements. (a) Rewrite the construction of $*_i G_i$ in a more detailed form.

(b) Verify the universal property of $*_i G_i$.

(c) The *free group* $G(X)$ on the set X comes equipped with a mapping $\eta \colon X \to |G(X)|$ (the insertion of the basis) with the usual universal property

(i) for every mapping $f \colon X \to |H|$ with values in a group H, there is a unique homomorphism $g \colon G(X) \to H$ such that $g\eta = f$.

The free group on the empty set is the trivial group, the free group on the singleton is the additive group $\mathbb{Z}$.

(d) In general, the free group $G(X)$ can be constructed as a free product of copies of the additive group $\mathbb{Z}$

$$G(X) = *_{x \in X} \mathbb{Z}. \tag{6.52}$$

The solution is obtained as a categorical sum of copies of the free object on the singleton, in the same way as the free R-module on a set, in (4.5). This is a general fact in category theory: cf. [G5], Exercise 4.1.3(e).

(e) In the alphabet $\Sigma_{x \in X} |\mathbb{Z}|$, it is convenient to replace the x-indexed copy $\mathbb{Z} \times \{x\}$ with the free multiplicative group $\{x^k \mid k \in \mathbb{Z}\}$ generated by the singleton $\{x\}$, whose elements are formal powers of x. The elements of $G(X)$ can thus be described as equivalence classes $[w] \in W/\sim$

$$[w] = x_1^{k_1} x_2^{k_2} ... x_p^{k_p} \qquad (x_j \in X, \, k_j \in \mathbb{Z}), \tag{6.53}$$

of words in this alphabet, where we add up the exponents of consecutive terms with the same base, and cancel all occurrences of the form x^0.

The free group $\mathbb{Z} * \mathbb{Z}$ on two elements is already non commutative.

6.4.2 Pushouts of groups

We have already considered pushouts of topological spaces, in 2.3.6. We are now interested in the same topic (a categorical colimit), in the category Gp of all groups.

There are given two homomorphisms f, g with the same domain G_0. Their *pushout* is a group G equipped with two maps $u_i \colon G_i \to G$ $(i = 1, 2)$ which form a commutative square $u_1 f = u_2 g$, in a universal way:

$$
\begin{array}{ccccc}
 & & G_1 & & \\
 & {\scriptstyle f}\nearrow & \big\downarrow{\scriptstyle u_1} & {\scriptstyle v_1}\searrow & \\
G_0 & & & G \; \xrightarrow{\;-w\;>\;} H & \\
 & {\scriptstyle g}\searrow & \big\uparrow{\scriptstyle u_2} & {\scriptstyle v_2}\nearrow & \\
 & & G_2 & &
\end{array}
\tag{6.54}
$$

(*) for every triple (H, v_1, v_2) such that $v_1 f = v_2 g$ there exists a unique homomorphism $w \colon G \to H$ such that $w u_1 = v_1$, $w u_2 = v_2$.

The solution always exists, and is determined up to isomorphism, as shown below.

Exercises and complements. (a) The elementary case where G_0 is the trivial group leads to the free product $G = G_1 * G_2$, with its canonical injections.

(b) In the general case, one builds a solution as a quotient $G = (G_1 * G_2)/K$ of the free product (much in the same way as in Top, in 2.3.6): K is the normal subgroup of $G_1 * G_2$ generated by all elements of the form $u_1 f(x)\, u_2 g(x^{-1})$, for $x \in G_0$.

(c) Prove the uniqueness of the solution up to isomorphism.

6.4.3 The van Kampen Theorem

Let X be a space; let X_1, X_2 be two subspaces, and $X_0 = X_1 \cap X_2$. Suppose that $X = \operatorname{int} X_1 \cup \operatorname{int} X_2$ and $x_0 \in X_0$.

Suppose also that X_0, X_1 and X_2 are path connected (whence X is also). The following diagram of groups and inclusions is a pushout in Gp

$$
\begin{array}{ccc}
\pi_1(X_0, x_0) & \xrightarrow{\;u_1\;} & \pi_1(X_1, x_0) \\
{\scriptstyle u_2}\big\downarrow & & \big\downarrow{\scriptstyle v_1} \\
\pi_1(X_2, x_0) & \xrightarrow[\;v_2\;]{} & \pi_1(X, x_0)
\end{array}
\tag{6.55}
$$

Proof This theorem, also named after Seifert and van Kampen, was proved around 1930. Here it will be deduced from a corollary of its version for groupoids, in 6.4.6. $\qquad\square$

6.4.4 Exercises and complements (Computations)

(a) Prove that $\pi_1(\mathbb{S}^n, e_1) = 0$, for $n > 1$.

(b) (*The figure-eight space*) Compute the fundamental group of the figure-eight space (in 2.5.1), using the same open cover used for its homology groups. This exercise makes precise what we have already 'experimented' in 6.2.6(b).

The result is easily extended to the fundamental group of a bouquet of circles $\mathbb{S}^1 \vee \ldots \vee \mathbb{S}^1$, or a bouquet of spheres.

(c) In a space X with a non-commutative fundamental group one can always find two loops, at the same basepoint, which are loop homotopic and not 2-homotopic. *Hints:* use Lemma 6.3.6.

(d) Compute the fundamental group of the space $P_n = \mathbb{D}^2/R_n$ defined in 2.5.8 (including the projective plane, for $n = 2$).

*(e) The fundamental group of the Klein bottle $\mathbb{K}$ is generated by two elements α, β under the relation $\alpha = \beta\alpha\beta$. The generators can be realised as $\alpha = [a]$, $\beta = [b]$, where a and b are the loops used in 2.6.4.

We recall that their homology classes $[a]_H$ and $[b]_H$ generate $H_2(\mathbb{K}) \cong \mathbb{Z} \oplus \mathbb{Z}_2$, under the relation $2[b]_H = 0$. The fundamental group of $\mathbb{K}$ is not commutative.

The reader should not find it difficult to complete the missing points, using the open cover (U, V) of Exercise 2.6.4(a).

6.4.5 The van Kampen Theorem for groupoids

Let X be a space; let X_1, X_2 be two subspaces, and $X_0 = X_1 \cap X_2$. Suppose that $X = \operatorname{int} X_1 \cup \operatorname{int} X_2$.

The following diagram of small groupoids and functors (induced by inclusions) is a pushout of groupoids (defined by the universal property 6.4.2(*), in the category Gpd*)*

$$
\begin{array}{ccc}
\Pi_1 X_0 & \xrightarrow{u_1} & \Pi_1 X_1 \\
{\scriptstyle u_2}\downarrow & & \downarrow{\scriptstyle v_1} \\
\Pi_1 X_2 & \xrightarrow{v_2} & \Pi_1 X
\end{array}
\tag{6.56}
$$

Proof This is the groupoid version of the van Kampen Theorem, by Ronald Brown [Br1] (also exposed in [Br2], [Br3]). It does not need the hypothesis of path-connectedness of the spaces X_0, X_1 and X_2.

The proof is written in 6.4.9. $\qquad\qquad\square$

6.4.6 Corollary

We assume the same hypotheses on X, X_1, X_2 and X_0, as in the previous theorem.

(a) Let $A \subset X$ be a representative *subset of X_0, X_1, X_2, in the sense that it meets all the path components of these subspaces, and therefore of X: see 6.1.9.*

Then, restricting all the groupoids of (6.56) to the vertices in A, we still have a pushout of groupoids, where $A_i = X_i \cap A$

$$
\begin{array}{ccc}
\Pi_1 X_0 | A_0 & \xrightarrow{u_1} & \Pi_1 X_1 | A_1 \\
{\scriptstyle u_2}\downarrow & & \downarrow{\scriptstyle v_1} \\
\Pi_1 X_2 | A_2 & \xrightarrow[v_2]{} & \Pi_1 X | A
\end{array}
\qquad (6.57)
$$

and we use the notation of 6.1.9: $\Pi_1 X | A$ is the full subgroupoid of $\Pi_1 X$ with vertices in A; and so on.

(b) If X_0, X_1 and X_2 are path connected (as well as X), and $x_0 \in X_0$, we can take $A = \{x_0\}$ as the representative set. Then the four groupoids fall down to groups, and our statement gives the ordinary van Kampen's Theorem 6.4.4.

Proof (a) We write $X_3 = X$ and $A_3 = A$.

In the given hypotheses, we have seen in 6.1.9 that there are deformation retracts (for $i = 0, ..., 3$)

$$
J_i \colon \Pi_1 X_i | A_i \rightleftarrows \Pi_1 X_i \colon P_i, \qquad P_i J_i = \mathrm{id}, \qquad \lambda_i \colon \mathrm{id} \cong J_i P_i, \qquad (6.58)
$$

determined by choosing a path $\lambda_x \colon x \simeq P(x)$ in X which, for $x \in X_i$, stays in X_i. (We begin by a choice for $x \in X_0 = X_1 \cap X_2$, and extend it to $X_1 \setminus X_0$ and $X_2 \setminus X_0$, covering also X.) If $x \in A$, we take the trivial path $\lambda_x = e_x \colon x \simeq x = P(X)$.

Forgetting the natural isomorphisms λ_i (which are not needed here), it is now sufficient to prove that 'a retract of a pushout is a pushout', a general fact of category theory proved in Lemma 6.4.8. (The argument is straightforward; an interested reader can state and prove independently this result.)

(b) A straightforward consequence of (a): a commutative square of groups is a pushout in Gp if (and only if) it is in Gpd. $\qquad\square$

6.4.7 The fundamental groupoid of the circle

The groupoid version of van Kampen's Theorem can be applied to the usual open arcs X_1 and X_2 of the circle, whose intersection has two connected components (note that the group version does not apply).

It is sufficient (and convenient) to take a representative set $A = \{p, q\}$ formed of two points, one in each connected component of $X_1 \cap X_2$

$$
\begin{aligned}
X_1 &= \mathbb{S}^1 \setminus \{\underline{s}\} \simeq \{*\}, \\
X_2 &= \mathbb{S}^1 \setminus \{\underline{n}\} \simeq \{*\}, \\
X_0 &= X_1 \cap X_2 \simeq A = \{p, q\}.
\end{aligned}
\tag{6.59}
$$

Then $G_1 = \Pi_1 X_1 | A$ and $G_2 = \Pi_1 X_2 | A$ are two copies of the indiscrete groupoid (see 1.3.1) on two vertices p, q, while $G_0 = \Pi_1 X_0 | A$ is the discrete groupoid on the same vertices.

The pushout $G = \Pi_1 \mathbb{S}^1 | A$ of G_1 and G_2 over G_0 is the groupoid on the objects p, q freely generated by two arrows $\lambda \colon p \to q$ and $\mu \colon q \to p$.

In particular, $G(p)$ is the free group generated by the endomorphism $\alpha = \lambda\mu \colon p \to p$, and is isomorphic to $\mathbb{Z}$; similarly $G(q)$ is the free group generated by the endomorphism $\beta = \mu\lambda \colon q \to q$.

We find again that the fundamental group of the circle is the infinite cyclic group. This proof does not use the theory of covering maps.

6.4.8 Lemma (Pushouts and retracts)

In a category C we have two commutative squares

$$
\begin{array}{ccc}
X_0 \xrightarrow{u_1} X_1 & \qquad & A_0 \xrightarrow{u_1'} A_1 \\
\downarrow{\scriptstyle u_2} \qquad \downarrow{\scriptstyle v_1} & \qquad & \downarrow{\scriptstyle u_2'} \qquad \downarrow{\scriptstyle v_1'} \\
X_2 \xrightarrow[v_2]{} X_3 & \qquad & A_2 \xrightarrow[v_2']{} A_3
\end{array}
\tag{6.60}
$$

We also have four retractions $j_i \colon A_i \rightleftarrows X_i \colon p_i$ $(0 \leqslant i \leqslant 3)$ consistent with the previous squares (with $i = 1, 2$)

$$
u_i j_0 = j_i u_i', \qquad v_i j_i = j_3 v_i', \qquad p_i u_i = u_i' p_0, \qquad p_3 v_i = v_i' p_i
\tag{6.61}
$$

which means that the sections $j_i \colon A_i \to X_i$ form a commutative cube, from the right square to the left, and the retractions $p_i \colon X_i \to A_i$ form a commutative cube, from the left square to the right.

Then, if the left square is a pushout, the right one is also.

**Note.* The 'same' holds for all colimits.

Proof There are given two morphisms $h_i\colon A_i \to Y$ ($i = 1, 2$) such that $h_1 u_1' = h_2 u_2'$. We define $f_i = h_i p_i\colon A_i \to Y$ and get a commutative square: $f_i u_i = h_i p_i u_i = h_i u_i' p_0$ does not depend on $i = 1, 2$.

There is thus a unique $f\colon X_3 \to Y$ such that $f v_i = f_i$ (for $i = 1, 2$), and a morphism $h = f j_3\colon A_3 \to Y$ satisfying the required conditions

$$hv_i' = f j_3 v_i' = f v_i j_i = f_i j_i = h_i p_i j_i = h_i,$$

for $i = 1, 2$. Moreover, if also $h'\colon A_3 \to Y$ gives $h' v_i' = h_i$ (for $i = 1, 2$), then $h' p_3 v_i = h' v_i' p_i = h_i p_i = f_i$, whence $h' p_3 = f$ and $h' = f j_3 = h$. $\qquad\square$

6.4.9 Proof of Theorem 6.4.5

We shall use the regular concatenation of consecutive paths (in (1.74)), written as $a_1 * \ldots * a_n$.

Let G be a small groupoid (with composition written in the same way) and take two functors $F_i\colon \Pi_1 X_i \to G$ which coincide on $\Pi_1 X_0$ ($F_1 u_1 = F_2 u_2$); we have to prove that they have a unique 'extension' $F\colon \Pi_1 X \to G$. On the objects this is obvious, since $X = X_1 \cup X_2$ and $X_0 = X_1 \cap X_2$.

Let then $a\colon x \simeq x'$ be a path in X. By the Lebesgue Number Lemma 1.4.9, there is a finite decomposition $0 < 1/n < 2/n < \ldots < 1$ of the standard interval such that each subinterval $[(i-1)/n, i/n]$ is mapped by a into X_1 or X_2 (possibly both); we call it a *suitable* decomposition for our data. Thus, a is a regular concatenation $a_1 * \ldots * a_n$ where each path $a_i\colon [0, 1] \to X$ is contained in X_{k_i} (with $k_i = 1, 2$).

We define

$$F[a] = F_{k_1}[a_1] * \ldots * F_{k_n}[a_n] \in G(F(x), F(x')). \qquad (6.62)$$

First, this morphism $F[a]$ does not depend on the choice of k_i: if $\mathrm{Im}\,(a_i) \subset X_1 \cap X_2 = X_0$, then $F_1 u_1 = F_2 u_2$ shows that $F_1[a_i] = F_2[a_i]$.

Second, $F[a]$ does not depend on the choice of n: if also m gives a suitable decomposition, one can use the regular decomposition in mn subintervals to prove that m and n give the same result.

Third, $F[a]$ does not depend on the representative path a. It is sufficient to show this for a second path $a'\colon x \to x'$, linked to the first by a 2-homotopy $H\colon a \simeq_2 a'$; in other words, $H\colon \mathbb{I}^2 \to X$ has degenerate 1-indexed faces, and 2-indexed faces a, a'.

Applying the Lebesgue Number Lemma to the compact metric square $[0, 1]^2$, there is some integer $n > 0$ such that each elementary square

$$[(i-1)/n, i/n] \times [(j-1)/n, j/n] \qquad (i, j = 1, \ldots, n),$$

is mapped by H into X_1 or X_2.

H can thus be obtained as a pasting of its n^2 restrictions to these squares, reparametrised as $H_{ij}\colon \mathbb{I}^2 \to X_{k(i,j)} \subset X$

$$H = (H_{11} *_1 H_{21} *_1 \ldots *_1 H_{n1}) *_2 \ldots *_2 (H_{1n} *_1 H_{2n} *_1 \ldots *_1 H_{nn}), \quad (6.63)$$

where $*_1$ denotes a horizontal concatenation (in the first coordinate) and $*_2$ a vertical one. Every 2-cube $K = H_{ij}$ yields, by Lemma 6.1.8, a 2-homotopy relation in $X_{k(i,j)}$

$$\partial_2^- K * \partial_1^+ K \simeq_2 \partial_1^- K * \partial_2^+ K. \quad (6.64)$$

Therefore, using the fact that the 1-indexed faces on the boundary, $\partial_1^- H_{1i}$ and $\partial_1^+ H_{ni}$, are constant paths, and the coincidence of faces between contiguous little squares, we can gradually move from the path a to a', following a lexicographic order

$$\begin{aligned}
F[a] &= F_{k(1,1)}[\partial_2^- H_{11}] * \ldots * F_{k(n,1)}[\partial_2^- H_{n1}] \\
&= F_{k(1,1)}[\partial_2^- H_{11}] * \ldots * F_{k(n,1)}[\partial_2^- H_{n1}] * F_{k(n,1)}[\partial_1^+ H_{n1}] \\
&= F_{k(1,1)}[\partial_2^- H_{11}] * \ldots * F_{k(n,1)}[\partial_1^- H_{n1}] * F_{k(n,1)}[\partial_2^+ H_{n1}] \\
&= F_{k(1,1)}[\partial_2^- H_{11}] * \ldots * F_{k(n,1)}[\partial_1^+ H_{n-1,1}] * F_{k(n,2)}[\partial_2^- H_{n2}] \quad (6.65) \\
&= \ldots = F_{k(1,2)}[\partial_2^- H_{12}] * \ldots * F_{k(n,2)}[\partial_2^- H_{n2}] \\
&= \ldots = F_{k(1,n)}[\partial_2^- H_{1n}] * \ldots * F_{k(n,n)}[\partial_2^- H_{nn}] \\
&= \ldots = F_{k(1,n)}[\partial_2^+ H_{1n}] * \ldots * F_{k(n,n)}[\partial_2^+ H_{nn}] = F[a'].
\end{aligned}$$

The figure shows the passage from the first line of formula (6.65), represented in the bottom row (with $n' = n - 1$), to the fourth line, where the path $\partial_2^- H_{n1} \simeq_2 \partial_2^- H_{n1} * \partial_1^+ H_{n1}$ has been replaced with the path

$$\partial_1^- H_{n1} * \partial_2^+ H_{n1} = \partial_1^+ H_{n'1} * \partial_2^- H_{n2},$$

(marked with arrows) applying (6.64) and contiguity with the near squares.

At the end of the process the first line of formula (6.65) is replaced with the last, represented in the top row of the figure. Now $F\colon \Pi_1 X \to G$ is also well defined on arrows.

To show that it preserves composition just note that, if two consecutive paths a, b have a suitable decomposition on n subintervals, then $a*b$ inherits a suitable decomposition $a * b = a_1 * \ldots * a_n * b_1 * \ldots * b_n$ which separates the original paths.

Finally, the uniqueness of the functor F is obvious. $\qquad\square$

6.5 Hints at the higher homotopy groups

We briefly sketch the definition and some properties of the higher homotopy groups $\pi_n(X, x_0)$, state the higher dimensional Hurewicz Theorem (without proof) and define n-connected spaces.

Literature. Homotopy groups can be studied in [Ha], Chapter 4, and [Hu].

6.5.1 Homotopy groups

Let (X, x_0) be a pointed space. For every $n \geqslant 0$ we consider the set of relative maps

$$\Omega_n(X, x_0) = \mathsf{Top}_2((\mathbb{I}^n, \underline{\partial}\mathbb{I}^n), (X, x_0)), \tag{6.66}$$

formed of all n-cubes $a\colon \mathbb{I}^n \to X$ that take the border $\underline{\partial}\mathbb{I}^n$ (in $\mathbb{R}^n$) to the basepoint x_0, called *relative cubes at x_0*. This set will be pointed at the map $u_n\colon \mathbb{I}^n \to X$, constant at x_0.

We define $\pi_n(X, x_0)$ as the quotient (pointed) set

$$\pi_n(X, x_0) = \Omega_n(X, x_0)/\!\simeq, \tag{6.67}$$

modulo relative homotopy: two relative cubes $a, a'\colon (\mathbb{I}^n, \underline{\partial}\mathbb{I}^n) \to (X, x_0)$ give the same class $[a] = [a']$ if there is a relative homotopy $h\colon a \simeq a'$ (see (3.1)). This is an ordinary homotopy $a \simeq a'$ with *fixed boundary*

$$h\colon \mathbb{I}^n \times \mathbb{I} \to X, \qquad h(s,0) = a(s), \quad h(s,1) = a'(s) \quad (s \in \mathbb{I}^n),$$

$$h(s,t) = x_0 \quad \text{for } s \in \underline{\partial}\mathbb{I}^n \text{ and } t \in \mathbb{I}. \tag{6.68}$$

The case $n = 0$ gives a mere pointed set, examined in Exercise 6.5.2(a).

For $n \geqslant 1$, the set $\Omega_n(X, x_0)$ is equipped with n operations, the concatenations $a *_i b$ in each direction $i = 1, \ldots, n$ (already considered for i-consecutive 2-cubes, in (6.13))

$$(a *_i b)(t_1, \ldots, t_n) = \begin{cases} a(t_1, \ldots, 2t_i \ldots, t_n), & \text{for } 0 \leqslant t_i \leqslant 1/2, \\ b(t_1, \ldots, 2t_i - 1, \ldots, t_n), & \text{for } 1/2 \leqslant t_i \leqslant 1. \end{cases} \tag{6.69}$$

These operations are easily seen to be well defined: cf. Exercise 6.5.2(b).

For $n = 1$ we have the single operation $a * b$ of path concatenation, which induces the group structure $[a][b] = [a * b]$ of $\pi_1(X, x_0)$.

For $n \geqslant 2$, all these operations induce the same operation $[a] + [b] = [a *_i b]$, making $\pi_n(X, x_0)$ into an abelian group: this is proved in Exercise 6.5.2(d).

6.5.2 Exercises and complements

(a) For $n = 0$ the previous definition gives again the pointed set $\pi_0(X, x_0) = (|X|/\simeq, [x_0])$ of path components of X, introduced in (1.70).

(b) For $n \geqslant 1$, each operation $*_i$ is well defined in the set $\Omega_n(X, x_0)$ and induces an operation $[a *_i b]$ on $\pi_n(X, x_0)$, making it into a group. In each case, the identity is the class $[u_n]$ of the constant map $\mathbb{I}^n \to X$ at x_0.

(c) For $n \geqslant 2$, two distinct operations $*_i$, $*_j$ of $\Omega_n(X, x_0)$ satisfy the *middle-four interchange* property

$$(a *_i b) *_j (c *_i d) = (a *_j c) *_i (b *_j d). \tag{6.70}$$

As a consequence, the induced operations in $\pi_n(X, x_0)$ also satisfy it.

(d) For $n \geqslant 2$, all the induced operations $[a] +_i [b] = [a *_i b]$ in $\pi_n(X, x_0)$ coincide, giving a unique operation $[a] + [b]$ which makes $\pi_n(X, x_0)$ into an abelian group. *Hints:* apply the exercise below.

(e) If a set A is equipped with two operations, say $+$ and $\bullet$, which have the same identity e and satisfy the middle-four interchange property

$$(x + y) \bullet (z + u) = (x \bullet z) + (y \bullet u), \qquad \text{for all } x, y, z, u \in A, \tag{6.71}$$

then these operations are commutative and coincide.

> *Note.* The associativity of the operations is not assumed. Nevertheless, this result does not apply to the operations of $\Omega_n(X, x_0)$, which do not have an identity.

*(f) Formally, the sequence of structures of all $\pi_n(X, x_0)$ is made clearer by these remarks:

(i) $\pi_0(X, x_0)$ is a pointed set,

(ii) $\pi_1(X, x_0)$ is a group, which is a group *in* the category Set$_\bullet$,

(iii) $\pi_2(X, x_0)$ is an abelian group, which is a group *in* the category Gp,

(iv) $\pi_n(X, x_0)$ is an abelian group, which is a group *in* the category Ab (for $n \geqslant 3$).

All this makes sense defining a *group in* a category C with finite products, as an object G equipped with mappings $m: G \times G \to G$ (*multiplication*), $i: G \to G$ (*inversion*), $e: G^0 \to G$ (*identity*) satisfying some axioms.

An interested reader can develop these issues (which are not difficult), or see [M2], page 3. Point (iii) follows from Exercise (e).

6.5.3 Homotopy functors

(a) Extending the case $n = 1$ (in 6.2.1), we have a sequence of functors defined on the category of pointed spaces

$$\pi_n \colon \mathsf{Top}_{\bullet} \to \mathsf{C}_n, \qquad (X, x) \mapsto \pi_n(X, x),$$

$$f_* \colon \pi_n(X, x) \to \pi_n(Y, y), \qquad f_*([a]) = [fa], \tag{6.72}$$

with values in the category C_n of pointed sets (for $n = 0$), or groups (for $n = 1$), or abelian groups (for $n \geqslant 2$). A pointed map $f \colon (X, x) \to (Y, y)$ is sent to the morphism $\pi_n(f) = f_*$ which takes the class of a relative cube $a \in \Omega_n(X, x)$ to the class of $fa \in \Omega_n(Y, y)$.

For $n = 0$, this mapping preserves the basepoint. For $n \geqslant 1$ it preserves the structural operation: at the level of cubes, we already have that $f(a *_i b) = fa *_i fb$, for $i = 1, ..., n$.

(b) We can now extend an issue already noted for $n = 1$ (in 6.2.2(c)): the functor π_n can be identified with the homotopy functor associated to the pointed space $(\mathbb{S}^n, e_1)$, defined in (1.71)

$$\pi_n(X, x_0) = [(\mathbb{S}^n, e_1), (X, x_0)]_{\bullet}. \tag{6.73}$$

In fact, a relative cube $a \colon \mathbb{I}^n \to X$ at x_0 amounts to a pointed map $\hat{a} \colon (\mathbb{S}^n, e_1) \to (X, x_0)$, and a homotopy $h \colon a \simeq b$ with fixed boundary amounts to a pointed homotopy $\hat{h} \colon \hat{a} \simeq \hat{b}$.

This also works for $n = 0$: a point $a \colon \{0\} \to X$ is the same as a pointed map $\hat{a} \colon (\mathbb{S}^0, 1) \to (X, x_0)$ with $\hat{a}(-1) = a(0)$, and the path component of $a(0)$ corresponds to the class of $\hat{a}$ up to pointed homotopy.

6.5.4 Complements

The rest of this section is meant to give some further information on the higher homotopy groups; easy proofs are only sketched, complex proofs are referred to.

(a) (*The transition isomorphism*) The pointed set $\pi_0(X, x_0) = (|X|/{\simeq}, [x_0])$ stays the same when x_0 varies in a path connected component of X.

For $n \geqslant 1$, a path $c \colon x_0 \simeq x_1$ in the space X gives a transition isomorphism

$$\hat{c}_n \colon \pi_n(X, x_0) \to \pi_n(X, x_1), \tag{6.74}$$

which extends the 1-dimensional case (in 6.3.1).

For a 0-connected space X, the symbol $\pi_n(X)$ can thus denote the isomorphism type of each $\pi_n(X, x_0)$, with due care (see the introduction to Section 6.3).

The case $n = 2$ is sufficient to show the general pattern: a relative cube $A\colon \mathbb{I}^2 \to X$ at x_0 is transformed into a relative cube at x_1, pasting around A various double homotopies built with the path c, as in the following picture

$$M'(s,t) = c(\max(1 - s, t)),$$
$$M(s,t) = c(\max(s, t)),$$
$$C_1(s,t) = c(t),$$
$$\dots \tag{6.75}$$
$$L(s,t) = c^\sharp(\min(s,t)),$$
$$L'(s,t) = c^\sharp(\min(1 - s, t)).$$

(b) (*Cartesian products*) The functor $\pi_n\colon \mathsf{Top}_\bullet \to \mathsf{C}_n$ preserves products, for all $n \geqslant 0$.

The case $n = 0$ is an easy consequence of the fact that a product $\prod_i X_i$ of non-empty spaces is path connected if and only if all the factors X_i are. The case $n = 1$ was proved in 6.2.4 (and deduced from the fundamental groupoid case in 6.1.6); for higher n, the proof is as in 6.1.6.

*(c) The higher homotopy groups $\pi_k(\mathbb{S}^1)$ are trivial, for $k \geqslant 2$.

This is proved in [Ha], Proposition 4.1, as a consequence of the lifting properties of the exponential map $p\colon \mathbb{R} \to \mathbb{S}^1$.

*(d) It follows that $\pi_k(\mathbb{T}^n)$ is trivial, for $n \geqslant 1$ and $k \geqslant 2$. On the other hand, we have seen that the homology groups $H_k(\mathbb{T}^n)$ are only trivial for $k > n$ (in Exercise 5.6.7(e)).

In particular, any map $\mathbb{S}^2 \to \mathbb{T}$ is homotopic to a constant: the homotopy groups do not detect the cavity of dimension 2 in the torus.

6.5.5 *Hurewicz Theorem* (in higher dimension)

Suppose that the 0-connected space X has

$$\pi_1(X) = \dots = \pi_{n-1}(X) = 0, \tag{6.76}$$

for $n \geqslant 2$. Then, for every $x_0 \in X$, there is a Hurewicz isomorphism

$$h_n\colon \pi_n(X, x_0) \to H_n(X), \qquad h_n[a] = [a]_H, \tag{6.77}$$

that takes the class of a relative n-cube $a\colon (\mathbb{I}^n, \partial\mathbb{I}^n) \to (X, x_0)$ to its homology class.

Proof We only remark that the cube a is a cycle, because its faces are constant cubes of degree $\geqslant 1$. The proof of the theorem can be found in [Ha], Theorem 4.32. $\qquad\square$

6.5.6 Higher connectedness

(a) Let X be a 1-connected space, which means that it is non-empty, path connected, and has $\pi_1(X) = 0$. By Theorem 6.5.5, $\pi_2(X) \cong H_2(X)$.

We say that X is *2-connected* if $\pi_2(X) = 0$, or equivalently $H_2(X) = 0$.

(b) More generally we say that a space X is *n-connected*, for $n \geqslant 1$, if it is non-empty, path connected, and satisfies the following equivalent conditions:

(i) $\pi_1(X) = \dots = \pi_n(X) = 0$,

(ii) $\pi_1(X) = 0, \qquad H_2(X) = \dots = H_n(X) = 0$.

Then $\pi_{n+1}(X) \cong H_{n+1}(X)$.

The equivalence readily follows from the Hurewicz Theorem, by induction. Let us note that, in (ii), we cannot replace π_1 with H_1: we have seen in 6.3.5(c) that $H_1(X)$ can be trivial when $\pi_1(X)$ is not.

(c) Let $n \geqslant 2$. We have seen in 6.4.4(a) that $\pi_1(\mathbb{S}^n) = 0$. Since $H_k(\mathbb{S}^n) = 0$ for $0 < k < n$, we deduce that

$$\pi_k(\mathbb{S}^n) = 0 \ \text{ for } 0 < k < n, \qquad \pi_n(\mathbb{S}^n) \cong \mathbb{Z}. \qquad (6.78)$$

For $n \geqslant 1$, the sphere $\mathbb{S}^n$ is $(n-1)$-connected and we have an isomorphism

$$h_n \colon \pi_n(\mathbb{S}^n, e_1) \to H_n(X),$$
$$h_n[a] = [a]_H \qquad (\text{for } a \colon (\mathbb{I}^n, \partial\mathbb{I}^n) \to (X, x_0)). \qquad (6.79)$$

*(d) The higher homotopy groups of $\mathbb{S}^n$ (far from vanishing!) are still the subject of research: there are infinitely many pairs (n, k) for which $\pi_k(\mathbb{S}^n)$ is known, and infinitely many for which it is not.

*(e) The group $\pi_3(\mathbb{S}^2, e_1)$ is infinite cyclic, generated by the *Hopf map* $p \colon \mathbb{S}^3 \to \mathbb{S}^2$: see [Hu], Section III.5 and Proposition V.6.5.

7

Complements on categories and topology

This is brief exposition of some issues that have already appeared above, either pertaining to category theory (limits, colimits and adjoint functors), or to general topology (the compact-open topology of the spaces of maps). These issues are related, because the compact-open topology comes out of an adjunction.

This sketch might perhaps be a stimulus to further explore these notions, in the appropriate sources.

Literature. All books on categories, e.g. [M2, Bo1, AHS, G5], have chapters on limits and adjunctions. Applications in Topology, Algebraic Topology and Homological Algebra can be found in all of them, and form a large part of [G5], in Chapters 5 and 6.

The compact-open topology is dealt with in most books on Topology.

7.1 An overview of categorical limits and adjoint functors

Universal properties lead to crucial developments: categorical limits and adjoint functors — the underlying layout of various aspects of Algebra and Topology.

In Category Theory, *limits* extend products and the classical 'projective limits' of Algebra and Topology, casting all of them in the same mould. Dually, *colimits* extend and unify sums and the classical 'inductive limits'; they also include pushouts, already used here to 'paste' spaces, groups or groupoids.

Adjunctions, a global way of presenting universal properties, were introduced in 1958 by D. Kan, a mathematician mainly working in homotopy theory [Ka]: a clear illustration of the dialectic relationship between Topology and Category Theory.

7.1.1 Limits and colimits

(a) Let I be a small category and $X \colon I \to C$ a functor, written in index notation (as in (1.86)), for $i \in I = \mathrm{Ob}\, I$ and $a \colon i \to j$ in I:

$$X \colon I \to C, \qquad i \mapsto X_i, \quad a \mapsto (X_a \colon X_i \to X_j). \tag{7.1}$$

A *cone* of X is an object A of C (the *vertex* of the cone) equipped with a family of maps $(f_i \colon A \to X_i)_{i \in I}$ in C, such that the following triangles commute, for $a \colon i \to j$ in I

$$X_a \cdot f_i = f_j. \tag{7.2}$$

The *limit* of $X \colon I \to C$ is a universal cone $(L, (u_i \colon L \to X_i)_{i \in I})$. This means a cone of X such that every cone $(A, (f_i \colon A \to X_i)_{i \in I})$ factorises uniquely through the former; in other words, there is a unique arrow $f \colon A \to L$ such that, for all $i \in I$, $u_i f = f_i$.

The solution is determined up to isomorphism. If it exists, the limit object L is denoted as $\mathrm{Lim}(X)$.

The uniqueness is proved in the usual way; it also follows from the presentation of limits as universal arrows, in 7.1.3.

(b) Dually, a *cocone* $(A, (f_i \colon X_i \to A)_{i \in I})$ of X satisfies the following condition, for $a \colon i \to j$ in I

$$f_j X_a = f_i. \tag{7.3}$$

The *colimit* of the functor X is a universal cocone $(L', (u_i \colon X_i \to L')_{i \in I})$: the universal property says now that for every cocone $(A, (f_i))$ of X there is a unique map $f \colon L' \to A$ such that $f u_i = f_i$ (for all i). The colimit object, if it exists, is denoted as $\mathrm{Colim}(X)$.

*(c) A *weak limit* (resp. a *weak colimit*) of X is a cone (resp. a cocone) that is only assumed to satisfy the existence part of the universal property.

These issues are important in Homotopy Theory. For instance, a homotopy pullback (resp. pushout) of topological spaces [Mat] gives a weak pullback (resp. pushout) in the homotopy category $\mathrm{ho}\mathsf{Top}$.

7.1.2 Examples and complements

(a) The product $\Pi\, X_i$ of a family $(X_i)_{i \in I}$ of objects of C is the limit of the corresponding functor $X \colon \mathsf{I} \to \mathsf{C}$, defined on the discrete category whose objects are the elements of I. The sum $\Sigma\, X_i$ is the colimit of the same functor.

(b) A *pushout* in C is the colimit of a functor $X \colon \mathsf{I} \to \mathsf{C}$ defined on the category $1 \leftarrow 0 \to 2$. It was already used in Top (see 2.3.6(a)), in Gp (see 6.4.2) and in Gpd (see 6.4.5). The dual notion, a *pullback* in C, is the limit of a functor $X \colon \mathsf{I}^{op} \to \mathsf{C}$ defined on the opposite category $1 \to 0 \leftarrow 2$.

(c) The *equaliser* in C of a pair of parallel morphisms $f, g \colon X \to Y$ is the limit of the obvious functor defined on the category $0 \rightrightarrows 1$ by these morphisms.

In Set, the equaliser of a pair of mappings $f, g \colon X \to Y$ is the subset E of X on which they coincide

$$E = \{ x \in X \mid f(x) = g(x) \}, \tag{7.4}$$

or, more precisely, the inclusion $E \to X$.

In Top the solution is the same, with the subspace topology on E. We have similar constructions in the usual categories of structured sets, putting on E the structure induced by X.

(d) The *coequaliser* in C of a pair of parallel morphisms $f, g \colon X \to Y$ is the colimit of the same functor. (Note that the domain category $0 \rightrightarrows 1$ is selfdual.)

In Set, the coequaliser of f, g is the quotient Y/R modulo the least equivalence relation containing all pairs $(f(x), g(x))$, for $x \in X$; more precisely, the colimit is the projection $Y \to Y/R$.

In Top we have the same solution, with the quotient topology on Y/R. There are similar constructions in Gp, Ab, $R\,\mathsf{Mod}$ ..., where R is the congruence of the current algebraic structure, generated by all pairs $(f(x), g(x))$.

(e) All limits in a category can be constructed with products and equalisers (provided they exist): an easy result, proved in every book on category theory. For instance, one can see [M2], Section V.2, Theorem 1, or [G5], Theorem 2.2.4.

Dually, all colimits can be constructed with sums and coequalisers. In particular, pushout can be constructed with binary sums and coequalisers, as we have seen in 2.3.6(c) for spaces and 6.4.2 for groups, following a pattern that can be easily extended to every category.

The usual categories of structured sets have all limits and colimits, which is expressed saying that they are *complete* and *cocomplete*.

7.1.3 Limits and colimits by universal arrows

Let a functor $X\colon \mathsf{I} \to \mathsf{C}$ be given, and consider the diagonal functor, already defined in (1.88) (with $i \in \mathrm{Ob}\,\mathsf{I}$ and $a \in \mathrm{Mor}\,\mathsf{I}$)

$$D\colon \mathsf{C} \to \mathsf{C}^{\mathsf{I}}, \qquad (DA)_i = A, \qquad (DA)_a = \mathrm{id}\,A. \tag{7.5}$$

A natural transformation $f\colon DA \to X$ is the same as a cone of X of vertex A.

The limit of X in C is the same as a universal arrow $(L, \varepsilon\colon DL \to X)$ from the diagonal functor D to the object X of C^{I} (cf. 1.6.6).

Dually, the colimit of X in C is the same as a universal arrow $(L', \eta\colon X \to DL')$ from the object X of C^{I} to the functor D.

7.1.4 Introducing adjunctions

We begin by two well known situations, which help us to introduce adjoint functors.

(a) The forgetful functor $U\colon \mathsf{Top} \to \mathsf{Set}$ has two well known 'extreme approximations' to an inverse which does not exist, the functors $D, C\colon \mathsf{Set} \to \mathsf{Top}$, where DX (resp. CX) is the set X with the discrete (resp. indiscrete) topology. In fact they provide the set X with the finest (resp. the coarsest) topology.

For every set X and every space Y we can identify the following hom-sets

$$\mathsf{Top}(DX, Y) = \mathsf{Set}(X, UY), \qquad \mathsf{Set}(UY, X) = \mathsf{Top}(Y, CX), \tag{7.6}$$

since every mapping $X \to Y$ becomes continuous if we put on X the discrete topology, and every mapping $Y \to X$ becomes continuous if we put on X the indiscrete one.

The first fact will say that D is left adjoint to U (and U right adjoint to D), which is written as $D \dashv U$. The second says that $U \dashv C$. The name comes from the analogy with adjoint operators between Hilbert spaces.

(b) In other cases it may be convenient to follow another approach, based on universal arrows.

To wit, let us start from the forgetful functor $U\colon R\,\mathsf{Mod} \to \mathsf{Set}$. We have seen, in (4.5), that, for every set X, there is a universal arrow from X to the functor U

$$F(X) = \bigoplus_{x \in X} R, \qquad \eta\colon X \to UF(X), \tag{7.7}$$

consisting of the free R-module $F(X)$ on the set X, with the insertion of the basis η.

The fact that this universal arrow exists *for every X in* Set allows us to construct a 'backward' functor $F\colon$ Set $\to R\,$Mod, left adjoint to U.

F is already defined on the objects. For a mapping $f\colon X \to Y$ in Set

$$
\begin{array}{ccccc}
X & \xrightarrow{\ \eta X\ } & UF(X) & & F(X) \\
{\scriptstyle f}\downarrow & & \ \downarrow{\scriptstyle U(Ff)} & & \ \downarrow{\scriptstyle Ff} \\
Y & \xrightarrow[\ \eta Y\ ]{} & UF(Y) & & F(Y)
\end{array}
\tag{7.8}
$$

we define $F(f)$ by the universal property of ηX, as the unique morphism $F(X) \to F(Y)$ in $R\,$Mod such that $U(Ff)$ makes the previous square commute. (In this example $F(f)$ is just the linear extension of f; but the general pattern is better seen if we proceed in a formal way.)

Further applications of the universal property readily show that F preserves composition and identities (as evident in our particular case). Let us note that we have extended the given object-function $F(X)$ in the only way that makes the family (η_X) into a natural transformation

$$\eta\colon 1 \to UF\colon \text{Set} \to \text{Set},$$

called the *unit* of the adjunction.

(c) The relationship between these functors U and F (assuming now that we have defined both) can be equivalently expressed as in (a), by a family of bijective mappings

$$\varphi_{XA}\colon R\,\text{Mod}(FX, A) \to \text{Set}(X, UA) \qquad (X \text{ in Set}, \ A \text{ in } R\,\text{Mod}), \tag{7.9}$$

which is natural in the variables X and A (on the category $\text{Set}^{\text{op}} \times R\,\text{Mod}$).

Concretely, $\varphi_{XA}(g\colon FX \to A)$ is the restriction of the R-homomorphism g to the basis X of $F(X)$; its bijectivity means that every mapping $f\colon X \to UA$ has a unique extension to a homomorphism $g\colon FX \to A$.

7.1.5 Main definitions

An *adjunction $F \dashv G$*, with a functor $F\colon \mathsf{C} \to \mathsf{D}$ *left adjoint* to a functor $G\colon \mathsf{D} \to \mathsf{C}$, can be equivalently presented in four forms.

(i) We assign two functors $F\colon \mathsf{C} \to \mathsf{D}$ and $G\colon \mathsf{D} \to \mathsf{C}$ together with a family of bijections

$$\varphi_{XY}\colon \mathsf{D}(FX, Y) \to \mathsf{C}(X, GY) \qquad (X \text{ in } \mathsf{C}, \ Y \text{ in } \mathsf{D}),$$

which is natural in X, Y. More formally, the family (φ_{XY}) is a functorial isomorphism

$$\varphi\colon \mathsf{D}(F(-), -) \longrightarrow \mathsf{C}(-, G(-))\colon \mathsf{C}^{\text{op}} \times \mathsf{D} \longrightarrow \text{Set}.$$

(ii) We assign a functor $G\colon \mathsf{D} \to \mathsf{C}$ and, for every object X in C, a universal arrow (see 1.6.6)

$$(F_0 X, \eta X\colon X \to GF_0 X), \qquad \text{from the object } X \text{ to the functor } G.$$

(ii*) We assign a functor $F\colon \mathsf{C} \to \mathsf{D}$ and, for every object Y in D, a universal arrow

$$(G_0 Y, \varepsilon Y\colon FG_0 Y \to Y), \qquad \text{from the functor } F \text{ to the object } Y.$$

(iii) We assign two functors $F\colon \mathsf{C} \to \mathsf{D}$ and $G\colon \mathsf{D} \to \mathsf{C}$, together with two natural transformations

$$\eta\colon \mathrm{id}\,\mathsf{C} \to GF \ (\text{the } unit), \qquad \varepsilon\colon FG \to \mathrm{id}\,\mathsf{D} \ (\text{the } counit),$$

which satisfy the *triangular identities*: $\varepsilon F.F\eta = \mathrm{id}\,F$ and $G\varepsilon.\eta G = \mathrm{id}\,G$

$$
F \xrightarrow{\ F s\eta\ } FGF \xrightarrow{\ \varepsilon F\ } F \underset{\mathrm{id}\,F}{} \qquad
G \xrightarrow{\ \eta G\ } GFG \xrightarrow{\ G\varepsilon\ } G \underset{\mathrm{id}\,F}{}
\tag{7.10}
$$

A proof of the equivalence can be found in [M2], Section IV.1, Theorem 2, or in [G5], Theorem 3.1.5. Essentially:

- given (i) one defines $\eta X = \varphi_{X,FX}(\mathrm{id}\,FX)\colon X \to GFX$ and $\varepsilon Y = (\varphi_{GY,Y})^{-1}(\mathrm{id}\,GY)\colon FGY \to Y$,

- given (ii) one defines $F(X) = F_0 X$, the morphism $F(f\colon X \to X')$ by the universal property of ηX, and the morphism $\varphi_{XY}(g\colon FX \to Y)$ as $Gg.\eta X\colon X \to GY$,

- given (iii) one defines the mapping $\varphi_{XY}\colon \mathsf{D}(F(X),Y) \to \mathsf{C}(X,G(Y))$ as above, and also a backward mapping $\psi_{XY}(f\colon X \to GY) = \varepsilon Y.Ff$; then one proves that these mappings are inverse to each other, by the triangular identities.

These forms have different features.

Form (i) is the classical definition of an adjunction, and is at the origin of the name.

Form (ii) is used when we start from a given functor G and want to construct its left adjoint (possibly less easy to define). Form (ii*) is used in a dual way.

The 'algebraic' form (iii) leads to the formal theory of adjunctions. *In fact, it makes sense in an abstract 2-category.*

7.1.6 Examples and comments

In the previous chapters there are various cases where the general existence of universal arrows ensures the existence of an adjoint.

(a) (*Free objects*) As we have seen in 1.6.7(b), given a functor $U\colon \mathsf{A} \to \mathsf{Set}$, the free A-object over a set X is an object $F(X)$ equipped with a universal arrow $\eta\colon X \to UF(X)$ from the object X to the functor U. The existence of all free objects in A amounts to the existence of the left adjoint $F \dashv U$.

In particular, we have constructed the free abelian group in 1.2.5, the free monoid in 1.6.7(c), the free R-module in (4.5), and the free group in 6.4.1. In fact, all algebraic structures defined by 'equational axioms' have free objects: the following point is an extension of this fact.

(b) Many interesting constructions in algebra arise as a left adjoint to a forgetful functor $U\colon \mathsf{A} \to \mathsf{B}$, between categories of algebraic structures of which the second is 'weaker' than the first — in the sense of having less structure, or less properties, or both.

Here we have considered the inclusion functor $U\colon \mathsf{Ab} \to \mathsf{Gp}$, whose left adjoint is the *abelianisation functor* $(-)^{\mathrm{ab}}\colon \mathsf{Gp} \to \mathsf{Ab}$, with unit the canonical projection $pG\colon G \to G^{\mathrm{ab}} = G/[G,G]$ (see 6.3.3(e), (f)).

A general result in this sense can be found in [G5], Theorem 4.4.5; several concrete instances are computed in Section 4.1, *loc. cit.*

This layout includes point (a) letting $\mathsf{B} = \mathsf{Set}$, the category of trivial algebraic structures.

(c) (*Limit functor as a right adjoint*) The description of the limit of a functor $X\colon \mathsf{I} \to \mathsf{C}$ as a universal arrow $(LX, \varepsilon_X\colon D(LX) \to X)$ from the diagonal functor $D\colon \mathsf{C} \to \mathsf{C}^{\mathsf{I}}$ to the object X of C^{I} (in 7.1.3) shows that the existence of all I-based limits in C amounts to the existence of a right adjoint to D, the *limit functor* $L\colon \mathsf{C}^{\mathsf{I}} \to \mathsf{C}$, with counit $\varepsilon\colon DL \to \mathrm{id}\,\mathsf{C}$.

In particular, if the category C has all products indexed by the set I, the diagonal functor $D\colon \mathsf{C} \to \mathsf{C}^I$ has a right adjoint, the functor $\Pi\colon \mathsf{C}^I \to \mathsf{C}$ *of I-indexed products*.

(d) Dually, if the category C has all I-based colimits, the diagonal functor $D\colon \mathsf{C} \to \mathsf{C}^{\mathsf{I}}$ has a left adjoint, the *colimit functor* $L'\colon \mathsf{C}^{\mathsf{I}} \to \mathsf{C}$, with unit $\eta\colon \mathrm{id}\,\mathsf{C} \to DL'$.

In particular, if the category C has all sums indexed by the set I, there is a functor $\Sigma\colon \mathsf{C}^I \to \mathsf{C}$ *of I-indexed sums*, left adjoint to the diagonal functor $D\colon \mathsf{C} \to \mathsf{C}^I$.

(e) As usual, the theory of adjunctions is selfdual, but the concrete examples, based on categories of structured sets, are not.

Thus, the forgetful functor $U\colon \mathsf{A} \to \mathsf{Set}$ of an equational algebraic structure has always a left adjoint, but generally lacks a right one. Comparing (c) and (d), we see that the existence of a left or a right adjoint to the diagonal functor D are independent facts.

7.1.7 Main properties of adjunctions

(a) (*Uniqueness and existence*) Given a functor, its left adjoint (or its right adjoint) is uniquely determined up to isomorphism: this follows from the uniqueness property of universal arrows, in 1.6.6.

A standard way of proving the existence (under suitable hypothesis) is the Adjoint Functor Theorem of P. Freyd: see [M2], Section V.6, Theorem 2, or [G5], Section 3.5.

(b) (*Composition*) As a consequence of the composition of universal arrows (in 1.6.6(d)), two consecutive adjunctions

$$F: \mathsf{C} \rightleftarrows \mathsf{D} : G, \qquad \eta: 1 \to GF, \quad \varepsilon: FG \to 1,$$
$$H: \mathsf{D} \rightleftarrows \mathsf{E} : K, \qquad \rho: 1 \to KH, \quad \sigma: HK \to 1, \tag{7.11}$$

give a composed adjunction from the first to the third category

$$HF: \mathsf{C} \rightleftarrows \mathsf{E} : GK,$$
$$(G\rho F).\eta: 1 \to GF \to GK\,HF, \tag{7.12}$$
$$\sigma.(H\varepsilon K): HF\,GK \to HK \to 1.$$

Accordingly, an adjunction can be written as an arrow (distinguished, for instance, by a dot)

$$(F, G, \eta, \varepsilon): \mathsf{C} \rightarrowtail \mathsf{D}, \tag{7.13}$$

here *directed as the left adjoint*. (According to the context, the other direction can be preferable.)

    There is a category AdjCat of small categories and adjunctions, with this composition and obvious identities.

(c) (*Duality*) Applying duality on categories, functors and transformations, an adjunction $(F, G, \eta, \varepsilon): \mathsf{C} \rightarrowtail \mathsf{D}$ is transformed into an adjunction

$$(G^{\mathrm{op}}, F^{\mathrm{op}}, \varepsilon^{\mathrm{op}}, \eta^{\mathrm{op}}): \mathsf{D}^{\mathrm{op}} \rightarrowtail \mathsf{C}^{\mathrm{op}}, \tag{7.14}$$

with left adjoint $G^{\mathrm{op}}: \mathsf{D}^{\mathrm{op}} \to \mathsf{C}^{\mathrm{op}}$ and unit $\varepsilon^{\mathrm{op}}: \mathrm{id}\,\mathsf{D}^{\mathrm{op}} \to F^{\mathrm{op}}G^{\mathrm{op}}$.

(d) (*Preservation properties*) A left adjoint preserves all the existing colimits, while a right adjoint preserves the existing limits. The proof is straightforward, cf. [G5], Exercise 6.3.4(g).

7.1.8 Equivalence and adjunctions

An *adjoint equivalence* is an adjunction $(F, G, \eta, \varepsilon): \mathsf{C} \rightarrowtail \mathsf{D}$ where $\eta: 1 \to GF$ and $\varepsilon: FG \to 1$ are functorial isomorphisms. Equivalently, it is an equivalence of categories where the isomorphisms $GF \cong \mathrm{id}\,\mathsf{C}$ and $FG \cong \mathrm{id}\,\mathsf{D}$ are bound by the triangular identities.

An important theorem states that the following conditions on a functor $F\colon \mathsf{C} \to \mathsf{D}$ are equivalent, forming a simple *characterisation of the equivalence of categories*:

(i) the functor F is an equivalence of categories,

(ii) the functor F can be completed to an adjoint equivalence $(F, G, \eta, \varepsilon)$,

(iii) the functor F is faithful, full and *essentially surjective on objects*.

The last property means that: for every object Y of D there exists some object X in C such that $F(X)$ is isomorphic to Y in D. The proof of the equivalence of these conditions requires the axiom of choice: see [M2], Section V.4, or [G5], Theorem 1.5.8.

An interested reader can write the proof, extending the solution of Exercise 6.1.9(b).

Examples and comments. (a) The (large) category of finite sets is equivalent to its full subcategory of finite cardinals, the sets $n = \{0, 1, ..., n-1\}$, which is small (and cannot be isomorphic to the former).

In fact, it is sufficient to apply characterisation (iii) to the embedding of the full subcategory of finite cardinals.

(b) Similarly, the category of finite dimensional vector fields on the field K is equivalent to its full subcategory on the objects K^n, for $n \geqslant 0$.

*The latter can be realised, up to categorical isomorphism, as a category $\mathrm{Mat}(K)$ whose objects are the natural numbers, and whose arrows $n \to m$ are the matrices of type $m \times n$, with entries in K (by the usual matrix representation of linear mappings); the composition is matrix product.

The only tricky, or amusing, point is that we have to introduce a zero matrix $0_{m,n}$ of type $m \times n$ (to represent the zero morphism $K^n \to K^m$) even when $m = 0$ or $n = 0$. This 'formal matrix' can hardly be written as a table, which should have no rows and some columns, or some rows and no columns, or nothing at all — all these cases being distinguished...*

7.2 The compact-open topology

We end with the exponential law in Top: for a locally compact space A, the functor $- \times A\colon \mathsf{Top} \to \mathsf{Top}$ has a right adjoint $(-)^A\colon \mathsf{Top} \to \mathsf{Top}$, where the space $Y^A = \mathsf{Top}(A, Y)$ has the compact-open topology.

In particular, the standard interval $\mathbb{I}$ gives the path functor $P(Y) = \mathsf{Top}(\mathbb{I}, Y)$, right adjoint to the cylinder functor $I(X) = X \times \mathbb{I}$. Homotopies of maps $X \to Y$ are equivalently represented by maps $I(X) \to Y$ or $X \to P(Y)$.

7.2.1 The exponential law, II

Let us come back to the exponential law of sets, dealt with in 4.1.7. For every set A there is a natural bijection in the variables X, Y

$$\varphi^A_{XY} : \mathsf{Set}(X \times A, Y) \longrightarrow \mathsf{Set}(X, Y^A),$$

$$(f : X \times A \to Y) \mapsto (g : X \to Y^A), \quad g(x) = f(x, -) : A \to Y, \tag{7.15}$$

where $Y^A = \mathsf{Set}(A, Y)$.

We can now say that this is a family of adjunctions, indexed by the set A: the functor $- \times A : \mathsf{Set} \to \mathsf{Set}$ has a right adjoint, the functor $(-)^A = \mathsf{Set}(A, -) : \mathsf{Set} \to \mathsf{Set}$.

Equivalently, for every set Y, there is a universal arrow $(Y^A, \varepsilon Y)$ from the functor $- \times A$ to the set Y itself

$$\varepsilon Y : Y^A \times A \to Y, \quad \varepsilon Y(h, a) = h(a) \quad (evaluation). \tag{7.16}$$

The universal property states that, for every map $f : X \times A \to Y$ there is a unique map $g : X \to Y^A$ such that

$$\varepsilon Y(g \times A) = f, \tag{7.17}$$

which means (again) that $g(x)(a) = f(x, a)$, for all $x \in X$ and $a \in A$.

7.2.2 Exponentiable objects and spaces

More generally, let C be a category with finite products. An object A is said to be *exponentiable* in C (for cartesian products) if the functor $- \times A : \mathsf{C} \to \mathsf{C}$ has a right adjoint. The latter is then written as $(-)^A : \mathsf{C} \to \mathsf{C}$.

Equivalently, A is exponentiable if, for every object Y, there is a universal arrow $(Y^A, \varepsilon Y)$ from the functor $- \times A$ to the object Y

$$\varepsilon Y : Y^A \times A \to Y, \tag{7.18}$$

which is still called *evaluation*.

The category C is said to be *cartesian closed* if all its objects are exponentiable.

Examples and complements. (a) We have seen that Set is cartesian closed, with $Y^A = \mathsf{Set}(A, Y)$.

(b) Top is not cartesian closed, but we show below that *every locally compact space A is exponentiable*: for every space Y, the space Y^A is the set of maps $\mathsf{Top}(A, Y)$ endowed with the compact-open topology.

*(c) As a partial converse to the previous result, every exponentiable space which is Hausdorff must be locally compact, as proved in [He], Section 7.3.18. In particular, the rational line $\mathbb{Q}$ is not exponentiable.

A topological space is exponentiable if and only if it satisfies a property called *core compactness*, or *quasi local compactness*: every nbd V of a point contains a smaller one W, which is *relatively compact under V*; this relation, denoted by $W \ll V$, means that any open cover of V contains a finite subcover of W.

This can be seen in [LoR], together with the topology of the exponential space in this case, and many related results for other 'topological categories'.

*(d) The category Cat is cartesian closed, with $Y^A = \mathsf{Cat}(A, Y)$. The exponential law is an extension of that of Set, that one can easily write down, or find in the solution of Exercise 3.4.7(a), in [G5].

*(e) $R\,\mathsf{Mod}$ (and Ab) is not cartesian closed. But we have seen, in 4.1.7(b), an exponential law based on the tensor product in $R\,\mathsf{Mod}$, instead of the cartesian product, giving the adjunction $- \otimes_R A \dashv \mathrm{Hom}_R(A, -)$.

This is a prime example in the theory of monoidal categories (equipped with a 'tensor product'), exponentiable objects therein, and monoidal closed categories: cf. [M2, EK, Kl].

7.2.3 The compact-open topology

The set of maps $\mathsf{Top}(X, Y)$ between two topological spaces can be given diverse topologies: the *compact-open topology* is likely the most well known.

Its open subsets are generated by the sets

$$W(K, V) = \{h \in \mathsf{Top}(X, Y) \mid h(K) \subset V\}, \tag{7.19}$$

where K is any compact subspace of X and V is any open subset of Y. A general open subset is a union of finite intersections of *distinguished* open sets, of the previous form. (The latter form a subbase of the topology.)

This is particularly important if A is *locally compact*, in the sense that every point has a fundamental system of compact neighbourhoods.

7.2.4 Theorem

Every locally compact space A is exponentiable: the functor

$$F \colon \mathsf{Top} \to \mathsf{Top}, \qquad F(X) = X \times A, \quad F(f) = f \times A, \tag{7.20}$$

has a right adjoint

$$G \colon \mathsf{Top} \to \mathsf{Top}, \qquad G(Y) = Y^A, \quad G(g) = g_\sharp \colon h \mapsto gh, \tag{7.21}$$

where Y^A is the set $\mathsf{Top}(A, Y)$, with the compact-open topology.

The counit of the adjunction is the evaluation map

$$\varepsilon_Y \colon FG(Y) = Y^A \times A \to Y, \qquad \varepsilon_Y(h,a) = h(a). \qquad (7.22)$$

Proof (a) We have defined the space $G(Y) = Y^A = \mathsf{Top}(A,Y)$, and the mapping ε_Y. We have to prove that the latter is continuous, and the pair (Y^A, ε_Y) is a universal arrow from the functor F to the object Y.

As in (7.17), this means that, for every map $f \colon X \times A \to Y$, there is a unique map $g \colon X \to Y^A$ such that $g(x)(a) = f(x,a)$, for all $x \in X$ and $a \in A$.

(b) First we prove that, for every Y, the mapping ε_Y is continuous at any point $(h_0, a_0) \in Y^A \times A$.

Let V be an open neighbourhood of $h_0(a_0)$ in Y. As A is locally compact, there is a compact neighbourhood K of a_0 in A such that $h_0(K) \subset V$. The distinguished open set $W(K,V) \times K$ of $Y^A \times A$ is a neighbourhood of (h_0, a_0), which the mapping ε_Y takes into V.

(c) Now, we prove that, for every map $f \colon X \times A \to Y$, the mapping

$$g \colon X \to Y^A, \qquad g(x) = f(x,-) \colon A \to Y, \qquad (7.23)$$

is well defined, i.e. $f(x,-)$ is continuous on A, for every $x \in X$.

We fix $x_0 \in X$ and prove the continuity of $h = g(x_0)$ at $a_0 \in A$. Let V be an open neighbourhood of $h(a_0) = f(x_0, a_0)$ in Y. As $f \colon X \times A \to Y$ is continuous, there is a neighbourhood $U \times K$ of (x_0, a_0) such that $f(U \times K) \subset V$, and $h(K) = f(\{x_0\} \times K) \subset V$.

(d) Finally, the mapping g is continuous at an arbitrary point $x_0 \in X$. (Its uniqueness under the condition $f = \varepsilon Y.(g \times A)$ is obvious, at the level of sets.)

It is sufficient to consider an open neighbourhood of $g(x_0) = f(x_0, -)$ of the form $W(K,V)$, where K is compact in A, V is open in Y, and $f(\{x_0\} \times K) \subset V$.

For any $a \in K$ there is a basic open neighbourhood $U_a \times A_a$ of (x_0, a) in $X \times A$ such that $f(U_a \times A_a) \subset V$. All A_a form an open cover of the compact subspace K of A, and we can extract a finite family (A_{a_i}) which covers K. Then the intersection $U = \cap U_{a_i}$ is an open neighbourhood of x_0, and $f(U \times K) \subset V$, because $(x,a) \in U \times K$ implies that a belongs to some A_{a_i} and $x \in U_{a_i}$. But $f(U \times K) \subset V$ means that $g(U) \subset W(K,V)$, and the proof is complete. $\qquad \square$

7.2.5 The path functor

In particular, the cylinder functor $I = -\times \mathbb{I}\colon \mathsf{Top} \to \mathsf{Top}$ has a right adjoint

$$P\colon \mathsf{Top} \to \mathsf{Top},$$

$$P(Y) = Y^{\mathbb{I}}, \qquad P(g)(a) = ga \quad (\text{for } g\colon Y \to Y', \, a\colon \mathbb{I} \to Y), \tag{7.24}$$

called the *cocylinder* or *path functor*. $P(Y)$ is the space of all paths $\mathbb{I} \to Y$, with the compact-open topology. The counit is the evaluation of paths

$$\varepsilon Y\colon P(Y) \times \mathbb{I} \to Y, \qquad (a,t) \mapsto a(t). \tag{7.25}$$

The universal property states that, for every map $f\colon X \times \mathbb{I} \to Y$ there is a unique map $g\colon X \to PY$ such that $f(x,t) = g(x)(t)$ (for all $x \in X$ and $t \in \mathbb{I}$).

In other words, it says that a homotopy $\varphi\colon f \to g$, can be equivalently described by a map $\hat{\varphi}\colon IX \to Y$ or a map $\check{\varphi}\colon X \to PY$.

The functor P inherits from the interval $\mathbb{I}$, contravariantly, a dual structure, which we write with the same symbols used for the cylinder functor, in (1.57) and (1.85):

$$\begin{aligned} \partial^\alpha\colon P(Y) \to Y, \qquad & \partial^\alpha(a) = a(\alpha) & (\textit{faces}), \\ e\colon Y \to P(Y), \qquad & e(y)(t) = y & (\textit{degeneracy}), \\ r\colon P(Y) \to P(Y), \qquad & r(a)(t) = a(1-t) & (\textit{reversion}). \end{aligned} \tag{7.26}$$

In this description, the faces of a homotopy $\check{\varphi}\colon X \to PY$ are defined as $\partial^\alpha \check{\varphi}\colon X \to Y$.

7.2.6 Complements

We can now prove a fact already used above: if A is a locally compact space (or, more generally, an exponentiable space), the functor $-\times A\colon \mathsf{Top} \to \mathsf{Top}$ preserves topological projections.

In fact, the topological projection $p\colon X \to X/R$ can be viewed as the coequaliser of two maps $f,g\colon R \rightrightarrows X$, where $R \subset X \times X$ has the subspace topology, and f,g are the restrictions of the cartesian projections $p_1, p_2\colon X \times X \to X$.

Now, the left adjoint $-\times A\colon \mathsf{Top} \to \mathsf{Top}$ preserves all colimits (by 7.1.7(d)), so that the coequaliser of the maps

$$f \times A,\, g \times A\colon R \times A \rightrightarrows X \times A,$$

is the map $p \times A\colon X \times A \to (X/R) \times A$. The latter is thus a topological projection, on the quotient of $X \times A$ modulo the equivalence relation $R \times A$.

8

Solution of the exercises

Easy exercises and exercises marked with * may be left to the reader.

8.1 Exercises of Chapter 1

8.1.1 Solutions of 1.1.1

(a) Let $a < b$ in $\mathbb{R}$. We want to prove that there are homeomorphisms, for the following forms of non-degenerate intervals:

$$\text{(i)} \qquad \mathbb{R} \cong \,]a, +\infty[\,\cong\,]-\infty, b[\,\cong\,]a, b[\,\cong\,]0, 1[\qquad\qquad (open),$$

$$\text{(ii)} \qquad [a, +\infty[\,\cong\,]-\infty, b] \,\cong\, [a, b[\,\cong\,]a, b] \,\cong\, [0, 1[\qquad\qquad (semiopen),$$

$$\text{(iii)} \quad [a, b] \cong [0, 1] \qquad\qquad\qquad\qquad\qquad\qquad\qquad\qquad\qquad (compact).$$

First let us note that, for every $a \in \mathbb{R}$ and $\lambda \neq 0$, the affine mapping

$$f \colon \mathbb{R} \to \mathbb{R}, \qquad\qquad f(x) = \lambda x + a, \qquad\qquad (8.1)$$

is a homeomorphism.

Point (iii) is already proved: the compact interval $[0, 1]$ is homeomorphic to its f-image $[a, a + \lambda]$ (with $\lambda > 0$), and therefore to any non-degenerate compact interval.

Similarly, the open interval $]0, 1[$ is homeomorphic to any bounded open interval $]a, b[$. But $\mathbb{R}$ is homeomorphic to the open interval $J = \,]-\pi/2, \pi/2[$, as shown by the invertible function $\tan \colon J \to \mathbb{R}$, and to the open interval $J' = \,]0, +\infty[$, as shown by the invertible function $\exp \colon \mathbb{R} \to J'$. To conclude point (i), J' is homeomorphic to any $]a, +\infty[$ by a translation, and the latter is homeomorphic to $]-\infty, -a[$ by $f(x) = -x$.

Point (ii) is proved in the same way. The affine mapping (8.1) proves that all the intervals $[a, b[$ are homeomorphic to each other, and all the intervals $]a, b]$ are also. But the former are homeomorphic to the latter,

299

since $[0, 1[\cong]0, 1]$, via $f(x) = 1 - x$. Taking into account that $[0, \pi/2[\cong [0, +\infty[$, via the tangent function, the last verifications are obvious.

(b) The idea is to distinguish the presence of the end-points by properties of connectedness.

Let X be a connected space. We say that a point $x \in X$ *disconnects* the space if $X \setminus \{x\}$ is no longer connected. Plainly, given a homeomorphism $f \colon X \to Y$, the point x disconnects X if and only if $f(x)$ disconnects Y.

We can say something more: f restricts to a bijection $f' \colon A \to B$ between the subsets of the points which disconnect their space; and of course it also restricts to a bijection between their complements, the subsets of points which do not disconnect the space. (If useful, we can also remark that we have two homeomorphisms between the corresponding subspaces.)

Now, in the interval $]0, 1[$ every point disconnects the space. The interval $[0, 1[$ has precisely one point which does not disconnect it, namely 0. The interval $[0, 1]$ has two such points.

(c) Trivially, a singleton space is only homeomorphic to a singleton. To prove that $\mathbb{R}$ is not homeomorphic to any higher $\mathbb{R}^n$ we proceed as in (b): any point $x \in \mathbb{R}$ disconnects the line, while each pierced space $\mathbb{R}^n \setminus \{x\}$ is path connected, for $n > 1$.

(d) A degenerate interval is not equipotent to the finite sphere $\mathbb{S}^0 = \{-1, 1\}$. Any non-degenerate interval J is disconnected by any interior point, while any sphere $\mathbb{S}^n$ stays path connected if we take out a point.

8.1.2 Solutions of 1.1.2

(a) A map $x \colon \{*\} \to X$ amounts to a point of X, and a homotopy $\varphi \colon x \simeq y \colon \{*\} \to X$ amounts to a path in X, from x to y. All this will be better analysed in Section 1.4.

(b) $\mathbb{R}^n$ is a topological vector space on the topological field $\mathbb{R}$, which means that the linear combination

$$c \colon \mathbb{R}^n \times \mathbb{R}^n \times \mathbb{R} \times \mathbb{R} \longrightarrow \mathbb{R}^n, \qquad c(v, w, \lambda, \mu) = \lambda v + \mu w,$$

is a continuous mapping.

The mapping φ can be written as a composite of continuous mappings

$$X \times \mathbb{I} \longrightarrow \mathbb{R}^n \times \mathbb{R}^n \times \mathbb{I} \times \mathbb{I} \longrightarrow \mathbb{R}^n,$$

$$(x, t) \mapsto (f(x), g(x), 1 - t, t) \mapsto (1 - t)f(x) + tg(x),$$

whose image, in the given hypotheses, is contained in Y.

8.1.3 Solutions of 1.1.3

(a) Using the $\mathbb{R}$-linear structure of $\mathbb{R}^n$, the origin is a deformation retract of $\mathbb{R}^n$

$$p\colon \mathbb{R}^n \to \{0\}, \qquad \varphi\colon ip \simeq \mathrm{id}\,\mathbb{R}^n, \quad \varphi(x,t) = tx. \tag{8.2}$$

This point is a particular case of the next.

(b) If $X \subset \mathbb{R}^n$ is starred with respect to $x_0 \in X$, the singleton $\{x_0\}$ is a deformation retract of X, by the affine homotopy $\varphi\colon mp \simeq \mathrm{id}\,X$ (see (1.3)) that travels on the line segments of $\mathbb{R}^n$, from x_0 to each $x \in X$

$$\begin{aligned}
p\colon X \rightleftarrows \{x_0\} : m, \qquad & pm_0 = \mathrm{id}, \qquad \varphi\colon mp \simeq \mathrm{id}\,X, \\
\varphi\colon X \times \mathbb{I} \to X, \qquad & \varphi(x,t) = (1-t)x_0 + tx.
\end{aligned} \tag{8.3}$$

(c) Plainly, (i) is equivalent to (ii):

- if X is contractible, there is a homotopy $\varphi\colon mp \simeq \mathrm{id}\,X$, as in (1.4),
- conversely, if $\mathrm{id}\,X$ is homotopic to a map $X \to X$ constant at $x_0 \in X$, then $\{x_0\}$ is a deformation retract of X.

The rest follows easily, by whisker composition of homotopies and maps (see (1.60)).

(d) The normalisation mapping N is obviously the identity on $\mathbb{S}^n$. Moreover we have an affine homotopy $\varphi\colon mN \simeq \mathrm{id}\,X$

$$\varphi(x,t) = (1-t)N(x) + tx \qquad ((x,t) \in X \times \mathbb{I}), \tag{8.4}$$

that moves on each line segment from $N(x)$ to x, staying in the pierced space X.

(e) If A is a deformation retract of X, any homotopy $\varphi\colon mp \simeq \mathrm{id}\,X$ satisfies the required conditions. Conversely, if these are satisfied, we have a map $p = \varphi(-,0)\colon X \to A$ that satisfies $p(a) = \varphi(a,0) = a$ (for $a \in A$); moreover $\varphi(x,0) = p(x) = mp(x)$ and $\varphi(x,1) = x$ (for $x \in X$).

An accurate reader will note a light abuse of notation in this argument: a more formal writing would distinguish the endomap $e = \varphi(-,0)\colon X \to X$ from its codomain-restriction $p\colon X \to A$, satisfying $e = mp$. According to the context, abuses of notation can be harmless or a source of errors; yet, an excessive formalisation would make mathematics boring and unreadable.

8.1.4 Solutions of 1.1.6

We begin by proving (b). First, $\mathbb{S}^n \subset \mathbb{R}^{n+1}$, but the group $H_n(\mathbb{S}^n) \cong \mathbb{Z}$ cannot be embedded in $H_n(\mathbb{R}^{n+1}) = 0$ (for $n > 0$).

Second, let us suppose that the inclusion $m \colon A \to X$ has a (continuous!) retraction

$$A \xrightarrow{\ m\ } X \xrightarrow{\ p\ } A \qquad\qquad pm = \mathrm{id}\,A. \qquad (8.5)$$

Then the functor H_n gives

$$H_n(A) \xrightarrow{\ m_*\ } H_n(X) \xrightarrow{\ p_*\ } H_n(A) \qquad p_* m_* = \mathrm{id}\,H_n(A), \qquad (8.6)$$

showing that the homomorphism m_* has a left inverse, and is injective.

Putting together these two facts, we have verified (a): $\mathbb{S}^n$ is not a retract of a contractible space, for $n > 0$.

For $n = 0$ the argument can be adapted: the group $H_0(\mathbb{S}^0) \cong \mathbb{Z}^2$ cannot be embedded in $H_0(\mathbb{R}) \cong \mathbb{Z}$; but this would be a circular argument, because the computation of the group $H_0(X)$ (in Exercise 2.2.5(d)) will depend on the partition of the space X in path components.

8.1.5 Solutions of 1.1.8

(a) The proof is obvious: if V is open in Y, $f^{-1}(V) = \bigcup_i f_i^{-1}(V)$.

(b) As in (a), using the preimage $f^{-1}(C) = \bigcup_i f_i^{-1}(C)$ of a closed subset of Y.

(c) By hypothesis, U is open in X if and only if every trace $u_i^{-1}(U)$ is open in X_i. If all the restrictions $f_i = f u_i \colon X_i \to Y$ are continuous and V is open in Y, then every $u_i^{-1} f^{-1}(V) = f_i^{-1}(V)$ is open in X_i, and $f^{-1}(V)$ is open in X.

The converse is trivial: all restrictions of continuous mappings are continuous.

8.1.6 Solutions of 1.2.2

(a) The euclidean line is homotopy equivalent to the singleton space, but its underlying set is infinite.

(b) Let $\varphi \colon X \times \mathbb{I} \to Y$ be a homotopy $f \simeq g \colon X \to Y$. For every $x \in X$, the mapping $\varphi(x, -) \colon \mathbb{I} \to Y$ is continuous and its image is a path connected subspace of Y containing $f(x)$ and $g(x)$.

Therefore, if A is the connected (resp. path connected) component of x in X, $f(A)$ and $g(A)$ are contained in the same connected (resp. path connected) component of Y; this means that $\mathrm{Cc}(f) = \mathrm{Cc}(g)$ and $\Pi_0(f) = \Pi_0(g)$.

The same holds for the functor $\mathcal{A} = \mathsf{Top}(-, \mathbb{Z})$, because a map $b \colon Y \to \mathbb{Z}$ is sent to the maps

$$\mathcal{A}(f)(b) = bf \colon X \to \mathbb{Z}, \qquad \mathcal{A}(g)(b) = bg \colon X \to \mathbb{Z},$$

which coincide: for every $x \in X$ the path $b\varphi(x, -) \colon \mathbb{I} \to \mathbb{Z}$ is constant, and its endpoints $bf(x)$, $bg(x)$ coincide.

8.1.7 Solutions of 1.2.3

(b) A map $a \colon X \to \mathbb{Z}$ gives such a decomposition, letting $X_k = a^{-1}\{k\}$. This correspondence is plainly bijective.

(c) If X is empty we only have the empty mapping $\emptyset \to \mathbb{Z}$. Otherwise, every constant mapping $X \to \mathbb{Z}$ is continuous, and $\mathcal{A}(X)$ is at least countable.

(d) We can assume that X is non-empty, because this is a consequence of all the conditions we are considering.

(i) $\Rightarrow$ (ii). If X is connected, each map $a \colon X \to \mathbb{Z}$ is constant, and any point $x \in X$ gives a canonical isomorphism of rings (independent of x)

$$\varphi \colon \mathcal{A}(X) \to \mathbb{Z}, \qquad \varphi(a) = a(x). \tag{8.7}$$

Obviously, (ii) $\Rightarrow$ (iii) and (iv). Conversely, if X is not connected, we have partition $X = U \cup V$ in disjoint non-empty open sets. This gives two non-zero maps $a, b \colon X \to \mathbb{Z}$

$$a(U) = \{1\}, \quad a(V) = \{0\}, \qquad b(U) = \{0\}, \quad a(V) = \{1\},$$

with $ab = 0$. Therefore $\mathcal{A}(X)$ is not an integral domain.

Choosing two elements $x \in U$, $x' \in V$, we have a surjective homomorphism $\mathcal{A}(X) \to \mathbb{Z} \times \mathbb{Z}$, $a \mapsto (a(x), a(x'))$ of rings (and of the underlying additive groups), which shows that $\mathcal{A}(X)$ is not a cyclic group.

(e) It is an extension of the previous points. We choose an element $x_i \in X_i$ in each connected component. This gives a homomorphism with values in the product ring $\mathbb{Z}^I$

$$\varphi \colon \mathcal{A}(X) \to \mathbb{Z}^I, \qquad \varphi(a) = (a(x_i))_{i \in I}. \tag{8.8}$$

This homomorphism is canonical and injective, because a is constant on each connected component. It is also surjective, in our hypothesis: any family $(\lambda_i)_{i \in I}$ of integers defines a mapping $a \colon X \to \mathbb{Z}$, letting $a(x) = \lambda_i$ when $x \in X_i$; this mapping is continuous, because the preimage of any $\lambda \in \mathbb{Z}$ is a union of (open) connected components of X.

*(f) The ring $\mathcal{A}(X)$ is isomorphic to the subring $R \subset \mathbb{Z}^{\mathbb{N}}$ formed of the sequences $(\lambda_n)_{n \in \mathbb{N}}$ of integers which are eventually constant: there exists some $k \in \mathbb{N}$ such that $\lambda_n = \lambda_k$ for all $n \geqslant k$.

In fact, we have a ring-homomorphism

$$h \colon \mathcal{A}(X) \to \mathbb{Z}^{\mathbb{N}}, \qquad h(a)_n = a(1/(n+1)), \tag{8.9}$$

which is injective: the sequence $\lambda_n = a(1/(n+1))$ converges to $\lambda_\infty = a(0)$ in the discrete space $\mathbb{Z}$, which means that it is eventually constant at this value. The image of h is the subring R of $\mathbb{Z}^\mathbb{N}$.

R is countable: its underlying additive group $|R|$ is the internal direct sum of two countable subgroups of $\mathbb{Z}^\mathbb{N}$

$$|R| = \left(\bigoplus_{n \in \mathbb{N}} \mathbb{Z}\right) \oplus \mathbb{Z}e,$$

namely the ideal of eventually null sequences and the subring generated by the unit element $e = (1, 1, ..., 1, ...)$ of $\mathbb{Z}^\mathbb{N}$.

R is thus the unital ring canonically associated to the non-unital ring $\bigoplus_{n \in \mathbb{N}} \mathbb{Z}$. We also note that the additive group $|R|$ is isomorphic to $\bigoplus_{n \in \mathbb{N}} \mathbb{Z}$. The polynomial ring $\mathbb{Z}[X]$ in one indeterminate has the 'same' underlying additive group, but is an integral ring, quite different from R.

*(g) Every irrational number ξ determines a subset

$$A_\xi = \;]-\infty, \xi[\cap \mathbb{Q} = \;]-\infty, \xi] \cap \mathbb{Q},$$

which is open and closed in $\mathbb{Q}$. Its characteristic function $\mathbb{Q} \to \mathbb{Z}$ is continuous.

This proves that $\mathcal{A}(\mathbb{Q})$ is at least a continuum. On the other hand, $\mathcal{A}(\mathbb{Q}) \subset \mathsf{Set}(\mathbb{Q}, \mathbb{Z})$ cannot have a greater cardinal.

8.1.8 Solutions of 1.2.4

Points (a) and (b) are obvious.

*(c) We have seen in Exercise 1.2.3(f) that the ring $\mathcal{A}(X)$ is countable. On the other hand, if D is an infinite discrete space, the ring $\mathcal{A}(D) \cong \mathbb{Z}^D$ is uncountable: $\mathsf{Set}(\mathbb{N}, \mathbb{Z})$ has already the cardinal of the continuum.

8.1.9 Solutions of 1.3.3

(a) In Set, if $x_0 \in A \subset X$, we can define a retraction $p\colon X \to A$ letting $p(x) = x$ for $x \in A$, and $p(x) = x_0$ otherwise. In $\mathsf{Set}_\bullet$ we proceed in the same way, using the basepoint.

(b) If the subgroup $A \subset X$ has a retraction $p\colon X \to A$ in Ab, the subgroup $B = \operatorname{Ker} p$ meets our conditions. In fact, if $x \in A \cap B$, then $x = p(x) = 0$; if $x \in X$, then $x = p(x) + (x - p(x)) \in A + B$.

Conversely, if there exists a subgroup $B \subset X$ such that $A \cap B = \{0\}$ and $A + B = X$, each element $x \in X$ can be uniquely written as $x = a + b$, with $a \in A$ and $b \in B$. This gives a canonical isomorphism $X \cong A \oplus B$, and a retraction $p\colon X \to A$.

(c) A cyclic group of prime order has no other subgroups than the necessary ones. As to the group $\mathbb{Z}$, any two non-trivial subgroups $h\mathbb{Z}$ and $k\mathbb{Z}$ meet in a non-trivial subgroup.

(d) We have to check that U is open in A (if and) only if it is the m-preimage of an open subset of X. In fact, $U = (pm)^{-1}(U) = m^{-1}(p^{-1}(U))$ is the m-preimage of $p^{-1}(U)$.

(e) We have to check that U is open in A if (and only if) its p-preimage is open in X. In fact, if $p^{-1}(U)$ is open in X, $U = m^{-1}(p^{-1}(U))$ is open in A.

8.1.10 Solutions of 1.4.2

(a) The mappings $a(2t)$ and $b(2t - 1)$ have the same outcome at $t = 1/2$, namely $a(1) = b(0)$; the mapping $a * b \colon \mathbb{I} \to X$ is thus well defined. Its continuity follows from the Finite Closed Cover Lemma, applied to the cover of $\mathbb{I}$ formed by the closed intervals $[0, 1/2]$ and $[1/2, 1]$.

In this case, one can also use left and right continuity of $a * b$, at $t = 1/2$.

(d) A consequence of the canonical homeomorphism $p' \colon \mathbb{I}/\partial\mathbb{I} \to \mathbb{S}^1$, $p'[t] = e^{2\pi i t}$, in (1.51).

8.1.11 Solutions of 1.4.3

(a) The mappings $\varphi(x, 2t)$ and $\psi(x, 2t - 1)$ take the same value when $t = 1/2$, namely $g(x)$. Moreover they are continuous on the subspaces $X \times [0, 1/2]$ and $X \times [1/2, 1]$, respectively, and the latter form a finite closed cover of IX.

Here, continuity cannot be reduced to left and right continuity, as in Exercise 1.4.2(a): the Finite Closed Cover Lemma is the appropriate tool to construct maps 'by cases', in homotopy theory.

(d) We have a homotopy equivalence between X and the singleton space

$$p \colon X \rightleftarrows \{*\} \colon m, \qquad pm = \mathrm{id}, \qquad \varphi \colon mp \simeq \mathrm{id}\, X,$$
$$\varphi \colon X \times \mathbb{I} \to X, \qquad \varphi(x, 0) = m(*), \qquad \varphi(x, 1) = x \quad (x \in X). \tag{8.10}$$

Therefore, for each $x \in X$, the map $\varphi(x, -) \colon \mathbb{I} \to X$ is a path from $x_0 = m(*)$ to x, and X is path connected.

We have seen that the singleton $\{x_0\}$ is a deformation retract of X. But in fact any point $x_1 \in X$ would do: if $m' \colon \{*\} \to X$ corresponds to the point x_1 and $a \colon \mathbb{I} \to X$ is a path from x_1 to x_0, the map $\psi(x, t) = a(t)$ is a homotopy $\psi \colon m'p \simeq mp \colon X \to X$ (between constant maps), and the concatenation $\psi * \varphi \colon m'p \simeq \mathrm{id}\, X$ shows that $\{x_1\}$ is also a deformation retract of X.

8.1.12 Solutions of 1.4.6

(a) A point x of the space X is identified with a map $x\colon \{*\} \to X$, and the path component $[x]_P$ is identified with the homotopy class of the map $x\colon \{*\} \to X$. In this way we identify the functor Π_0 with the hom-functor $[\{*\}, -]\colon \mathsf{hoTop} \to \mathsf{Set}$.

More precisely, or pedantically, we have defined a canonical isomorphism $\Pi_0(X) \to [\{*\}, X]$ between these functors, see 1.5.2.

(b) A point x of the pointed space X is identified with the pointed map $\hat{x}\colon \mathbb{S}^0 \to X$ which reaches x (and takes the basepoint 1 to 0_X); the path component $[x]_P$ is identified with the pointed homotopy class of the map $\hat{x}$.

In this way we identify the functor π_0 with the hom-functor $[\mathbb{S}^0, -]_{\bullet}\colon \mathsf{hoTop}_{\bullet} \to \mathsf{Set}_{\bullet}$. (Again, these functors are canonically isomorphic.)

8.1.13 Solutions of 1.4.7

(a) The (convex) space $\mathbb{I}$ is contractible (by 1.1.3(b)), whence a path $a\colon \mathbb{I} \to X$ is homotopic to a constant path $e_x\colon \mathbb{I} \to X$ in the same path component (by 1.1.3(c)). Plainly, two constant paths $e_x, e_y\colon \mathbb{I} \to X$ are homotopic if and only if the points x, y are, which means that they belong to the same path component of X.

(b) The first point is proved as for ordinary homotopies of general maps: trivial homotopies, reversed homotopies and homotopy concatenation (defined in 1.4.3) can be restricted to path homotopies with fixed endpoints.

Moreover, if $h\colon a \simeq_2 b\colon x \simeq y$, then $k(s,t) = h(1-s,t)$ is a 2-homotopy $a^{\sharp} \simeq_2 b^{\sharp}\colon y \simeq x$.

(c) The functions $\mathrm{id}\,\mathbb{I}$ and f are paths in $\mathbb{I}$, with the same endpoints: 0 and 1. As a natural choice, we can use the affine homotopy

$$h\colon \mathrm{id}\,\mathbb{I} \simeq_2 f\colon \mathbb{I} \to \mathbb{I}, \quad h(s,t) = (1-t).s + t.f(s) \quad (s,t \in \mathbb{I}), \tag{8.11}$$

which is a 2-homotopy, because our paths have the same endpoints. By whisker composition, we have a 2-homotopy $ah\colon a \simeq_2 af\colon \mathbb{I} \to X$.

(d) Using the previous point, it is sufficient to use a strictly increasing reparametrisation $f\colon \mathbb{I} \to \mathbb{I}$ that transforms the regular decomposition $0 < 1/n < \ldots < 1$ into an arbitrary decomposition $0 < t_1 < \ldots < 1$.

The simplest choice is the piecewise affine function

$$f(t) = (nt - k)(t_k - t_{k-1}) + t_k, \tag{8.12}$$

where $k = 1, \ldots, n$ and $t \in [(k-1)/n, k/n]$.

8.1.14 Solutions of 1.5.1

(a) For a scalar $\lambda \in R$, every R-homomorphism $f \colon A \to B$ gives a commutative square

$$
\begin{array}{ccc}
A & \xrightarrow{\ \lambda\ } & A \\
{\scriptstyle f}\downarrow & & \downarrow{\scriptstyle f} \\
B & \xrightarrow[\ \lambda\ ]{} & B
\end{array}
\qquad\qquad \lambda f(x) = f(\lambda x) \quad (\text{for } x \in A). \tag{8.13}
$$

(b) The R-homomorphism $\varphi_R \colon R \to R$ is the multiplication by a scalar $\lambda = \varphi(1)$. For every R-module A, the naturality of φ on the homomorphism $f \colon R \to A$ that sends 1 to an element $x \in A$ gives the relation $\varphi_A(x) = \varphi_A(f(1)) = f(\varphi_R(1)) = f(\lambda) = \lambda x$.

8.1.15 Solutions of 1.5.2

(d) The claim follows from the naturality of $\psi \colon H \to K$ on the general component $\varphi X \colon FX \to GX$

$$
\begin{array}{ccc}
HFX & \xrightarrow{\ \psi FX\ } & KFX \\
{\scriptstyle H(\varphi X)}\downarrow & & \downarrow{\scriptstyle K(\varphi X)} \\
HGX & \xrightarrow[\ \psi GX\ ]{} & KGX
\end{array}
\tag{8.14}
$$

8.1.16 Solutions of 1.5.3

(a) A map $f \colon X \to Y$ gives squares of maps

$$
\begin{array}{ccc}
X \xrightarrow{\ \partial^\alpha\ } IX \\
\end{array}
\qquad
\begin{array}{ccc}
IX \xrightarrow{\ e\ } X \\
\end{array}
\qquad
\begin{array}{ccc}
IX \xrightarrow{\ r\ } IX \\
\end{array}
\tag{8.15}
$$

$$
\begin{array}{ccc}
X & \xrightarrow{\ \partial^\alpha\ } & IX \\
{\scriptstyle f}\downarrow & & \downarrow{\scriptstyle If} \\
Y & \xrightarrow[\ \partial^\alpha\ ]{} & IY
\end{array}
\qquad
\begin{array}{ccc}
IX & \xrightarrow{\ e\ } & X \\
{\scriptstyle If}\downarrow & & \downarrow{\scriptstyle f} \\
IY & \xrightarrow[\ e\ ]{} & Y
\end{array}
\qquad
\begin{array}{ccc}
IX & \xrightarrow{\ r\ } & IX \\
{\scriptstyle If}\downarrow & & \downarrow{\scriptstyle If} \\
IY & \xrightarrow[\ r\ ]{} & IY
\end{array}
$$

which are easily seen to be commutative.

(b) We have a natural isomorphism $\varphi \colon F \to G$. If $f \simeq g \colon X \to Y$ in Top, the naturality condition of φ on f and g gives two commutative squares

$$
\begin{array}{ccc}
FX & \xrightarrow{\ \varphi X\ } & GX \\
{\scriptstyle Ff}\downarrow\downarrow{\scriptstyle Fg} & & {\scriptstyle Gf}\downarrow\downarrow{\scriptstyle Gg} \\
FY & \xrightarrow[\ \varphi Y\ ]{} & GY
\end{array}
\tag{8.16}
$$

By hypothesis, $Ff = Fg$ and the components of φ are invertible in D. It follows that $Gf = Gg$.

8.1.17 Solutions of 1.5.4

(c) A natural transformation $\varphi\colon F \to G\colon \mathsf{C} \to \mathsf{D}$ gives a functor

$$\Phi\colon \mathsf{C} \times \mathbf{2} \to \mathsf{D}, \qquad \Phi(-,0) = F, \qquad \Phi(-,1) = G,$$
$$\Phi(f\colon X \to X', \iota\colon 0 \to 1) = \varphi(f)\colon F(X) \to G(X'), \tag{8.17}$$

where $\iota\colon 0 \to 1$ is the non-identity arrow of $\mathbf{2}$ and $\varphi(f)$ is defined in (1.80). Similarly, we have a functor

$$\Phi'\colon \mathsf{C} \to \mathsf{D}^{\mathbf{2}}, \qquad \Phi'(X) = \varphi_X\colon F(X) \to G(X),$$
$$\Phi'(f\colon X \to X') = (Ff, Gf)\colon \varphi_X \to \varphi_{X'}. \tag{8.18}$$

Both procedures are invertible.

8.1.18 Solutions of 1.5.5

(a) Any object X_0 of C gives a constant functor $G\colon \mathbf{1} \to \mathsf{C}$ which is easily seen to be weak inverse to F: we have $FG = \mathrm{id}\,\mathbf{1}$ and $GF \simeq \mathrm{id}\,\mathsf{C}$.

8.1.19 Solutions of 1.6.4

(a) First (i) implies (ii): the relation $up + vq = \mathrm{id}\,C$ is ensured by $p(up + vq) = p$ and $q(up + vq) = q$. Conversely, to show that (ii) implies (i), take two R-homomorphisms $f\colon X \to A$ and $g\colon X \to B$. If $h\colon X \to C$ satisfies $ph = f$ and $qh = g$, then $h = (up + vq)h = uf + vg$; conversely, the morphism $h = uf + vg\colon X \to C$ does have f, g as components.

One proves in the same way that (i*) is equivalent to (ii).

8.1.20 Solutions of 1.6.5

(a) The equivalence of these conditions on the object Z is obvious. If there is such an object, any morphism $A \to B$ that factorises through Z coincides with the additive identity of the group $\mathsf{C}(A, B)$, by the distributivity of composition.

(b) Any full subcategory of Ab inherits a preadditive structure; taking all non-trivial groups, any object A has at least two distinct endomorphisms: the trivial one and the identity.

(c) In Rng the polynomial ring $\mathbb{Z}[X]$ on one indeterminate is the initial object, while the trivial ring $\{0\}$ is the terminal object. In Rng' the trivial ring $\{0\}$ is initial and terminal.

(d) Using the description by components and co-components, the morphisms

$$f_1 = (1,0)\colon X \to X \times Y, \qquad f_2 = (0,1)\colon Y \to X \times Y,$$

give a morphism $f = [f_1, f_2]\colon X + Y \to X \times Y$ as required. It is not invertible in $\mathsf{Set}_\bullet$ and $\mathsf{Top}_\bullet$ (generally).

(e) In the same way as a category on a single object amounts to a monoid (cf. 1.3.2(c)).

*(f) A sum of sets $(X, u_i\colon X_i \to X)$ in Set is (easily proved to be) still a sum in the category of relations of sets. In the latter, by selfduality, X is also the product of the family (X_i), with projections given by the reversed relations $u_i^\sharp\colon X \to X_i$.

8.1.21 Solutions of 1.6.7

(c) For a set X, we view an n-tuple $w = (x_1, x_2, ..., x_n) \in X^n$ as an n-ary *word* in the alphabet X. Two words are multiplied by *juxtaposition* (or *concatenation*)

$$(x_1, ..., x_p)(y_1, ..., y_q) = (x_1..., x_p, y_1, ..., y_q). \tag{8.19}$$

We are also considering the empty word $e = (\ ) \in X^0$, of length 0, which acts as a unit for this operation: $ew = w = we$, for every word w.

The disjoint union $M(X) = \sum_{n \in \mathbb{N}} X^n$ of all sets X^n is thus a monoid. It is actually the *free monoid* on the set X, by the mapping 'of unary words'

$$\eta\colon X \to |M(X)|, \qquad \eta(x) = (x). \tag{8.20}$$

It is easy to verify that $(M(X), \eta)$ is a universal arrow from the set X to the forgetful functor $U\colon \mathsf{Mon} \to \mathsf{Set}$ of the category of monoids.

Also here we can identify the element $x \in X$ with the unary word $\eta(x) = (x)$, so that each element of $M(X)$ can be written in a unique way as a finite product $x_1 x_2 \ldots x_n$ of elements of the basis (including the empty product $e = \prod_{i \in \emptyset} x_i$).

8.2 Exercises of Chapter 2

8.2.1 Solutions of 2.1.2

(a) First we prove that property (i) determines its solution, up to a coherent isomorphism (as any universal property). In fact, if we have two of them, $k_i\colon K_i \to A$ $(i = 1, 2)$, there are unique morphisms u, v such that

$$u\colon K_1 \to K_2, \qquad v\colon K_2 \to K_1, \qquad k_2 u = k_1, \qquad k_1 v = k_2,$$

and $vu\colon K_1 \to K_1$ is the identity, because $k_1(vu) = k_2 u = k_1 = k_1(\mathrm{id}\, K_1)$. Similarly $uv = \mathrm{id}\, K_2$, so that u, v are isomorphisms, inverse to each other.

Now, it is sufficient to remark that the embedding $\ker f\colon \mathrm{Ker}\, f \rightarrowtail A$ does satisfy (i): if $fh = 0$ the image of h is contained in $\mathrm{Ker}\, f$, and we can (and must) define u as the codomain-restriction of h.

(b) The uniqueness of the solution to (i*), up to isomorphism, is proved as above. (One can just reverse all arrows, including f, and start from two solutions $c_i\colon B \to C_i$.)

Then we remark that the projection $\mathrm{cok}\, f\colon B \twoheadrightarrow B/f(A)$ satisfies (i*): if $hf = 0$ the kernel of h contains $f(A)$, and we can (and must) define u as the homomorphism induced by h.

(c) The existence and uniqueness of the morphism u' comes from the universal property of $\ker g$: in fact the composite $u(\ker f)$ annihilates g, because $gu(\ker f) = vf(\ker f) = 0$.

Similarly, the existence and uniqueness of the morphism v' comes from the universal property of $\mathrm{cok}\, f$.

(d) We apply (c), taking into account that $\mathrm{Coim}\, f = \mathrm{Cok}\,(\ker f)$ and $\mathrm{Im}\, f = \mathrm{Ker}\,(\mathrm{cok}\, f)$.

(f) In **Set.** and **Top.** the kernel of a morphism is the preimage of the basepoint, while the cokernel is the quotient of the codomain which collapses the image.

8.2.2 *Solutions of 2.1.3*

(a)–(f) Obvious.

(g) The canonical factorisations $f = mp$ and $g = nq$ of two consecutive homomorphisms give a commutative diagram (2.12), where $\mathrm{Ker}\, q = \mathrm{Ker}\, g$.

In this commutative diagram, the pair (f, g) is exact if and only if (m, q) is exact, p is epi and n is mono; adding $\mathrm{Ker}\, p$ and $\mathrm{Cok}\, n$, these three conditions are expressed by short exact sequences.

(h) In this exact sequence, f is epi if and only if $\mathrm{Im}\, f = \mathrm{Ker}\, g$ is B, if and only if $g = 0$. Similarly h is mono if and only if $\mathrm{Im}\, g = \mathrm{Ker}\, h$ is 0, if and only if $g = 0$. The rest is a consequence

8.2.3 *Solutions of 2.1.6*

(a) A straightforward consequence of Exercise 2.1.2(d). More concretely, one can apply condition (i') of the Induction Lemma 2.1.4: the formula $\partial_n f_n(x) = f_{n-1}\partial_n(x)$ says that:

- a cycle $x \in Z_n(A)$ is sent to $f_n(x) \in Z_n(B)$,

- a boundary $\partial_n(x) \in B_{n-1}(A)$ is sent to the boundary $\partial_n f_n(x)$ in B.

(b) It is a straightforward application of the Induction Lemma 2.1.4: $B_n(f)$ is a restriction of $Z_n(f)$, and the latter induces $H_n(f)$ on the homology quotients.

8.2.4 Solutions of 2.1.7

(a) The reflexivity property $f \simeq f$ comes from the sequence of zero homomorphisms $0: A_n \to B_{n+1}$. For the symmetry property, we take the opposite components $-\varphi_n: A_n \to B_{n+1}$. For transitivity, we add the components of two homotopies $\varphi: f \to g$ and $\psi: g \to h$

$$\varphi + \psi: f \to h, \qquad (\varphi + \psi)_n = \varphi_n + \psi_n. \tag{8.21}$$

Consistence with composition can be readily verified on the whisker composition of $\varphi: f \to g: A \to B$ with two chain morphisms $h: A' \to A$, $k: B \to B'$. Then the sequence of homomorphisms

$$\psi_n = k_{n+1}\varphi_n h_n: A'_n \to B'_{n+1} \tag{8.22}$$

gives a homotopy $\psi: kfh \to kgh$

$$\begin{aligned}
\partial_{n+1}\psi_n + \psi_{n-1}\partial_n &= \partial_{n+1}(k_{n+1}\varphi_n h_n) + (k_n\varphi_{n-1}h_{n-1})\partial_n \\
&= (k_n\partial_{n+1}\varphi_n h_n) + (k_n\varphi_{n-1}\partial_n h_n) \\
&= k_n(\partial_{n+1}\varphi_n + \varphi_{n-1}\partial_n)h_n = k_n(g_n - f_n)h_n.
\end{aligned}$$

(b) We have a homotopy $\varphi: f \to g: A \to B$. For every cycle $z \in Z_n(A)$, the homology class $[z] \in H_n(A)$ gives

$$H_n(g)[z] - H_n(f)[z] = [g_n(z) - f_n(z)] = [\partial_{n+1}\varphi_n(z) + \varphi_{n-1}\partial_n(z)] = 0,$$

because z is an n-cycle and $\partial_{n+1}\varphi_n(z)$ is an n-boundary.

8.2.5 Solutions of 2.1.9

(a)–(c). Omitting degrees, we have:

$$\begin{aligned}
\partial\partial(a, x, b) &= \partial(\partial a - x, -\partial x, \partial b + x) \\
&= (\partial\partial a - \partial x + \partial x, \partial\partial x, \partial\partial b + \partial x - \partial x) = 0,
\end{aligned}$$

$$\begin{aligned}
\partial(If)(a, x, b) &= (\partial f(a) - f(x), -\partial f(x), \partial f(b) + f(x)) \\
&= (f\partial a - f(x), -f\partial(x), f\partial b + f(x)) = (If)\partial(a, x, b),
\end{aligned}$$

$$\partial^-\partial(a) = (\partial a, 0, 0) = \partial\partial^-(a), \quad \partial^+\partial(a) = (0, 0, \partial a) = \partial\partial^+(a),$$

$$e\partial(a, x, b) = e(\partial a - x, -\partial x, \partial b + x) = \partial a + \partial b = \partial e(a, x, b),$$

$$r\partial(a, x, b) = r(\partial a - x, -\partial x, \partial b + x) = (\partial b + x, \partial x, \partial a - x)$$
$$= \partial(b, -x, a) = \partial r(a, x, b).$$

(d) We take a general chain morphism $\Phi = [f, \varphi, g] \colon IA \to B$, where the component of degree n

$$\Phi_n = [f_n, \varphi_{n-1}, g_n] \colon (IA)_n \to B_n,$$
$$(a, x, b) \mapsto f_n(a) + \varphi_{n-1}(x) + g_n(b), \tag{8.23}$$

is written as in 1.6.2, by its co-components in **Ab**

$$f_n \colon A_n \to B_n, \quad \varphi_{n-1} \colon A_{n-1} \to B_n, \quad g_n \colon A_n \to B_n.$$

We now analyse the condition $\partial_{n+1}\Phi_{n+1} = \Phi_n\partial_{n+1}$ on a general element $(a, x, b) \in (IA)_{n+1}$

$$\partial_{n+1}\Phi_{n+1}(a, x, b) = \partial_{n+1}(f_{n+1}a + \varphi_n x + g_{n+1}b)$$
$$= \partial_{n+1}f_{n+1}a + \partial_{n+1}\varphi_n x + \partial_{n+1}g_{n+1}b,$$

$$\Phi_n\partial_{n+1}(a, x, b) = \Phi_n(\partial_{n+1}a - x, -\partial_n x, \partial_{n+1}b + x)$$
$$= f_n\partial_{n+1}a - f_n x - \varphi_{n-1}\partial_n x + g_n\partial_{n+1}b + g_n x.$$

Their equality is equivalent to saying that:

- $\partial_{n+1}f_{n+1}(a) = f_n\partial_{n+1}(a)$,
- $\partial_{n+1}\varphi_n(x) + \varphi_{n-1}\partial_n(x) = g_n(x) - f_n(x)$,
- $\partial_{n+1}g_{n+1}(b) = g_n\partial_{n+1}(b)$,

for all $a \in A_{n+1}$, $x \in A_n$ and $b \in A_{n+1}$.

Finally, giving the chain morphism Φ is equivalent to giving two chain morphisms $f, g \colon A \to B$ and a chain homotopy $\varphi \colon f \simeq g$.

8.2.6 *Solutions of 2.2.1*

(a) Letting $j \leqslant i$ and applying $\delta_j^\beta \delta_i^\alpha$ to $(t_1, ..., t_{n-1})$ we get:

$$\delta_j^\beta \delta_i^\alpha(t_1, ..., t_{n-1}) = \delta_j^\beta(t_1, ..., t_{i-1}, \alpha, ..., t_{n-1})$$
$$= (t_1, ..., t_{j-1}, \beta, ..., t_{i-1}, \alpha, ..., t_{n-1}),$$

where, in the last step, the inserted α has been shifted to position $i + 1$. This proves that $\delta_j^\beta \delta_i^\alpha = \delta_{i+1}^\alpha \delta_j^\beta$.

The other verifications are similar.

8.2.7 Solutions of 2.2.3

(a) We form a diagram as in (2.14)

$$
\begin{array}{ccccc}
\mathbb{Z}\mathrm{Deg}_n K & \rightarrowtail & \mathbb{Z}K_n & \twoheadrightarrow & C_n(K) \\
{\scriptstyle f'}\downarrow & & {\scriptstyle f}\downarrow & & {\scriptstyle f''}\downarrow \\
\mathbb{Z}\mathrm{Deg}_{n-1} K & \rightarrowtail & \mathbb{Z}K_{n-1} & \twoheadrightarrow & C_{n-1}(K)
\end{array}
\qquad (8.24)
$$

where $f = \partial_n$ is defined on the canonical basis as

$$\partial_n(a) = \Sigma_{i,\alpha}\,(-1)^{i+\alpha}\,\partial_i^\alpha a.$$

This homomorphism has a restriction f', because of (2.45). By the Induction Lemma 2.1.4, there is a unique homomorphism $f'' \colon C_n(K) \to C_{n-1}(K)$ that makes diagram (8.24) commutative.

(b) For a generator $a \in K_{n+1}$ we have

$$\partial_n \partial_{n+1}(a) = \Sigma_{i,j}\,(-1)^{i+j+\alpha+\beta}\,\partial_i^\alpha \partial_j^\beta(a), \qquad (8.25)$$

where α, β take the values $0, 1$, and $i = 1, ..., n$, $j = 1, ..., n+1$.

This sum can be split in two parts, for $j \leqslant i$ and $j > i$, respectively

$$
\begin{aligned}
\partial_n \partial_{n+1}(a) = {}& \Sigma_{j \leqslant i}\,(-1)^{i+j+\alpha+\beta}\,\partial_i^\alpha \partial_j^\beta(a) \\
& + \Sigma_{j > i}\,(-1)^{i+j+\alpha+\beta}\,\partial_i^\alpha \partial_j^\beta(a).
\end{aligned}
$$

Working on the left addendum, we apply the cubical relations of the faces $(\partial_i^\alpha \partial_j^\beta = \partial_j^\beta \partial_{i+1}^\alpha$ for $j \leqslant i)$, and then a change of variables on the indices (replacing j with i, and i with $j - 1$)

$$
\begin{aligned}
\Sigma_{j \leqslant i}\,(-1)^{i+j+\alpha+\beta}\,\partial_i^\alpha \partial_j^\beta(a) = {}& \Sigma_{j \leqslant i}\,(-1)^{i+j+\alpha+\beta}\,\partial_j^\beta \partial_{i+1}^\alpha(a) \\
= {}& - \Sigma_{i < j}\,(-1)^{i+j+\alpha+\beta}\,\partial_i^\alpha \partial_j^\beta(a).
\end{aligned}
$$

The left addendum is thus the additive inverse of the right one, and $\partial_n \partial_{n+1}(a) = 0$.

(c) If $A = B \cup C$ is a disjoint union of sets, $\mathbb{Z}A$ is the internal direct sum of two subgroups, which can be identified with $\mathbb{Z}B$ and $\mathbb{Z}C$, so that $\mathbb{Z}A/\mathbb{Z}B \cong \mathbb{Z}C$. This is applied to the set $A = K_n$, its subset $B = \mathrm{Deg}_n K$ and the complement $C = A \setminus B$.

8.2.8 Solutions of 2.2.5

(a) There are no mappings $\mathbb{I}^n \to \emptyset$. Therefore $C_+(\emptyset)$ is the trivial chain complex: all its components are 0, and all homology groups are 0.

For the singleton, there is a unique mapping $a_n \colon \mathbb{I}^n \to \{*\}$, for $n \geqslant 0$.

The differential of the *non-normalised* chain complex always involves an even number of summands (the cube $\mathbb{I}^n$ has $2n$ faces)

$$\partial_n(a_n) = \Sigma_{i,\alpha}(-1)^{i+\alpha}a_{n-1} = 0,$$

and all its homology groups are isomorphic to $\mathbb{Z}$. But we are interested in the normalised chain complex, and every generator a_n is degenerate in positive degree: $a_n = e_i(a_{n-1})$, for $n > 0$ and any i.

Thus the cubical chain complex $C_+\{*\}$ of the singleton is isomorphic to

$$... \; 0 \to 0 \to 0 \to \mathbb{Z} \tag{8.26}$$

and its homology is $\mathbb{Z}$ in degree 0, and trivial elsewhere.

(b) Here it is convenient to use the 'augmented' chain complex

$$... \; C_2(X) \xrightarrow{\partial_2} C_1(X) \xrightarrow{\partial_1} C_0(X) \xrightarrow{\tilde{\partial}_0} \mathbb{Z} \tag{8.27}$$

where we have added the homomorphism $\tilde{\partial}_0(\Sigma \lambda_i x_i) = \Sigma \lambda_i$. Plainly $\tilde{\partial}_0 \partial_1 = 0$, because the boundary of a 1-cube is the formal difference of two points, with coefficients 1 and -1.

The homomorphism $\tilde{\partial}_0$ is surjective, because $X \neq \emptyset$; moreover $\operatorname{Ker} \tilde{\partial}_0 = \operatorname{Im} \partial_1$, because X is path-connected. Therefore $\tilde{\partial}_0$ induces an isomorphism on $H_0(X) = C_0(X)/\operatorname{Im} \partial_1$

$$H_0(X) \to \mathbb{Z}, \qquad [x] \mapsto 1_{\mathbb{Z}}. \tag{8.28}$$

(c) A cube $a\colon \mathbb{I}^n \to X$ has a path-connected image, contained in one path component X_i; its faces 'belong' to the same component. We also note that a is degenerate in X if and only if it is as a cube of X_i.

All this proves the canonical isomorphism of chain complexes in (2.58). Applying the functor H_n we get an isomorphism

$$H_n(\oplus_{i\in I} C_+(X_i)) \to H_n(X), \qquad [(z_i)_{i\in I}] \mapsto \Sigma_{i\in I}\,[z_i]. \tag{8.29}$$

We also have the isomorphism (2.29)

$$H_n(\oplus_{i\in I} C_+(X_i)) \to \oplus_{i\in I} H_n(C_+(X_i)), \qquad [(z_i)_{i\in I}] \mapsto ([z_i])_{i\in I}, \tag{8.30}$$

and composing the inverse of the latter with (8.29) we obtain the isomorphism we are looking for, sending the essentially finite family $([z_i])_{i\in I}$ to the homology class $\Sigma_{i\in I}\,[z_i]$ of $H_n(X)$.

(d) A consequence of (b) and (c). Without using (c), one can extend the argument of (b) using the homomorphism

$$\tilde{\partial}_0\colon C_0(X) \to F\Pi_0(X), \qquad x \mapsto [x]_P,$$

(whose kernel is the image of $\partial_1 \colon C_0(X) \to C_1(X)$), and its naturality in the variable X.

(e) A consequence of (a), (b) and (c).

8.2.9 Solutions of 2.2.8

(b) Let $h \colon \mathbb{I}^2 \to X$ be a homotopy with fixed endpoints, from a to b, as in (1.73). Then the boundary of the 2-cube h is the 1-chain $a - b$

$$\partial_2(h) = -e_x + e_y + a - b = a - b, \tag{8.31}$$

as the degenerate paths $e_x = e_1(x)$ and $e_y = e_1(y)$ are zero in $C_1(X)$.

(c) Similarly, we take a loop homotopy $h \colon \mathbb{I}^2 \to X$ from a to b, as in (1.78). The boundary of the 2-cube h is again the 1-chain $a - b$

$$\partial_2(h) = -c + c + a - b = a - b, \tag{8.32}$$

but here a and b are loops, and we have proved that $[a] = [b]$.

(d) A consequence of (c), as a constant loop is a degenerate 1-cube.

(e) It is sufficient to construct a 2-cube $h \colon \mathbb{I}^2 \to X$ with the following faces, where e_x is the constant path at $x = a(0) = c(0)$

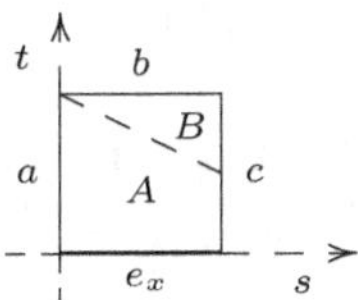

For instance, using the Closed Cover Lemma 1.1.8(b), we can define h on the subspaces $A, B \subset \mathbb{I}^2$ represented above:

$$A = \{(s, t) \in \mathbb{I}^2 \mid t \leqslant 1 - s/2\}, \qquad B = \{(s, t) \in \mathbb{I}^2 \mid t \geqslant 1 - s/2\},$$

using the map a on A, and the map b on B:

$$h(s, t) = \begin{cases} a(t/(1 - s/2)), & \text{for } (s, t) \in A, \\ b(s + 2t - 2), & \text{for } (s, t) \in B. \end{cases}$$

In fact, these expressions coincide on the segment $t = 1 - s/2$, taking the constant value $a(1) = b(0)$. Finally:

$$\partial_2(h) = -a + c + e_x - b = c - a - b. \tag{8.33}$$

(f) The 2-cube $h\colon \mathbb{I}^2 \to X$, $h(s,t) = a(\min(s, 1-t))$

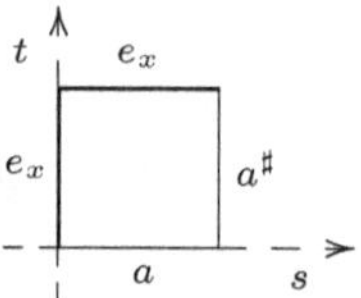

has the faces represented above. Therefore $\partial_2 h = a^\sharp + a$ is a boundary, and $[a + a^\sharp] = 0$.

8.2.10 Solutions of 2.3.4

(a) The differential of a small chain is small: for a chain $c \in C_n(X)$, we have seen that $\mathrm{supp}(\partial_n c) \subset \mathrm{supp}(c)$, in (2.53).

(b) If $W \cap U_i$ is open in U_i, its trace is open in $\mathrm{int}(U_i)$ and therefore in X. If this holds true for every $i \in I$, it follows that $W = \bigcup_i (W \cap \mathrm{int}(U_i))$ is open in X.

8.2.11 Solutions of 2.3.6

(a) The property holds in **Set**: the conditions $f = hu$ and $g = hv$ require that $h(x) = f(x)$ for $x \in U$, and $h(x) = g(x)$ for $x \in V$. On the other hand, defining h in this way is legitimate, because f and g coincide on $A = U \cap V$.

For **Top**, we only have to add that the mapping h is continuous; this follows from the Open Cover Lemma (in 1.1.8(a)), since h coincides with f on $\mathrm{int}\, U$, and with g on $\mathrm{int}\, V$.

(b) It is the usual proof of uniqueness, up to 'isomorphism', of something determined by a universal property. Here we suppose that the triple (Y, f, g) is also a solution. After $h\colon X \to Y$ we also have a unique map $k\colon Y \to X$ such that $u = kf$ and $v = kg$. Therefore

$$(kh)u = kf = u = (\mathrm{id}\, X)u, \qquad (kh)v = kg = v = (\mathrm{id}\, X)v,$$

and $kh = \mathrm{id}\, X$. Similarly $hk = \mathrm{id}\, Y$, and $h\colon X \to Y$ is a homeomorphism, inverse to k.

8.2.12 Solutions of 2.3.8

(a) The composite $k'_{*n} h'_{*n}\colon H_n(A) \to H_n(C)$ is obviously zero, because $k'_n h'_n\colon A_n \to C_n$ already is.

Suppose now that $b \in Z_n(B)$ and $k'_{*n}[b] = [k'_n(b)] = 0$. Thus $k'_n(b) = \partial_{n+1}(\hat{c})$, for some $\hat{c} \in C_{n+1}$, which can be lifted to $\hat{b} \in B_{n+1}$. We have formed the right part of the following diagram

$$
\begin{array}{ccccc}
 & & & \hat{b} \longmapsto \hat{c} & (n+1) \\
 & & & \downarrow & \\
\hat{a} \longmapsto b - \partial\hat{b} \longmapsto 0 & & & b \longmapsto c & (n) \\
\downarrow \qquad\quad \downarrow & & & \downarrow & \\
\partial\hat{a} \longmapsto \partial b = 0 & & & 0 & (n-1)
\end{array}
$$

The cycle $b - \partial_{n+1}\hat{b}$ belongs to $\operatorname{Ker} k'_n = \operatorname{Im} h'_n$, and can be lifted to an element $\hat{a} \in A_n$; the commutative left square shows that $k'_{n-1}(\partial_n \hat{a}) = \partial_n(b - \partial_{n+1}\hat{b}) = 0$; therefore $\partial_n \hat{a} = 0$ (as k'_{n-1} is mono), and the cycle $\hat{a}$ works as expected

$$h'_{*n}[\hat{a}] = [b - \partial_{n+1}\hat{b}] = [b].$$

(b) The composite $D_n k'_{*n} \colon H_n(B) \to H_{n-1}(A)$ is zero, as shown by the following diagram, where we are assuming that $b \in B_n$ is a cycle, so that $D_n k'_{*n}[b] = D_n[c] = [0]$

$$
\begin{array}{ccc}
b \longmapsto c & & (n) \\
\downarrow \qquad \downarrow & & \\
0 \longmapsto 0 \longmapsto 0 & & (n-1)
\end{array}
$$

Suppose now that $c \in Z_n(C)$ and $D_n[c] = [a] = 0$. Thus $a = \partial_{n+1}(\hat{a})$, for some $\hat{a} \in A_{n+1}$

$$
\begin{array}{ccc}
\hat{a} \qquad\quad b \longmapsto c & & (n) \\
\downarrow \qquad\quad \downarrow \qquad \downarrow & & \\
a \longmapsto \partial b \longmapsto 0 & & (n-1)
\end{array}
$$

It follows that $b - h'_n(\hat{a})$ is a cycle, and

$$k'_{*n}[b - h'_n(\hat{a})] = [k'_n(b) - k'_n h'_n(\hat{a})] = [k'_n(b)] = [c].$$

8.2.13 Solutions of 2.4.5

(a) Apply the MV-sequence, with $H_n(U) = H_n(V) = 0$ for $n \geqslant 1$.

(b) For $H_1(\Sigma X)$ we use the following portion of the MV-sequence, where $A \simeq X$ is path connected and h_0 is mono

$$0 \to H_1(\Sigma X) \xrightarrow{D_1} H_0(A) \xrightarrow{h_0} H_0(U) \oplus H_0(V).$$

8.2.14 Solutions of 24.9

(a) For $n = 1$, the space S^1 is the perimeter of the standard square, covered by the faces of $f_1 = \mathrm{id}\, \mathbb{I}^2$

$$z_1 = \partial_2 f_1 = -a + a' + b - b'.$$

Applying Exercises 2.2.8(f) and 2.2.8(e)

$$[z_1] = [a' + b'^\sharp + a^\sharp + b] = [a' * b'^\sharp * a^\sharp * b],$$

we find a simple loop in S^1, that corresponds to the simple loop $t \mapsto e^{2\pi it}$ ($t \in \mathbb{I}$) of the circle, used in Exercise 2.4.1(d) as a generator of $H_1(S^1)$.

(b) For $n \geqslant 2$ we remark that the lower face ∂_{n+1}^-

$$\partial_{n+1}^- f_n \colon \mathbb{I}^n \to \mathbb{I}^{n+1}, \qquad (t_1, ..., t_n) \mapsto (t_1, ..., t_n, 0), \qquad (8.34)$$

coincides with $f_{n-1} = \mathrm{id}\, \mathbb{I}^n$, under the usual nesting of all $\mathbb{R}^n$.

We suppose that $[\partial_n f_{n-1}]$ is a generator of $H_{n-1}(S^{n-1})$ and prove that the isomorphism of the MV-sequence used in Theorem 2.4.2

$$D_n \colon H_n(\mathbb{S}^n) \to H_{n-1}(A), \qquad D_n'[\partial_{n+1} f_n] = [\partial_n f_{n-1}],$$

takes $[\partial_{n+1} f_n]$ to a generator of $H_{n-1}(A) \cong H_{n-1}(S^{n-1})$.

In fact the cycle $z_n = \partial_{n+1} f_n \in C_n(S^n)$, a linear combination of n-cubes, can be split leaving $\partial_{n+1}^- f_n = f_{n-1}$ in $C_n(V)$ and all the others in $C_n(U)$

$$\partial_{n+1} f_n = c - c',$$

$$c = \sum_{i \leqslant n} (-1)^{i+\alpha} \partial_i^\alpha f_n + (-1)^n \partial_{n+1}^+ f_n \in C_n(U),$$

$$c' = (-1)^{n+1} f_{n-1} \in C_n(V).$$

Finally $D_n[\partial_{n+1} f_n] = [\partial_n c'] = (-1)^{n+1} [\partial_n f_{n-1}] = (-1)^{n+1} [z_{n-1}]$.

(c) Working for simplicity in dimension n (with $n > 1$), we form the composite of two homeomorphisms:

$$h = h'' h' \colon \mathbb{I}^n \to [-1, 1]^n \to \mathbb{D}^n,$$

$$h'(x) = 2x - (1, 1, ..., 1), \qquad h''(x) = \begin{cases} (||x||_\infty / ||x||)x, & \text{for } x \neq 0, \\ 0 & \text{for } x = 0. \end{cases}$$

Here $||x||_\infty = \max(x_1, ..., x_n)$ is the l_∞-norm of $\mathbb{R}^n$, which determines the cube

$$[-1, 1]^n = \{x \in \mathbb{R}^n \mid ||x||_\infty \leqslant 1\}.$$

The map $h'' \colon [-1, 1]^n \to \mathbb{D}^n$ takes every point x of the border of $[-1, 1]^n$ to its normalised point $x/||x|| \in \mathbb{S}^{n-1}$, and is linear on every line through the origin.

8.2.15 Solutions of 2.5.2

(a) If X is empty, $H_0(X) = 0$ and there is nothing to prove.

Otherwise $H_0(X)$ is the infinite cyclic group generated by the homology class $[x]_X$ of a point $x \in X$; this is sent by $f_{*0} \colon H_0(X) \to H_0(Y)$ to the homology class $[f(x)]_Y$, an element of the canonical basis of $H_0(Y)$ (by Exercise 2.2.5(d)). This proves that f_{*0} is a monomorphism.

(b) For any $x \in X$, the homology class $f_{*0}([x]_X) = [f(x)]_Y$ generates $H_0(Y)$.

(c) The first claim follows from (a): the components of $h_0 = (i_{*0}, j_{*0})$ are injective. The rest is a consequence (as in 2.1.3(h)) of the exact sequence

$$H_1(U) \oplus H_1(V) \xrightarrow{k_1} H_1(X) \xrightarrow{D_1} H_0(A) \xrightarrow{h_0} H_0(U) \oplus H_0(V).$$

8.2.16 Solutions of 2.5.3

(a) After the homology groups of the circle and the figure-eight space, we can easily guess that $H_1(C_n)$ is isomorphic to $\mathbb{Z}^n$, with a basis formed by the homology classes of n simple loops, one in each circle, while the higher homology groups of C_n are trivial.

This can be proved by induction on $n \geqslant 1$. The initial step being granted, we assume that our claim holds for $n - 1$ and prove it for $n \geqslant 2$, using the MV-sequence for the following open cover (U, V) (drawn for $n = 3$)

$$
\begin{aligned}
U &\simeq C_{n-1}, \\
V &\simeq \mathbb{S}^1, \\
A &= U \cap V \simeq \{*\},
\end{aligned}
\tag{8.35}
$$

where U is the union of C_{n-1} with a contractible open nbd of x in C_n, and V is the union of the last circle of C_n with a contractible open nbd of x.

All homology groups $H_k(C_n)$ are trivial for $k > 1$, as in 2.5.1(a).

For $H_1(C_n)$ we proceed as in 2.5.1(b), with the exact sequence

$$H_1(A) \xrightarrow{h_1} H_1(U) \oplus H_1(V) \xrightarrow{k_1} H_1(C_n).$$

Again, $H_1(A) = 0$ and k_1 is mono. But k_1 is also epi, because A is path connected (we apply Exercise 2.5.2(c)). Finally, using the inductive hypothesis, we have an isomorphism

$$H_1(C_n) \;\cong\; H_1(C_{n-1}) \oplus H_1(\mathbb{S}^1) \;\cong\; \mathbb{Z}^{n-1} \oplus \mathbb{Z},$$

that also proves our claim on the basis of $H_1(C_n)$.

(b) One can easily find an open cover (U, V) of X with:

$$U \simeq C_2, \qquad V \simeq \mathbb{S}^1, \qquad A = U \cap V \simeq \{*\},$$

that gives the same results as we found for the space C_3, namely:

$$H_1(X) \;\cong\; \mathbb{Z}^3, \qquad H_k(X) = 0 \quad \text{for } k > 1.$$

To prove that C_3 and X are not homeomorphic it is sufficient to note that C_3 has a single point that disconnects it, while X has two of them.

8.2.17 Solutions of 2.5.4

(a) Let us note that, for $j \neq i$, the subset $u_j^{-1}(u_i(U_i))$ is $\{0_{X_j}\}$ in case (i) and $\emptyset$ in case (ii).

A subset V of form (i) or (ii) is always open in X, because in both cases we have, for all j: $u_j^{-1}(V) = \bigcup_i u_j^{-1}(u_i(U_i)) = U_j$.

Conversely, every open subset V of X can be written as:

$$V = \bigcup_i V \cap \operatorname{Im} u_i = \bigcup_i u_i(U_i),$$

where each $U_i = u_i^{-1}(V)$ is open in X_i. Thus V is either of form (i) (when it contains the basepoint), or of form (ii) (when it does not).

(b) Let U be open in X_i and $V = u_i(U)$. If $0_{X_i} \notin U$, we have seen in (a) that the subset V is open in X (of form (ii)), and also open in $u_i(X_i)$.

Otherwise $0_{X_i} \in U$ and $0_X \in V$. The subset $V' = V \cup (\bigcup_{j \neq i} u_j(X_j))$ is an open subset of X of form (i). Now $V = V' \cap u_i(X_i)$ is just open in $u_i(X_i)$.

(c) A consequence of (b): the pointed sum X is the union of its subspaces $u_i(X_i)$, which meet at the basepoint 0_X; each of them is homeomorphic to X_i, and path-connected (resp. connected).

8.2.18 Solutions of 2.5.7

(a) Applying Theorem 2.5.6, the pointed sum $\mathbb{S}^1 \vee \mathbb{S}^2 \vee \ldots \vee \mathbb{S}^n$ satisfies our condition.

(b) One can take the pointed sum of β_1 copies of $\mathbb{S}^1$, β_2 copies of $\mathbb{S}^2$, ..., and β_n copies of $\mathbb{S}^n$.

8.2.19 Solutions of 2.5.8

(a) We use the following open cover (U, V) of P_n

$$U = P_n \setminus \{0\} \simeq \mathbb{S}^1, \qquad V = \text{int}\,\mathbb{D}^2 \simeq \{*\},$$
$$A = U \cap V = \text{int}\,\mathbb{D}^2 \setminus \{0\} \simeq \mathbb{S}^1. \tag{8.36}$$

The only non-obvious claim is the relation $P_n \setminus \{0\} \simeq \mathbb{S}^1$. We begin by observing that the homotopy equivalence $\mathbb{D}^2 \setminus \{0\} \simeq \mathbb{S}^1$ we used in 1.1.3(d) does not move the points of $\mathbb{S}^1$; therefore it induces a homotopy equivalence $P_n \setminus \{0\} \simeq \mathbb{S}^1/\mathbb{Z}_n$, with the orbit space of the action of $\mathbb{Z}_n$ on $\mathbb{S}^1$.

Now, this space is homeomorphic to $\mathbb{S}^1$: the surjective endomap $f(z) = z^n$ (a power in $\mathbb{C}$) induces a bijective map $\mathbb{S}^1/\mathbb{Z}_n \to \mathbb{S}^1$, which is a homeomorphism (from a compact space to a Hausdorff one).

Loosely speaking, winding the circle on itself, n times, we still get the circle.

To determine $H_1(P_n)$ we use the following exact sequence (omitting the trivial group $H_1(V)$)

$$H_1(A) \xrightarrow{h_1} H_1(U) \xrightarrow{k_1} H_1(P_n)$$

where k_1 is epi (by Exercise 2.5.2(c)), and therefore a cokernel of h_1.

We have thus to compute the homomorphism h_1. For $H_1(A) \cong H_1(\mathbb{S}^1)$ we choose the generator $\zeta = [a]_A$, where a is the simple loop drawn below, at the left

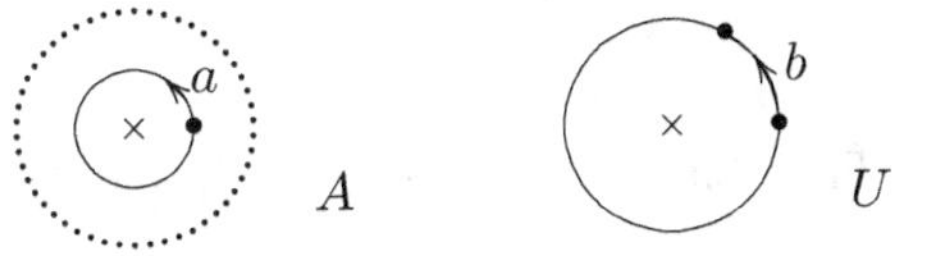

(8.37)

For $H_1(U)$ we choose the generator $\zeta' = [b]_U$, where b is a simple loop of U, the image of a path in the border of $\mathbb{D}^2$, from 1 to $e^{2\pi i/n}$. (The right figure is drawn for $n = 6$.)

Now $h_1[a]_A = [a]_U = n[b]_U$, because in U the loop a is loop homotopic to the image in P_n of the simple loop around the border of $\mathbb{D}^2$, which is the concatenation of n loops equal to b.

The homomorphism h_1 is thus, essentially, the multiplication by n in an infinite cyclic group. Its cokernel can be realised as $\mathbb{Z}_n$

$$H_1(A) \xrightarrow{h_1} H_1(U) \xrightarrow{p} \mathbb{Z}_n$$

by the projection p taking $\zeta' = [b]_U$ to the class $\bar{1}$: an epimorphism whose kernel is generated by $n[b]_U = h_1[a]_A$.

This proves that $H_1(P_n) \cong \mathbb{Z}_n$, with generator $[b]_{P_n}$.

(b) For $i > 1$ we have the exact sequence

$$H_i(U) \xrightarrow{\ k_i\ } H_i(P_n) \xrightarrow{\ D_i\ } H_{i-1}(A) \xrightarrow{\ h_{i-1}\ } H_{i-1}(U) \oplus H_{i-1}(V)$$

where $H_i U = 0$. We distinguish two cases, each of them giving $H_i(P_n) = 0$

- if $i > 2$: $H_{i-1}(A) = 0$, and $H_i(P_n) = 0$,

- if $i = 2$: $h_{i-1} = h_1$ has been computed above, and is a monomorphism (the multiplication by n in an infinite cyclic group); therefore $H_2(P_n) = \operatorname{Im} D_2 = \operatorname{Ker} h_1 = 0$.

8.2.20 Solutions of 2.5.9

(a) We can use the pointed sum

$$X = P_{T_1} \vee P_{T_2} \vee \ldots \vee P_{T_k}. \tag{8.38}$$

In fact, the origin has a contractible nbd in P_n, the open disc $\operatorname{int} \mathbb{D}^2$. Applying Theorem 2.5.7 and Exercise 2.5.8(a) we get, for $n > 1$

$$H_1(X) \cong \mathbb{Z}_{T_1} \oplus \mathbb{Z}_{T_2} \oplus \ldots \oplus \mathbb{Z}_{T_k}, \qquad H_n(X) = 0. \tag{8.39}$$

(b) One can use the suspension $Y = \Sigma^n X$ of the previous space. Applying (2.122) and (2.124):

$$H_k(Y) \cong H_{k-n}(X), \quad \text{for } k \geqslant n+1, \qquad H_k(Y) = 0, \quad \text{for } 0 < k \leqslant n.$$

(c) Applying again Theorem 2.5.7, the spaces P_4 and $P_2 \vee P_2$ have

$$H_1(P_4) \cong \mathbb{Z}_4, \qquad H_1(P_2 \vee P_2) \cong \mathbb{Z}_2 \oplus \mathbb{Z}_2,$$

with torsion coefficients 4 and $2, 2$ in degree 1 (and no others).

(d) Always applying Theorem 2.5.7, the spaces P_6 and $P_2 \vee P_3$ have indeed torsion coefficient 6 in degree 1

$$H_1(P_6) \cong \mathbb{Z}_6, \qquad H_1(P_2 \vee P_3) \cong \mathbb{Z}_2 \oplus \mathbb{Z}_3 \cong \mathbb{Z}_6,$$

and no other.

These two spaces are not homeomorphic: the basepoint disconnects $P_2 \vee P_3$, but no point of P_6 disconnects it.

One can form diverse path-connected spaces X having $H_1(X) \cong \mathbb{Z}_{30}$ (with torsion coefficient 30) and trivial homology in higher degree, like:

$$P_{30}, \qquad P_2 \vee P_{15}, \qquad P_3 \vee P_{10}, \qquad P_5 \vee P_6, \qquad P_2 \vee P_3 \vee P_5.$$

8.2.21 Solutions of 2.6.3

(a) We use an open cover (U, V) of the space $\mathbb{T}$, where y is the image of the centre of $\mathbb{I}^2$

$$U = \mathbb{T} \setminus \{y\}, \qquad V = \operatorname{int} \mathbb{I}^2 \simeq \{*\}, \qquad A = U \cap V \simeq \mathbb{S}^1, \tag{8.40}$$

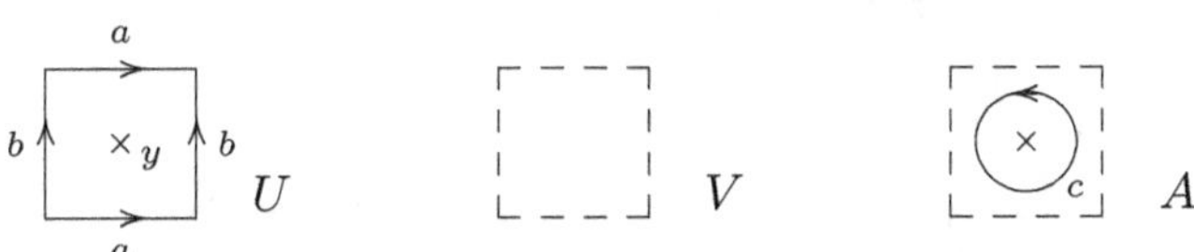

The space $U = \mathbb{T} \setminus \{y\}$ is homotopy equivalent to its retract formed by $\operatorname{Im} a \cup \operatorname{Im} b$ (the union of the equator and the meridian subspaces), which is the figure-eight space $\mathbb{S}^1 \vee \mathbb{S}^1$ examined in 2.5.1, with $H_1(\mathbb{S}^1 \vee \mathbb{S}^1) \simeq \mathbb{Z}^2$.

The following portion of the MV-sequence shows that $H_i(\mathbb{T}) = 0$ for $i > 2$

$$H_i(U) \xrightarrow{h_i} H_i(\mathbb{T}) \xrightarrow{D_i} H_{i-1}(A).$$

To compute $H_1(\mathbb{T})$ and $H_2(\mathbb{T})$ we use the exact sequence

$$0 \to H_2(\mathbb{T}) \xrightarrow{D_2} H_1(A) \xrightarrow{h_1} H_1(U) \xrightarrow{k_1} H_1(\mathbb{T}) \tag{8.41}$$

where $H_1(V) = 0$, and we know that k_1 is epi (by Exercise 2.5.2(c)).

The homomorphism h_1 is trivial, as shown by its value on a generator

$$h_1 = i_{*1} \colon H_1(A) \to H_1(U),$$
$$h_1([c]) = [c]_U = [a] + [b] - [a] - [b] = 0. \tag{8.42}$$

Here c is realised as a simple loop in A around the missing point y, as in the figure above; in U, c is loop homotopic to the loop $a * b * a^{\sharp} * b^{\sharp}$, whose homology class is trivial.

Coming back to the exact sequence (8.41), D_2 and k_1 are isomorphisms, and our claim is proved.

(b) The generators of $H_1(\mathbb{T})$ are the images of the two obvious generators of $H_1(\mathbb{S}^1 \vee \mathbb{S}^1)$, namely the homology classes $[a]$, $[b]$ of the equator and meridian loops.

As a generator of $H_2(\mathbb{T})$, our intuition suggests to use the canonical projection $p \colon \mathbb{I}^2 \to \mathbb{T}$, which enfolds (and detects) the 2-dimensional cavity of the torus. This 2-cube is readily seen to be a cycle

$$\partial_2 p = -b + b + a - a = 0.$$

To prove that $[p]$ is a generator we are to prove that $D_2[p]$ is a generator of $H_1(A)$, rewriting $[p] = [c + c']$ in a form on which we can compute D_2

$$c = -u + u' + v - v' \in C_2(U), \qquad c' \in C_2(V), \qquad (8.43)$$

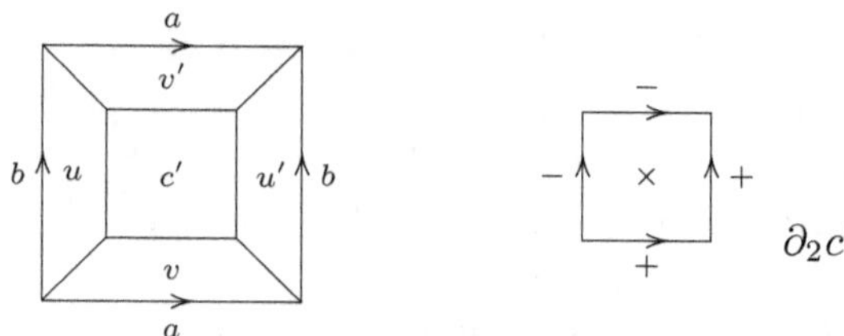

To make this precise, we start from a 3-cube $q\colon \mathbb{I}^3 \to \mathbb{T}$, whose image is represented in the figure above: first we squeeze the standard 3-cube on $\mathbb{I}^2$, as if we were orthogonally projecting a truncated pyramid on its basis, then we apply the projection $p\colon \mathbb{I}^2 \to \mathbb{T}$.

In this way:

- the lateral faces $u = \partial_1^- q$, $u' = \partial_1^+ q$, $v = \partial_2^- q$, $v' = \partial_2^+ q$ are homeomorphisms onto the quadrilaterals represented above, and form the chain $c \in C_2(U)$,

- the upper basis $c' = \partial_3^+ q$ is a homeomorphism onto the square $[1/4, 3/4]^2$, contained in $V = \operatorname{int} \mathbb{D}^2$,

- the lower basis $\partial_3^- q$ is the projection $p\colon \mathbb{I}^2 \to \mathbb{T}$ we are interested in (its image is covered by the other faces).

The 2-chain $\partial_3 q = c + c' - p$ is a boundary of the torus. The 2-cycle p is thus homologous to the cycle $c + c' = p + \partial_3 q$, on which we know how to compute D_2

$$D_2[p] = D_2[c + c'] = [\partial_2 c] = -[\partial_2 c'].$$

The proof is now complete: $\partial_2 c'$ is a cycle of $A = V \setminus \{*\}$, and $[\partial_2 c']$ is indeed a generator of $H_1(A) \cong H_1(\mathbb{S}^1)$, as follows from the Exercises of 2.2.8 (or shown in 2.4.9).

8.2.22 Solutions of 2.6.4

(a) We use the MV-sequence of the Klein bottle $\mathbb{K}$, with an open cover (U, V) similar to that used for the torus

$$U = \mathbb{T} \setminus \{y\}, \qquad V = \operatorname{int} \mathbb{I}^2 \simeq \{*\}, \qquad A = U \cap V \simeq \mathbb{S}^1, \qquad (8.44)$$

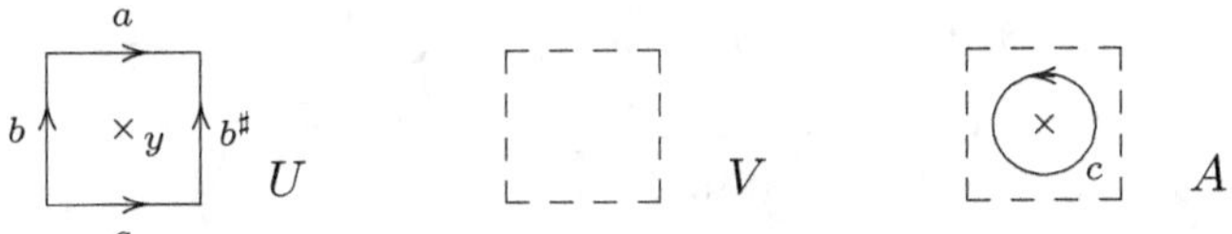

Again, $U \simeq \mathbb{S}^1 \vee \mathbb{S}^1$: the homotopy types of U, V and A are characterised as in 8.2.21. We find in the same way that $H_i(\mathbb{K}) = 0$ for $i > 2$.

Then we use the exact sequence

$$0 \to H_2(\mathbb{K}) \xrightarrow{D_2} H_1(A) \xrightarrow{h_1} H_1(U) \xrightarrow{k_1} H_1(\mathbb{K}) \to 0 \qquad (8.45)$$

where again $H_1(V) = 0$ and k_1 is epi.

Here the homomorphism $h_1 \colon H_1(A) \to H_1(U)$ is not trivial: for a simple loop c in A around the missing point y, we have

$$h_1([c]) = [c]_U = [a] + [b^\sharp] - [a] - [b] = -2[b]. \qquad (8.46)$$

Therefore h_1 is a monomorphism and $H_2(\mathbb{K}) = 0$. Moreover $\operatorname{Im} h_1$ is the subgroup of $\mathbb{Z}[a] \oplus \mathbb{Z}[b]$ generated by $2[b]$, and

$$H_1(\mathbb{K}) \cong \operatorname{cok} h_1 \cong \mathbb{Z} \oplus \mathbb{Z}_2,$$

with generators $[a]_\mathbb{K}$ (of infinite order) and $[b]_\mathbb{K}$ (of order 2).

8.2.23 Solution of 2.6.6

The spaces U, V and A can be described, up to homotopy equivalence, as

$$U \simeq \mathbb{T} \setminus \{*\}, \qquad V \simeq \mathbb{T} \setminus \{*\}, \qquad A \simeq \mathbb{S}^1, \qquad (8.47)$$

and we have already seen, in 8.2.21(a), that the pierced torus $\mathbb{T} \setminus \{*\}$ is homotopy equivalent to $\mathbb{S}^1 \vee \mathbb{S}^1$, the figure-eight space.

Moreover, as in (8.42):

$$i_{*1} = 0 \colon H_1(A) \to H_1(U), \qquad j_{*1} = 0 \colon H_1(A) \to H_1(V).$$

To compute $H_1(X)$ and $H_2(X)$ we use the exact sequence

$$0 \to H_2(X) \xrightarrow{D_2} H_1(A) \xrightarrow{0} H_1(U) \oplus H_1(V) \xrightarrow{k_1} H_1(X)$$

where we have proved that $h_1 = (i_{*1}, j_{*1}) = 0$, and we know that k_1 is epi, by Exercise 2.5.2(c).

Thus $D_2 \colon H_2(X) \cong H_1(A) \cong \mathbb{Z}$ and $k_1 \colon \mathbb{Z}^4 \cong H_1(X)$.

As generators of $H_1(X)$ we can take loops corresponding to two generators for each of the two tori.

8.2.24 Solutions of 2.7.3

(b) We start from an element $c \in C$ with $w(c) = 0$, and we want to prove that $c = 0$. As v' is mono, $h(c) = 0$ in D

$$\begin{array}{ccc}
c & \longmapsto & h(c) \\
\downarrow w & & \downarrow v' \\
0 & \longmapsto & 0
\end{array}
\qquad
\begin{array}{ccccccc}
a & & b & \longmapsto & c & \longmapsto & 0 \\
\downarrow u & & \downarrow v & & \downarrow w & & \downarrow v' \\
a' & \longmapsto & b' & \longmapsto & 0 & \longmapsto & 0
\end{array}
\qquad (8.48)$$

We move to the right diagram above; by the exactness of the upper row, there is some $b \in B$ such that $g(b) = c$; then $b' = v(b) \in B'$ goes to $w(c) = 0$ in C'.

By the exactness of the lower row, there is some $a' \in A'$ such that $f'(a') = b'$; but u is epi, and there is some $a \in A$ such that $u(a) = a'$.

By the commutativity of the left square, $vf(a) = f'u(a) = b' = v(b)$; but v is mono, and $f(a) = b$. Finally $c = gf(a) = 0$.

(c) We start from an element $c' \in C'$ and we want to prove that $c' \in \operatorname{Im} w$.

We form the left diagram below, where $d' = h'(c') \in D' = \operatorname{Im} v'$ and there is some $d \in D$ such that $d' = v'(d)$; moreover, $k(d) = 0$ in E, because u' is mono. By the exactness of the upper row, there is some $c \in C$ such that $h(c) = d$

$$\begin{array}{ccccc}
c & \longmapsto & d & \longmapsto & 0 \\
& & \downarrow v' & & \downarrow u' \\
c' & \longmapsto & d' & \longmapsto & 0
\end{array}
\qquad
\begin{array}{ccc}
b & \longmapsto & g(b) \\
\downarrow v & & \downarrow w \\
b' & \longmapsto & c' - w(c) & \longmapsto & 0
\end{array}
\qquad (8.49)$$

Moving to the right diagram above, the element $c' - w(c)$ goes to 0 in D' (because $h'(c') = d' = v'h(c) = h'w(c)$); there is thus some $b' \in B'$ such that $g'(b') = c' - w(c)$. As v is epi, there is some $b \in B$ such that $v(b) = b'$, and we have an element $g(b) \in C$ such that $wg(b) = c' - w(c)$. Finally we have found an element $c + g(b) \in C$ such that

$$w(c + g(b)) = w(c) + c' - w(c) = c'.$$

8.2.25 Solutions of 2.7.4

(a) Obviously, (iii) implies (i) and (ii).

(i) $\Rightarrow$ (iii). For $b \in B$, the element $b - nq(b)$ belongs to $\operatorname{Ker} q = \operatorname{Im} m$, and we define $p(b) \in A$ as the unique element such that $mp(b) = b - nq(b)$.

This mapping p satisfies the relation $mp + nq = \operatorname{id} B$; it is a homomorphism, because mp is, and m is injective. The relation $pm = \operatorname{id} A$ holds, because $mpm = m - nqm = m$ and m is injective.

(ii) $\Rightarrow$ (iii). For $c \in C$, there is some $b \in B$ such that $q(b) = c$, and we let $n(c) = b - mp(b)$; this does not depend on the choice of b: if $q(b') = c$, then $b - b' \in \operatorname{Ker} q = \operatorname{Im} m$, and $b - b' = m(a)$ for (a unique) $a \in A$, so that

$$(b - mp(b)) - (b' - mp(b')) = m(a) - mpm(a) = 0.$$

This mapping n satisfies the relation $mp + nq = \operatorname{id} B$; it is a homomorphism, because nq is, and q is surjective. The relation $qn = \operatorname{id} C$ holds, because $qnq = q - qmp = q$ and q is surjective.

(b) From (iii) it follows easily that $\operatorname{Im} m + \operatorname{Im} n = B$ and $\operatorname{Im} m \cap \operatorname{Im} n = 0$.

(c) Let $(c_i)_{i \in I}$ be a basis of C. For every $i \in I$ we choose some $b_i \in B$ such that $q(b_i) = c_i$, and define $n \colon C \to B$ as the unique homomorphism such that $n(c_i) = b_i$. This gives $qn = \operatorname{id} C$.

In general there are different possible choices. For instance, if C is infinite cyclic, with generator c, and $q(b) = c$, any element b' of the lateral $m(A) + b$ can be chosen as $n(c)$. This lateral is equipotent to A.

8.2.26 Solutions of 2.7.5

(a) Assuming 2.7.5(*), the morphism $f = up + vq - \operatorname{id} C$ gives $pf = 0$ and factorises through $v = \ker p$, as $f = vh$. It follows that $h = qvh = qf = 0$ and $f = 0$.

Conversely, if (2.179) is satisfied, let us prove that $v = \ker p$. We take $f \colon X \to C$ such that $pf = 0$; then $f = (up + vq)f = vqf$, and we have proved that f factorises (uniquely) through the monomorphism v.

(b) The composite (2.180) is computed as follows (where $(f, g) \colon A \to B \times B$)

$$\nabla (f \oplus g)\Delta = \nabla (f \times g)\Delta = \nabla (f, g) = f + g.$$

8.2.27 Solutions of 2.8.2

(b) The epimorphism $\rho X \colon \tilde{C}(X) \twoheadrightarrow C_+(X)$ in (2.188) gives a (natural) short exact sequence of unbounded chain complexes

$$\operatorname{Ker} \rho \overset{\iota X}{\rightarrowtail} \tilde{C}(X) \overset{\rho X}{\twoheadrightarrow} C_+(X) \qquad (\operatorname{Ker} \rho)_{-1} = \tilde{C}_{-1}(X) = \mathbb{Z}, \qquad (8.50)$$

where $\operatorname{Ker} \rho \subset \tilde{C}(X)$ has one non-trivial component, in degree -1, and $H_0(\operatorname{Ker} \rho) = (\operatorname{Ker} \rho)_{-1} = \mathbb{Z}$.

The derived exact homology sequence

$$0 \to \tilde{H}_0(X) \overset{\rho X}{\longrightarrow} H_0(X) \overset{DX}{\longrightarrow} \mathbb{Z} \overset{\rho X}{\longrightarrow} \tilde{H}_{-1}(X) \to 0$$

is natural on Top.

Its connecting morphism DX (in degree -1) and $pX = \iota_{*-1}X$ are computed as:

$$DX\colon H_0(X) \to \mathbb{Z}, \qquad (DX)[\textstyle\sum_i \lambda_i x_i] = [\tilde{\partial}_0(\textstyle\sum_i \lambda_i x_i)] = \textstyle\sum_i \lambda_i,$$

$$pX\colon H_{-1}(\operatorname{Ker}\rho) \to \tilde{H}_{-1}(X), \qquad (pX)(k) = [\iota X(k)] = [k].$$

(d) We have only to check that the homomorphism (2.192) is an isomorphism. In fact an element of $H_0(X)$ can be uniquely written as a sum

$$\textstyle\sum_i \lambda_i[x_i] = \textstyle\sum_{i\neq j} \lambda_i[x_i - x_j] + \textstyle\sum_i \lambda_i[x_j],$$

where the second addendum belongs to $\tilde{H}_0(X_j) \cong \mathbb{Z}$ and the first to $\tilde{H}_0(X)$.

(e) A consequence of (d): the trivial group has an empty basis.

*(g) For $n > 0$: $f_{*n} = 0$ and $\tilde{H}_n(X) = H_n(X)$. Otherwise f_{*0} is computed as:

$$f_{*0}[\textstyle\sum \lambda_i x_i] = \textstyle\sum \lambda_i[*] = \textstyle\sum \lambda_i \in H_0(\{*\}) = \mathbb{Z}, \tag{8.51}$$

and coincides with the homomorphism $DX\colon H_0(X) \to \mathbb{Z}$ of the exact sequence (2.191). The kernel and cokernel of the latter are indeed $\tilde{H}_0(X)$ and $\tilde{H}_{-1}(X)$.

8.2.28 Solutions of 2.8.6

(a) For the singleton, there is a unique mapping $a_n\colon \Delta^n \to \{*\}$, for every $n \geqslant 0$. The differential of the chain complex involves $n+1$ summands

$$\partial_n(a_n) = \textstyle\sum_i (-1)^i \partial_i a = \textstyle\sum_i (-1)^i a_{n-1},$$

and is 0 for n odd, and an isomorphism for n even.

The simplicial chain complex $S_+(\{*\})$ of the singleton is isomorphic to

$$\ldots\ \mathbb{Z} \xrightarrow{1} \mathbb{Z} \xrightarrow{0} \mathbb{Z} \xrightarrow{1} \mathbb{Z} \xrightarrow{0} \mathbb{Z}. \tag{8.52}$$

Its homology is $\mathbb{Z}$ in degree 0, and trivial elsewhere.

8.3 Exercises of Chapter 3

8.3.1 Solutions of 3.1.2

(a) The embedding $C_n(A) \rightarrowtail C_n(X)$ comes from an inclusion of canonical bases $c_n(A) \subset c_n(X)$ where

- $c_n(X)$ is the set of all non-degenerate n-cubes $a\colon \mathbb{I}^n \to X$,

- $c_n(A) = \{a \in c_n(X) \mid \operatorname{Im} a \subset A\}$.

Therefore $C_n(X, A)$ is the free abelian group generated by the complement $c_n(X, A) = c_n(X) \setminus c_n(A)$.

(b) The short exact sequence $C_n(A) \rightarrowtail C_n(X) \twoheadrightarrow C_n(X, A)$ certainly splits, because $C_n(X, A)$ is a free abelian group (see Exercise 2.7.4(c)). Here we can say something more: we have a canonical section $C_n(X, A) \rightarrowtail C_n(X)$ given by the inclusion $c_n(X, A) \subset c_n(X)$ of canonical bases, as in (a).

8.3.2 Solutions of 3.1.6

(a) In the exact sequence of the pair (X, A)

$$H_n A \xrightarrow{u_{*n}} H_n X \xrightarrow{v_{*n}} H_n(X, A) \xrightarrow{D_n} H_{n-1} A \xrightarrow{u_{*n-1}} H_{n-1} X$$

the morphisms u_{*n} and u_{*n-1} are isomorphisms, and $H_n(X, A) = 0$.

(b) The exact homology sequence of the pair shows that $H_0(X, A)$ is the cokernel of the homomorphism $H_0(A) \to H_0(X)$ induced by the inclusion.

(d) The exact sequence

$$H_n(\{x\}) \to H_n(X) \to H_n(X, x) \to H_{n-1}(\{x\})$$

gives a canonical isomorphism $H_n(X) \to H_n(X, x)$ for $n > 1$. In lower degree we have:

$$0 \to H_1 X \xrightarrow{k_1} H_1(X, x) \xrightarrow{D_1} H_0\{x\} \xrightarrow{h_n} H_0 X \xrightarrow{h_0} H_0(X, x) \to 0$$

where h_0 is injective, $D_1 = 0$ and k_1 is an isomorphism. On the right hand, the same sequence shows that $H_0(X, x) \cong H_0(X)/\mathrm{Im}\, h_0$, where $\mathrm{Im}\, h_0$ is generated by $[x] \in H_0(X)$.

Therefore $H_0(X, x)$ is freely generated by the path components of X, excluding that of x, which makes it isomorphic to $\tilde{H}_0(X)$, by Exercise 2.8.2(d).

8.3.3 Solutions of 3.1.7

(b) The commutative diagram of (unbounded) chain complexes

$$
\begin{array}{ccccc}
\tilde{C}(A) & \rightarrowtail & \tilde{C}(X) & \twoheadrightarrow & C_+(X, A) \\
\downarrow{\scriptstyle \rho A} & & \downarrow{\scriptstyle \rho X} & & \| \\
C_+(A) & \rightarrowtail & C_+(X) & \twoheadrightarrow & C_+(X, A)
\end{array}
\qquad (8.53)
$$

has short exact rows, and induces in homology the commutative diagram (3.17), with exact rows.

8.3.4 Solutions of 3.4.3

(a) Suppose that the inclusion $A \to X$ is a homotopy equivalence. We have the exact sequence

$$H_n(A) \to H_n(X) \to H_n(X, A) \to H_{n-1}(A) \to H_{n-1}(X)$$

where, as usual, we let $H_k(-) = 0$ on negative degrees. The group $H_n(X, A)$ is surrounded by two isomorphisms, and is trivial.

(c) A space X has a unique map $f\colon X \to \{*\}$ with values in the singleton. We define

$$\tilde{H}_n(X) = \mathrm{Ker}\,(f_{*n}\colon H_n(X) \to H_n(\{*\})) \quad (n \geqslant 0),$$
$$\tilde{H}_{-1}(X) = \mathrm{Cok}\,(f_{*0}\colon H_0(X) \to H_0(\{*\})).$$
$$(8.54)$$

Because of the dimension axiom, we readily have that:

$$\tilde{H}_n(X) = H_n(X), \quad \text{for } n > 0, \tag{8.55}$$

so that we are only interested in $\tilde{H}_0(X)$ and $\tilde{H}_{-1}(X)$. Moreover, by (b)

$$\tilde{H}_0(\emptyset) = 0, \qquad \tilde{H}_{-1}(\emptyset) = H_0(\{*\}) = G,$$
$$\tilde{H}_{-1}(X) = 0, \quad \text{for } X \neq \emptyset.$$
$$(8.56)$$

For the last point, it is sufficient to show that $f_{*0}\colon H_0(X) \to H_0(\{*\})$ is an epimorphism, for $X \neq \emptyset$. In fact, we have a retraction $\{*\} \to X \to \{*\}$, with induced homomorphisms $(G \to H_0(X) \to G) = \mathrm{id}\,G$.

Many other properties of reduced homology can now be proved, following the outline of 2.8.2.

8.3.5 Solutions of 3.5.3

(a) We use, as suggested, the following projections

$$\varepsilon_i\colon X^n \to X^{n-1}, \qquad \varepsilon_i(x_1, ..., x_n) = (x_1, ..., \hat{x}_i, ..., x_n),$$
$$\varepsilon_i \varepsilon_j = \varepsilon_j \varepsilon_{i+1}, \quad \text{for } j \leqslant i.$$

In this way:

$$d^{n-1}\lambda = \sum_{1 \leqslant i \leqslant n+1} (-1)^i \lambda \varepsilon_i,$$
$$d^n d^{n-1}\lambda = \sum_{i \leqslant n+1, j \leqslant n+2} (-1)^i \lambda \varepsilon_i \varepsilon_j$$
$$= \sum_{i<j} (-1)^i \lambda \varepsilon_i \varepsilon_j + \sum_{j \leqslant i} (-1)^i \lambda \varepsilon_i \varepsilon_j = 0.$$

In fact, the second addendum is the opposite of the first, by the usual change of variables:

$$\sum_{j \leqslant i} (-1)^i \lambda \varepsilon_i \varepsilon_j = \sum_{j \leqslant i} (-1)^i \lambda \varepsilon_j \varepsilon_{i+1} = -\sum_{i<j} (-1)^i \lambda \varepsilon_i \varepsilon_j.$$

8.3.6 Solutions of 3.5.5

(a) For the singleton X, a function $g\colon X^{n+1} \to G$ amounts to its value, and is locally zero if and only if it is zero. We identify all components $\overline{C}^n(X;G) = C^n(X;G) = G$.

The differential of the cochain complex $\overline{C}^+(X;G) = C^+(X;G)$

$$d^n\colon C^n(X;G) \to C^{n+1}(X;G), \qquad d^n g = \Sigma_{1\leqslant i\leqslant n+2}\,(-1)^i\, g, \qquad (8.57)$$

is zero for n even $\geqslant 0$ and $-\operatorname{id} G$ for n odd.

Therefore our cochain complex is exact in positive degree, and $\overline{H}^n(X;G) = 0$, for $n > 0$. Moreover we have got a canonical isomorphism

$$\overline{H}^0(X;G) = \operatorname{Ker}(d^0\colon C^0(X;G) \to C^1(X;G)) = C^0(X;G) \cong G,$$

that sends the function $\lambda\colon X \to G$ to $\lambda(*) \in G$.

8.3.7 Solutions of 3.5.8

(a) The relative cochain complex $\overline{C}^+(X, A; G)$ has an augmentation

$$0 \;\to\; \mathcal{A}(X, A; G) \;\xrightarrow{\eta}\; \overline{C}^0(X, A; G) \;\xrightarrow{d^0}\; \overline{C}^1(X, A; G) \qquad (8.58)$$

where a locally constant functions $\lambda\colon X \to G$ vanishing on A is sent to the class $\overline{\lambda} \in \overline{C}^0(X, A; G)$.

It is sufficient to prove that this sequence is exact, so that η embeds $\mathcal{A}(X, A; G)$ as $\operatorname{Ker} d^0 = \overline{H}^0(X, A; G)$.

First, η is injective: if $\eta(\lambda) = 0$, the (locally constant) function λ is locally zero, and everywhere 0. Secondly, $\operatorname{Ker} d^0 = \operatorname{Im} \eta$ because, for any function $\lambda\colon X \to G$ vanishing on A, the function

$$d^0(\lambda)\colon X^2 \to G, \qquad d^0(\lambda)(x_1, x_2) = \lambda(x_2) - \lambda(x_1),$$

is locally zero (at the diagonal) if and only if λ is locally constant.

(d) If the connected components of the space X are open, a function $\lambda\colon X \to G$ is locally constant if and only if it is constant on each connected component; this amounts to a family $(g_i) \in G^I$.

8.3.8 Solutions of 3.6.3

(a) A consequence of formula (3.94), which ensures that:

- the product of two cocycles is a cocycle,

- the product of a coboundary and a cocycle, in any order, is a coboundary:

$$dy = 0 \;\Rightarrow\; (dx)y = d(xy), \qquad dx = 0 \;\Rightarrow\; x(dy) = \pm d(xy).$$

8.3.9 Solutions of 3.6.4

(a) We have to verify that, for $\lambda \in C^p(X;G)$ and $\mu \in C^q(X;G)$

$$d^n(\lambda \bullet \mu) = d^p(\lambda) \bullet \mu + (-1)^p \lambda \bullet d^q(\mu). \tag{8.59}$$

Computing the two summands on the right, we single out a term in each of them:

$$(d^p\lambda \bullet \mu)(x_1, ..., x_{n+2})$$
$$= \Sigma_{1 \leqslant i \leqslant p+2} \, (-1)^i \, \lambda(x_1, ..., \hat{x}_i, ..., x_{p+2}) \, \mu(x_{p+2}, ..., x_{n+2})$$
$$= \Sigma_{1 \leqslant i \leqslant p+1} \, (-1)^i \, \lambda(x_1, ..., \hat{x}_i, ..., x_{p+2}) \, \mu(x_{p+2}, ..., x_{n+2})$$
$$+ (-1)^p \, \lambda(x_1, ..., x_{p+1}) \, \mu(x_{p+2}, ..., x_{n+2}),$$

$$(-1)^p(\lambda \bullet d^q\mu)(x_1, ..., x_{n+2})$$
$$= \Sigma_{1 \leqslant i \leqslant q+2} \, (-1)^{p+i} \, \lambda(x_1, ..., x_{p+1}) \, \mu(x_{p+1}, ..., \hat{x}_{p+i}, ..., x_{n+2})$$
$$= (-1)^{p+1} \lambda(x_1, ..., x_{p+1}) \, \mu(x_{p+2}, ..., x_{n+2})$$
$$+ \Sigma_{2 \leqslant i \leqslant q+2} \, (-1)^{p+i} \, \lambda(x_1, ..., x_{p+1}) \, \mu(x_{p+1}, ..., \hat{x}_{p+i}, ..., x_{n+2}).$$

Adding these results, the two terms we have put apart cancel, and we have:

$$(d^p(\lambda) \bullet \mu + (-1)^p \, \lambda \bullet d^q(\mu))(x_1, ..., x_{n+2})$$
$$= \Sigma_{1 \leqslant i \leqslant p+1} \, (-1)^i \, \lambda(x_1, ..., \hat{x}_i, ..., x_{p+2}) \, \mu(x_{p+2}, ..., x_{n+2})$$
$$+ \Sigma_{p+2 \leqslant i \leqslant n+2} \, (-1)^i \, \lambda(x_1, ..., x_{p+1}) \, \mu(x_{p+1}, ..., \hat{x}_i, ..., x_{n+2}).$$

The statement is thus proved:

$$d^n(\lambda \bullet \mu)(x_1, ..., x_{n+1})$$
$$= \Sigma_{1 \leqslant i \leqslant n+2} \, (-1)^i \, (\lambda \bullet \mu)(x_1, ..., \hat{x}_i, ..., x_{n+2})$$
$$= \Sigma_{1 \leqslant i \leqslant p+1} \, (-1)^i \, (\lambda \bullet \mu)(x_1, ..., \hat{x}_i, ..., x_{n+2})$$
$$+ \Sigma_{p+2 \leqslant i \leqslant n+2} \, (-1)^i \, (\lambda \bullet \mu)(x_1, ..., \hat{x}_i, ..., x_{n+2})$$
$$= \Sigma_{1 \leqslant i \leqslant p+1} \, (-1)^i \, \lambda(x_1, ..., \hat{x}_i, ..., x_{p+2}) \, \mu(x_{p+2}, ..., x_{n+2})$$
$$+ \Sigma_{p+2 \leqslant i \leqslant n+2} \, (-1)^i \, \lambda(x_1, ..., x_{p+1}) \, \mu(x_{p+1}, ..., \hat{x}_i, ..., x_{n+2}).$$

(b) The graded homomorphism $f^\sharp : C^*(X;R) \to C^*(Y;R)$ already preserves the product:

$$(f^{\sharp n}(\lambda \bullet \mu))(y_1, ..., y_{n+1}) = \lambda(fy_1, ..., fy_{p+1}) \, \mu(fy_{p+1}, ..., fy_{n+1})$$
$$= ((f^{\sharp p}\lambda) \bullet (f^{\sharp q}\mu))(y_1, ..., y_{n+1}).$$

8.3.10 Solutions of 3.7.3

(a) Cancelling the zero terms (given by repeated variables) and reordering the exterior products, we get:

$$\omega \wedge \omega' = \lambda_1\mu_1\, dx \wedge dydz + \lambda_2\mu_2\, dy \wedge dxdz + \lambda_3\mu_3\, dz \wedge dxdy$$
$$= (\lambda_1\mu_1 - \lambda_2\mu_2 + \lambda_3\mu_3)\, dxdydz. \tag{8.60}$$

(b) The properties (i)–(iii) of 3.6.2 are easily verified: consistency and distributivity hold by definition. Associativity can be verified on the canonical bases, and is obvious using formula (3.109), based on the extended symbols $dx^{i_1}dx^{i_2}\dots dx^{i_p}$.

With these symbols, it is also obvious that the products

$$(dx^{i_1}\dots dx^{i_p}) \wedge (dx^{j_1}\dots dx^{j_q}) = dx^{i_1}\dots dx^{i_p}dx^{j_1}\dots dx^{j_q},$$
$$(dx^{j_1}\dots dx^{j_q}) \wedge (dx^{i_1}\dots dx^{i_p}) = dx^{j_1}\dots dx^{j_q}dx^{i_1}\dots dx^{i_p}, \tag{8.61}$$

are related by a factor $(-1)^{pq}$.

8.3.11 Solutions of 3.7.4

(b) Beginning with the product $\rho = \lambda\mu$ of two C^∞-functions, we get the expected 1-form

$$d(\lambda\mu) = d(\textstyle\sum_{1\leqslant k\leqslant n} D_k\rho\, dx^k)$$
$$= \textstyle\sum_{1\leqslant k\leqslant n} ((D_k\lambda)\mu + \lambda(D_k\mu))\, dx^k = (d\lambda)\mu + \lambda(d\mu).$$

Now we apply definition (3.112) to a product $\omega \wedge \omega' = \sum_{i,j} \lambda_i\mu_j(dx^i \wedge dx^j)$, taking into account that each term $dx^i \wedge dx^j$ is either 0 or $\sigma(i,j)dx^k$, which is a basic differential, up to sign

$$d^{p+q}(\omega \wedge \omega') = \textstyle\sum_k d^0(\lambda_i\mu_j) \wedge (\sum_{i,j} dx^i \wedge dx^j)$$
$$= ((d\lambda_i)\mu_j + \lambda_i d(\mu_j)) \wedge (\textstyle\sum_{i,j} dx^i \wedge dx^j)$$
$$= (d\lambda_i)\mu_j \wedge (\textstyle\sum_{i,j} dx^i \wedge dx^j) + \lambda_i d(\mu_j)) \wedge (\sum_{i,j} dx^i \wedge dx^j)$$
$$= (d\lambda_i)\mu_j \wedge (\textstyle\sum_i dx^i) \wedge (\sum_j \mu_j\, dx^j) + (-1)^p (\sum_i \lambda_i\, dx^i) \wedge d(\mu_j) \wedge (\sum_j dx^j)$$
$$= d^p\omega \wedge \omega' + (-1)^p \omega \wedge d^q\omega'.$$

(c) The double differential of a C^∞-function is zero, by Schwartz theorem on the commutation of partial derivatives

$$d^1 d^0(\lambda) = d^1\left(\textstyle\sum_{1\leqslant i\leqslant n} \frac{\partial\lambda}{\partial x^i}\, dx^i\right) = \textstyle\sum_{i,j} \frac{\partial^2\lambda}{\partial x^i\partial x^j}\, dx^j dx^i$$
$$= \textstyle\sum_{j<i}\left(\frac{\partial^2\lambda}{\partial x^i\partial x^j} - \frac{\partial^2\lambda}{\partial x^j\partial x^i}\right) dx^j dx^i = 0.$$

The general case is a consequence, applying (b) and the vanishing of the differential on every dx^i:

$$d^{p+1} d^p \left(\sum_i \lambda_i \, dx^i \right) = d^{p+1} \left(\sum_i d^0(\lambda_i) \wedge dx^i \right)$$

$$= \sum_i \left(d^1 d^0(\lambda_i) \wedge dx^i - d^0(\lambda_i) \wedge d^p(dx^i) \right) = 0.$$

8.3.12   Solutions of 3.7.5

(a) The differential $d^0(\lambda) = \sum_i D_i \lambda \, dx^i$ is zero if and only if all the partial derivatives of λ vanish everywhere, which means that λ is locally constant, i.e. constant on every connected component of U. Thus $H^0_R(U)$ can be identified with the product algebra $\mathbb{R}^{Cc(U)}$, where $Cc(U)$ is the set of connected components of U.

    We have seen that Alexander–Spanier cohomology with coefficients in $\mathbb{R}$ gives the same algebra.

(b) The differential $d^1 \omega$ is computed 'as' in (3.102), for $n = 3$.

8.3.13   Solutions of 3.7.6

(a) Beginning with the product $\rho = \lambda\mu$ of two C^∞-functions, we obviously have $f^{\sharp 0}(\lambda\mu) = (\lambda\mu)f = (\lambda f)(\mu f) = f^{\sharp 0}(\lambda).f^{\sharp 0}(\mu)$.

Now we apply definition (3.114) to a product

$$\omega \wedge \omega' = \left(\sum_i \lambda_i \, dx^{i_1} \dots dx^{i_p} \right) \wedge \left(\sum_j \mu_j \, dx^{j_1} \dots dx^{j_q} \right)$$

$$= \sum_{i,j} \lambda_i \mu_j \, dx^{i_1} \dots dx^{i_p} dx^{j_1} \dots dx^{j_q},$$

where each term $dx^{i_1} \dots dx^{i_p} dx^{j_1} \dots dx^{j_q}$ is either 0 or a basic form, up to sign change

$$f^{\sharp p+q}(\omega \wedge \omega') = \sum_{i,j} \left((\lambda_i \mu_j) \circ f \right) d^0 f^{i_1} \dots d^0 f^{i_p} d^0 f^{j_1} \dots d^0 f^{j_q}$$

$$= \left(\sum_i (\lambda_i \circ f) \, d^0 f^{i_1} \dots d^0 f^{i_p} \right) \left(\sum_j (\mu_j \circ f) \, d^0 f^{j_1} \dots d^0 f^{j_q} \right)$$

$$= f^{\sharp p}(\omega) \wedge f^{\sharp q}(\omega').$$

(b) The statement is obvious for a 0-form λ on U:

$$d^0 f^{\sharp 0}(\lambda) = d^0(\lambda \circ f) = \sum_j \frac{\partial(\lambda \circ f)}{\partial y^j} \, dy^j$$

$$= \sum_{j,i} \left(\frac{\partial \lambda}{\partial x^i} \circ f \right) \frac{\partial f^i}{\partial y^j} \, dy^j = \sum_i \left(\frac{\partial \lambda}{\partial x^i} \circ f \right) d^0 f^i,$$

$$f^{\sharp 1} d^0(\lambda) = f^{\sharp 1} \left(\sum_i \frac{\partial \lambda}{\partial x^i} \, dx^i \right) = \sum_i \left(\frac{\partial \lambda}{\partial x^i} \circ f \right) d^0 f^i.$$

The general case is a consequence, taking into account point (a), and the fact that the differential vanishes on every differential:

$$d^p f^{\sharp p}(\textstyle\sum_i \lambda_i dx^{i_1}\dots dx^{i_p}) = d^p(\textstyle\sum_i (\lambda_i\circ f)d^0 f^{i_1} \wedge \dots \wedge d^0 f^{i_p})$$
$$= \textstyle\sum_i d^0(\lambda_i\circ f) \wedge d^0 f^{i_1} \wedge \dots \wedge d^0 f^{i_p},$$

$$f^{\sharp p+1}d^p(\textstyle\sum_i \lambda_i\, dx^{i_1}\dots dx^{i_p}) = f^{\sharp p+1}(\textstyle\sum_i d^0\lambda_i \wedge dx^{i_1}\dots dx^{i_p})$$
$$= \textstyle\sum_i f^{\sharp 1}(d^0\lambda_i) \wedge f^{\sharp 1}(dx^{i_1}) \dots f^{\sharp 1}(dx^{i_p})$$
$$= \textstyle\sum_i d^0(\lambda_i\circ f) \wedge d^0 f^{i_1} \wedge \dots \wedge d^0 f^{i_p}.$$

8.4 Exercises of Chapter 4

8.4.1 *Solutions of 4.1.8*

(a) Let $f\colon A \to B$ be an injective K-homomorphism. We take a basis $(a_i)_{i\in I}$ of A, and let $b_i = f(a_i)$. Then the family $(b_i)_{i\in I}$ is linearly independent in B, and can be extended to a basis $(b_i)_{i\in J}$ of B (with $I \subset J$). We define $g\colon B \to A$ as the unique linear mapping such that

$$g(b_i) = a_i, \text{ for } i \in I, \qquad g(b_i) = 0, \text{ for } i \in J \setminus I.$$

Then $gf\colon A \to A$ is the identity on each a_i, and therefore on A. (We are splitting B in the internal direct sum of $f(A) \cong A$ and $\operatorname{Ker} g$, another vector subspace of B.)

(b) If $f\colon A \to B$ is a surjective K-homomorphism and (b_i) is a basis of B, we choose an element $a_i \in f^{-1}\{b_i\}$ for every $i \in I$, and let $g\colon B \to A$ be the unique linear mapping such that $g(b_i) = a_i$ (for all i).

Then $fg\colon B \to B$ is the identity on each b_i, and therefore on B. (We are splitting A in the internal direct sum of $\operatorname{Ker} f$ and $g(B)$, a vector subspace of A isomorphic to B.)

(d) The mapping (for $\alpha\colon A \to K$)

$$\varphi\colon A \times B \to \operatorname{Hom}_K(A^*, B), \qquad \varphi(a,b)(\alpha) = \alpha(a).b,$$

is K-bilinear and induces our homomorphism $u\colon A\otimes_K B \to \operatorname{Hom}_K(A^*, B)$.

(This also works for R-modules, but the sequel would not: note that, for $\mathbb{Z}$-modules, $\operatorname{Hom}(A,\mathbb{Z}) = 0$ for every torsion abelian group A.)

Now, we choose a basis $(a_i)_{i\in I}$ of A. For every $i \in I$, we define a homomorphism α_i on this basis

$$\alpha_i\colon A \to K, \quad \alpha_i(a_i) = 1_K, \quad \alpha_i(a_j) = 0_K \quad (\text{for } j \neq i \text{ in } I),$$

which takes every vector $a = \sum_i \lambda_i a_i$ to the scalar λ_i. Note that $a = 0$ if and only if all homomorphisms α_i vanish on it.

To prove that each u_{AB} is injective, let $u(a \otimes b) = 0$, so that $\alpha_i(a)b = 0$, for all $i \in I$. Now, either $a = 0$ or there is some index i such that $\alpha_i(a) \in K^*$, and then $b = 0$. In both cases $a \otimes b = 0$.

If A and B are finitely generated, the domain and codomain of u have the same finite dimension on K, namely the product $\dim(A)\dim(B)$, and therefore u is an isomorphism.

8.4.2 Solutions of 4.2.1

(a) The trivial R-homomorphism $0_{AB} \colon A \to B$ is the identity of the group $\mathrm{Hom}_R(A, B)$. The trivial R-module X is characterised by the property $\mathrm{id}\, X = 0_{XX}$.

(b) A consequence of the characterisation 1.6.4(ii).

(c) The characterisation 2.7.4(iii) is obviously preserved by additive functors. The particular case is a consequence of Exercise 2.7.4(c).

8.4.3 Solutions of 4.2.7

(a) The first claim is a consequence of 4.1.5(e). The second is obvious: in diagram (4.39), one can choose a basis (x_i) in P, and choose $k(x_i) \in A$ so that $pk(x_i) = h(x_i)$, for all indices i.

(b) Plainly, a direct sum of projective modules is projective. Let X be a direct summand of the projective module P, with $n \colon X \rightleftarrows P \colon q$ and $qn = \mathrm{id}\, X$.

Given $p \colon A \twoheadrightarrow B$ and $h \colon X \to B$, there is a commutative square $hq = pk$, because P is projective

$$
\begin{array}{ccc}
P & \xrightarrow{\ q\ } & X \\
{\scriptstyle k}\big\downarrow & {\scriptstyle k'}\nearrow & \big\downarrow{\scriptstyle h} \\
A & \xrightarrow[\ p\ ]{} & B
\end{array}
\qquad\qquad (8.62)
$$

and the homomorphism $k' = kn \colon X \to A$ gives $pk' = pkn = hqn = h$.

The part on injective modules is similar.

(c) By (a) and (b), any summand of a free module is projective. Conversely, if P is projective, we choose an epimorphism $p \colon F \twoheadrightarrow P$ where F is a free module (one can always take the free module generated by the set $|P|$, or by any set of generators of P).

Then property (i) gives a homomorphism $k\colon P \to F$ such that $pk = \mathrm{id}\, P$

$$\tag{8.63}$$

8.4.4 Solutions of 4.2.8

(a) We start from the left short exact sequence in Ab

$$\mathbb{Z} \xrightarrow{\ i\ } \mathbb{Z} \xrightarrow{\ p\ } \mathbb{Z}_m \qquad\qquad \mathbb{Z}_n \xrightarrow{\ i\otimes\mathbb{Z}_n\ } \mathbb{Z}_n \xrightarrow{\ p\otimes\mathbb{Z}_n\ } \mathbb{Z}_m \otimes \mathbb{Z}_n \to 0$$

where the monomorphism i is the multiplication by $m \geqslant 2$. The tensor functor $-\otimes \mathbb{Z}_n$ gives the exact sequence on the right hand, where $\mathbb{Z}_m \otimes \mathbb{Z}_n$ is the cokernel of $i \otimes \mathbb{Z}_n$.

Now $\mathrm{Im}\,(i \otimes \mathbb{Z}_n) = m\mathbb{Z}_n = (m\mathbb{Z} \vee n\mathbb{Z})/n\mathbb{Z} = d\mathbb{Z}/n\mathbb{Z}$, where d is the greatest common divisor of m, n. By a Noether isomorphism

$$\mathbb{Z}_n/\mathrm{Im}\,(i \otimes \mathbb{Z}_n) = (\mathbb{Z}/n\mathbb{Z})/(d\mathbb{Z}/n\mathbb{Z}) \cong \mathbb{Z}/d\mathbb{Z} = \mathbb{Z}_d.$$

(b) A homomorphism $f\colon \mathbb{Z}_m \to A$ amounts to an element $a = f(\bar 1)$ such that $ma = 0$. We can identify $\mathrm{Hom}(\mathbb{Z}_m, A) = \{a \in A \mid ma = 0\}$.

Letting $n = n'd$ we also have that

$$\mathrm{Hom}(\mathbb{Z}_m, \mathbb{Z}_n) = \{\bar x \in \mathbb{Z}_n \mid mx \in n\mathbb{Z}\}$$
$$= \{\bar x \in \mathbb{Z}_n \mid x \in n'\mathbb{Z}\} = (n'\mathbb{Z})/(n\mathbb{Z}) \cong \mathbb{Z}/d\mathbb{Z}.$$

(c) Let $-\otimes A$ be exact. We take $a \in A$ with $na = 0$ and $n > 0$. Then the monomorphism $i\colon \mathbb{Z} \to \mathbb{Z}$ of multiplication by n is taken by $-\otimes A$ to a monomorphism $A \to A$ which sends a to 0, proving that $a = 0$.

8.4.5 Solutions of 4.2.9

(b) A consequence of (a) and 4.2.7(b).

(c) We have seen in (2.12) that a sequence of abelian groups can be decomposed in sequences of three objects, so that the given sequence is exact if and only if all the derived ones are short exact.

If the original exact sequence is upper unbounded, each additional object is a subgroup of an original one, and therefore a free abelian group. All the derived short exact sequences split (by 2.7.4(c)), and are preserved by any additive functor (by 4.2.1(c)).

8.4.6 Solutions of 4.3.3

(a) The chain complex $C_+(\{*\}; G)$ is isomorphic to

$$\ldots\ 0 \to 0 \to 0 \to 0 \to G. \tag{8.64}$$

Its homology is G in degree 0, and trivial elsewhere.

(b) Applying the right exact functor $- \otimes G$ to the augmented chain complex used in the solution to Exercise 2.2.5(b), we get an augmented chain complex

$$\ldots\ C_2(X; G) \xrightarrow{\partial_2} C_1(X; G) \xrightarrow{\partial_1} C_0(X; G) \xrightarrow{\tilde{\partial}_0} G \to 0 \tag{8.65}$$

exact at $C_0(X; G)$. Therefore the homomorphism $\tilde{\partial}_0((\sum \lambda_i\, x_i) \otimes g) = \sum \lambda_i g$ induces an isomorphism on $H_0(X; G) = C_0(X; G)/\operatorname{Im} \partial_1$

$$H_0(X; G) \to G, \qquad [x \otimes g] \mapsto g, \tag{8.66}$$

where x is any point of the 0-connected space X.

(c) The canonical isomorphism $\bigoplus_{i \in I} C_+(X_i) \to C_+(X)$ in (2.58) is preserved by the functor $- \otimes G$ (by (4.18)), giving the canonical isomorphism (4.50).

Applying the chain homology functor we get the isomorphism (4.51), as in Exercise 2.2.5(c).

(d) Let (Y_j) be the partition of Y in path components. By the previous points, f_{*0} amounts to embedding G in $\bigoplus_j G$, as the summand corresponding to the path component which contains $f(X)$.

8.4.7 Solutions of 4.3.8

(a) This is proved as in 2.4.1–2.4.2, with the present MV-sequence 4.3.6.

(b) For the space P_n we use the MV-sequence 4.3.6, with the open cover (U, V) used in the solution to Exercise 2.5.8(a):

$$U = P_n \setminus \{0\} \simeq \mathbb{S}^1, \qquad V = \operatorname{int} \mathbb{D}^2 \simeq \{*\}.$$
$$A = U \cap V = \operatorname{int} \mathbb{D}^2 \setminus \{0\} \simeq \mathbb{S}^1. \tag{8.67}$$

To determine $H_1(P_n; G)$ we use the following exact sequence (omitting $H_1(V; G) = 0$)

$$H_1(A; G) \xrightarrow{h_1} H_1(U; G) \xrightarrow{k_1} H_1(P_n; G) \xrightarrow{D_1}$$

$$H_0(A; G) \xrightarrow{h_0} H_0(U; G) \oplus H_0(V; G)$$

where h_0 is injective, by 4.3.3(d), whence $D_1 = 0$ and k_1 is epi: a cokernel of h_1.

As in 2.5.8(a), the homomorphism h_1 amounts to the multiplication by n in the group G, proving that $H_1(P_n; G) \cong G/nG$.

For $i > 1$ we have the exact sequence

$$0 \to H_i(P_n; G) \xrightarrow{D_i} H_{i-1}(A; G) \xrightarrow{h_{i-1}} H_{i-1}(U; G) \oplus H_{i-1}(V; G)$$

where $H_i(U; G) = H_i(V; G) = 0$. We distinguish two cases:

- if $i = 2$: $h_{i-1} = h_1$ has been computed above, as the multiplication by n in the group G; therefore

$$H_2(P_n; G) \cong \operatorname{Im} D_2 = \operatorname{Ker} h_1 = \{g \in G \mid ng = 0\},$$

- if $i > 2$: $H_{i-1}(A; G) = 0$ and $H_i(P_n; G) = 0$.

(c) The previous open cover (U, V) of P_n gives an open cover (U', V) for $\mathbb{S}^2$, where $U' = p(U)$ is the pierced sphere, a contractible space.

The projection $p \colon P_n \to \mathbb{S}^2$ agrees with these open covers: $p(U) = U'$ and $p(V) = V$. By the naturality of the corresponding MV-sequences, we have a commutative diagram with exact rows

$$
\begin{array}{ccccccccc}
0 & \longrightarrow & 0 & \longrightarrow & H_2(P_n; G) & \longrightarrow & H_1(A; G) & \longrightarrow & H_1(U; G) \\
& & & & \downarrow{\scriptstyle p_{*2}} & & \Big\| & & \downarrow{\scriptstyle p_{*1}} \\
0 & \longrightarrow & 0 & \longrightarrow & H_2(\mathbb{S}^2; G) & \longrightarrow & H_1(A; G) & \longrightarrow & 0
\end{array}
$$

By the Five Lemma 2.7.3(b), $p_{*2} \colon H_2(P_n; G) \to H_2(\mathbb{S}^2; G)$ is a monomorphism. Taking $G = \mathbb{Z}_n$, both groups are isomorphic to $\mathbb{Z}_n$, and p_{*2} is automatically an isomorphism.

Note. Generally, p_{*2} need not be epi: it is not the case with $G = \mathbb{Z}$.

8.4.8 Solutions of 4.4.3

(a) The cochain complex has a component G in degree zero, and is trivial elsewhere; the same holds for the graded group $H^+(\{*\}; G)$.

(b) $H^0(X; G) = Z^0(X; G)$ is the group of constant function $\lambda \colon X \to G$, because X is path connected; it can be identified with G, because $X \neq \emptyset$.

(c) Obvious, because any cube of X has image contained in some path component, and its faces as well.

(d) From (b) and (c).

8.5 Exercises of Chapter 5

8.5.1 Solutions of 5.1.3

(a) The identity $\operatorname{id} A$ of a module can be lifted to $\operatorname{id} A_+$, and gives $F_n(\operatorname{id} A) = \operatorname{id} F_n(A)$. A composite $gf: A \to B \to C$ can be lifted to $g_+ f_+: A_+ \to C_+$. An addition $f + g: A \to B$ can be lifted to $f_+ + g_+: A_+ \to B_+$.

(b) Obvious, because F is right exact: $F\varepsilon: FA_0 \to FA$ is a cokernel of $F\partial_1: FA_1 \to FA_0$, which 'is' $H_0(FA_+)$.

(c) Suppose we have a second choice of projective resolutions $\varepsilon': A'_+ \to A$, for each R-module. Then $\operatorname{id} A$ lifts to a homotopy equivalence $f_+: A_+ \to A'_+$ of chain complexes (by Exercise 5.1.2(d)). This gives an isomorphism $\rho_n A = H_n(f_+): F_n(A) \to F'_n(A)$ which only depends on the choice of resolutions. The naturality on a homomorphism $A \to B$ is easily checked.

8.5.2 Solutions of 5.1.6

(a) Let F be exact; for each R-module A, a projective resolution $A_+ \twoheadrightarrow A$ gives an exact chain complex FA_+, and $H_n(FA_+) = 0$ for $n > 0$. This property implies $F_1 = 0$, and the latter implies the exactness of F (using the exact sequence of derived functors).

(b) The projective R-module P has a projective resolution $A_+ \twoheadrightarrow P$ where $A_0 = P$ and $A_n = 0$ for $n > 0$. Therefore $F_n(P) = H_n(A_+) = 0$ for $n > 0$.

In particular, the right exact functor $F = - \otimes_R P$ is exact, by (a).

8.5.3 Solutions of 5.2.1

(a) Obvious: saying that B is flat means that the functor $- \otimes B$ is exact, or equivalently that its left derived functors are trivial (as already remarked in Exercise 5.1.6(a)).

(b) The free resolution $(m, p): \mathbb{Z} \rightarrowtail \mathbb{Z} \twoheadrightarrow \mathbb{Z}_m$ (where m stands for the multiplication by m) says that $\operatorname{Tor}(\mathbb{Z}_m, \mathbb{Z}_n)$ is the kernel of the homomorphism $m \otimes \mathbb{Z}_n: \mathbb{Z} \otimes \mathbb{Z}_n \to \mathbb{Z} \otimes \mathbb{Z}_n$, which is the multiplication $\overline{m}: \mathbb{Z}_n \to \mathbb{Z}_n$. This is easily seen to be isomorphic to $\mathbb{Z}_d$.

Recalling that $\mathbb{Z}_m \otimes \mathbb{Z}_n$ is also isomorphic to $\mathbb{Z}_d$ suggests an alternative proof. We have an exact sequence

$$0 \to \operatorname{Tor}(\mathbb{Z}_m, \mathbb{Z}_n) \to \mathbb{Z} \otimes \mathbb{Z}_n \xrightarrow{f} \mathbb{Z} \otimes \mathbb{Z}_n \to \mathbb{Z}_m \otimes \mathbb{Z}_n \to 0$$

where $\mathbb{Z}_m \otimes \mathbb{Z}_n \cong \mathbb{Z}_d$ has d elements, whence $\operatorname{Im} f$ has n/d elements (by Lagrange's Theorem on finite groups). Therefore $\operatorname{Coim} f$ has also n/d elements, and $\operatorname{Tor}(\mathbb{Z}_m, \mathbb{Z}_n)$ must have d elements (by the same argument); being a cyclic group, as a subgroup of $\mathbb{Z}_n$, it is isomorphic to $\mathbb{Z}_d$.

8.5.4 Solutions of 5.2.7

(d) The homology groups of $\mathbb{T}$ and $\mathbb{K}$ with integral coefficients are determined in 2.6.3 and 2.6.4. Then we apply the following known relations

$$\mathbb{Z} \otimes \mathbb{Z}_2 \cong \mathbb{Z}_2, \qquad \operatorname{Tor}(\mathbb{Z}, \mathbb{Z}_2) = 0,$$

$$\mathbb{Z}_2 \otimes \mathbb{Z}_2 \cong \mathbb{Z}_2, \qquad \operatorname{Tor}(\mathbb{Z}_2, \mathbb{Z}_2) \cong \mathbb{Z}_2.$$

(e) As above, using the homology groups of $\mathbb{P}^n$ with integral coefficients, determined in (2.170).

8.5.5 Solutions of 5.3.1

(a) $\operatorname{Ext}(A, -) = 0$ means that the functor $\operatorname{Hom}(A, -)$ is exact (by 5.1.8(a)); this is equivalent to saying that A is a projective $\mathbb{Z}$-module, i.e. a free abelian group (by 4.2.9(b)).

(b) Since $\operatorname{Hom}(\mathbb{Z}_m, \mathbb{Z}_n) \cong \mathbb{Z}_d$ (by Exercise 4.2.8(b)), we can give a proof similar to the second argument in 5.2.1(b). The exact sequence

$$0 \to \operatorname{Hom}(\mathbb{Z}_m, \mathbb{Z}_n) \to \operatorname{Hom}(\mathbb{Z}, \mathbb{Z}_n) \to \operatorname{Hom}(\mathbb{Z}, \mathbb{Z}_n) \to \operatorname{Ext}(\mathbb{Z}_m, \mathbb{Z}_n) \to 0$$

proves that $\operatorname{Ext}(\mathbb{Z}_m, \mathbb{Z}_n)$ too has d elements. Being a cyclic group, as a quotient of $\operatorname{Hom}(\mathbb{Z}, \mathbb{Z}_n) \cong \mathbb{Z}_n$, it is isomorphic to $\mathbb{Z}_d$.

8.5.6 Solutions of 5.4.1

(a) The expression $\partial_p a \otimes b + (-1)^p a \otimes \partial_q b$ is $\mathbb{Z}$-linear in both variables, a and b. Moreover, on a generator $a \otimes b \in A_p \otimes B_q$, with $n = p + q$

$$\partial_{n-1} \partial_n (a \otimes b) = \partial_{n-1}(\partial_p a \otimes b + (-1)^p a \otimes \partial_q b)$$

$$= \partial_{p-1} \partial_p a \otimes b + (-1)^{p-1} \partial_p a \otimes \partial_q b$$

$$+ (-1)^p \partial_p a \otimes \partial_q b + (-1)^p a \otimes \partial_{q-1} \partial_q b = 0.$$

(b) The canonical isomorphism of abelian groups, in (4.14)

$$\varphi \colon A \otimes B \to B \otimes A, \qquad a \otimes b \mapsto b \otimes a,$$

is extended to chain complexes, letting

$$\varphi_{pq} \colon A_p \otimes B_q \to B_q \otimes A_p, \qquad a \otimes b \mapsto (-1)^{pq} b \otimes a. \tag{8.68}$$

This is consistent with the differential:

$$\partial_n \varphi_n (a \otimes b) = (-1)^{pq} \partial_n (b \otimes a) = (-1)^{pq} \partial_q b \otimes a + (-1)^{pq+q} b \otimes \partial_p a,$$

$$\varphi_{n-1} \partial_n (a \otimes b) = \varphi_{n-1}(\partial_p a \otimes b + (-1)^p a \otimes \partial_q b)$$

$$= (-1)^{pq-q} b \otimes \partial_p a + (-1)^{p+pq-p} \partial_q b \otimes a.$$

Similarly, the isomorphisms (4.15) and (4.16) of abelian groups are extended to chain complexes

$$A_p \otimes \mathbb{Z} \to A_p, \qquad a \otimes \lambda \mapsto \lambda a. \tag{8.69}$$

$$A_p \otimes (B_q \otimes C_r) \to (A_p \otimes B_q) \otimes C_r, \qquad a \otimes (b \otimes c) \mapsto a \otimes (b \otimes c). \tag{8.70}$$

8.5.7 Solutions of 5.4.2

(b) For all p, q we have an exact sequence

$$\mathrm{Tor}(A_p'', B_q) \to A_p' \otimes B_q \to A_p \otimes B_q \to A_p'' \otimes B_q \to 0$$

where $\mathrm{Tor}(A_p'', B_q) = 0$, giving a short exact sequence $(f_p \otimes B_q, g_p \otimes B_q)$. We conclude applying (a).

8.5.8 Solutions of 5.4.3

(b) In these hypotheses on B, a tensor product $A \otimes B$ breaks into the direct sum of a sequence of complexes

$$A \otimes B_q = ((A_{n-q} \otimes B_q)_{n \geqslant 0}, (\partial_{n-q} \otimes B_q)_{n \geqslant 0}), \tag{8.71}$$

$$
\begin{array}{ccccccc}
\ldots A_2 \otimes B_0 & \xrightarrow{\partial_2 \otimes B_0} & A_1 \otimes B_0 & \xrightarrow{\partial_1 \otimes B_0} & A_0 \otimes B_0 & \longrightarrow & 0 \\[4pt]
\ldots A_1 \otimes B_1 & \xrightarrow{\partial_1 \otimes B_1} & A_0 \otimes B_1 & \longrightarrow & 0 & \longrightarrow & 0 \\[4pt]
\ldots A_0 \otimes B_2 & \longrightarrow & 0 & \longrightarrow & 0 & \longrightarrow & 0 \\[4pt]
\ldots\ \ldots\ \ldots
\end{array}
$$

Its homology is the direct sum of the homology of each row. Moreover each functor $- \otimes B_q \colon \mathrm{Ch}_+\mathrm{Ab} \to \mathrm{Ch}_+\mathrm{Ab}$ is exact, and preserves homology.

8.5.9 Solutions of 5.5.2

(d) To verify the consistency of the differential with the product, the procedure is similar to that of 3.6.4(a), for Alexander–Spanier's cochains.

We use the following identities, between maps $\Delta^p \to \Delta^{n+1}$ or $\Delta^q \to \Delta^{n+1}$

$$\tau_{p+1}^- \delta_{p+1,i} = \delta_{n+1,i}\, \tau_p^-, \qquad \tau_q^+ = \delta_{n+1,i}\, \tau_q^+ \qquad (0 \leqslant i \leqslant p = n - q), \tag{8.72}$$

$$\tau_p^- = \delta_{n+1,p+i}\tau_p^-, \qquad \tau_{q+1}^+ \delta_{q+1,i} = \delta_{n+1,p+i}\, \tau_q^+ \qquad (0 \leqslant i \leqslant q = n-p), \tag{8.73}$$

$$\tau_{p+1}^- \delta_{p+1,p+1} = \tau_p^-, \qquad \tau_q^+ = \tau_{q+1}^+ \delta_{q+1,0}. \tag{8.74}$$

On a simplex $a\colon \Delta^{n+1} \to X$ we have:

$$(d^p\lambda \smile \mu)(a) = (d^p\lambda)(a\tau^-_{p+1}).\mu(a\tau^+_q)$$
$$= \Sigma_{0\leqslant i\leqslant p+1}\,(-1)^i\,\lambda(a\tau^-_{p+1}\,\delta_{p+1,i})\,\mu(a\tau^+_q)$$
$$= \Sigma_{0\leqslant i\leqslant p}\,(-1)^i\lambda(a\tau^-_{p+1}\,\delta_{p+1,i})\,\mu(a\tau^+_q)$$
$$+ (-1)^{p+1}\lambda(a\tau^-_{p+1}\,\delta_{p+1,p+1})\,\mu(a\tau^+_q), \tag{8.75}$$

$$(-1)^p(\lambda \smile d^q\mu)(a) = (-1)^p\,\lambda(a\tau^-_p)\,(d^q\mu)(\tau^+_{q+1})$$
$$= \Sigma_{0\leqslant i\leqslant q+1}\,(-1)^{p+i}\,\lambda(a\tau^-_p)\,\mu(\tau^+_{q+1}\,\delta_{q+1,i})$$
$$= (-1)^p\,\lambda(a\tau^-_p)\,\mu(\tau^+_{q+1}\,\delta_{q+1,0})$$
$$+ \Sigma_{1\leqslant i\leqslant q+1}\,(-1)^{p+i}\,\lambda(a\tau^-_p)\,\mu(\tau^+_{q+1}\,\delta_{q+1,i}). \tag{8.76}$$

The last addendum in (8.75) and the first in (8.76) cancel, by (8.74). The sum of the remaining terms coincides with $d^n(\lambda \smile \mu)(a)$, applying (8.72) and (8.73):

$$(d^n(\lambda \smile \mu))(a) = \Sigma_{0\leqslant i\leqslant n+1}\,(-1)^i\,(\lambda \smile \mu)(a\delta_{n+1,i})$$
$$= \Sigma_{0\leqslant i\leqslant n+1}\,(-1)^i\,\lambda(a\delta_{n+1,i}\,\tau^-_p)\,\mu(a\delta_{n+1,i}\,\tau^+_q)$$
$$= \Sigma_{0\leqslant i\leqslant p}\,(-1)^i\,\lambda(a\delta_{n+1,i}\,\tau^-_p)\,\mu(a\delta_{n+1,i}\,\tau^+_q)$$
$$+ \Sigma_{1\leqslant i\leqslant q+1}\,(-1)^{p+i}\,\lambda(a\delta_{n+1,p+i}\,\tau^-_p)\,\mu(a\delta_{n+1,p+i}\,\tau^+_q)$$
$$= \Sigma_{0\leqslant i\leqslant p}\,(-1)^i\,\lambda(a\tau^-_{p+1}\,\delta_{p+1,i})\,\mu(a\tau^+_q)$$
$$+ \Sigma_{1\leqslant i\leqslant q+1}\,(-1)^{p+i}\,\lambda(a\tau^-_p)\,\mu(a\tau^+_{q+1}\,\delta_{q+1,i}).$$

8.5.10 Solutions of 5.5.6

(a) We use again the identities (8.72)–(8.74). We also note that $a\tau^-_0$ and $a\tau^+_0$ are 0-simplices and $\partial_0(a\tau^-_0) = \partial_0(a\tau^+_0) = 0$. Now

$$\partial_n \tau_n(a) = \partial_n(\Sigma_{p+q=n}\,a\tau^-_p \otimes a\tau^+_q)$$
$$= \Sigma_{p+q=n,p>0}\,(\partial_p(a\tau^-_p) \otimes a\tau^+_q)$$
$$+ \Sigma_{p+q=n,q>0}\,(-1)^p\,(a\tau^-_p \otimes \partial_q(a\tau^+_q))$$
$$= \Sigma_{p+q=n,p>0}\,\Sigma_{0\leqslant i\leqslant p}\,(-1)^i\,(a\tau^-_p\delta_i \otimes a\tau^+_q) \tag{8.77}$$
$$+ \Sigma_{p+q=n,q>0}\,\Sigma_{0\leqslant i\leqslant q}\,(-1)^{p+i}\,(a\tau^-_p \otimes a\tau^+_q\delta_i)$$
$$= \Sigma_{p+q=n}\,\Sigma_{0\leqslant i<p}\,(-1)^i\,(a\tau^-_p\delta_i \otimes a\tau^+_q) + A$$
$$+ B + \Sigma_{p+q=n}\,\Sigma_{1\leqslant i\leqslant q}\,(-1)^{p+i}\,(a\tau^-_p \otimes a\tau^+_q\delta_i).$$

We have set apart two summands, called A (for $i = p$) and B (for $i = 0$).

Using (8.74), we see that A and B cancel each other:

$$\begin{aligned}
A &= \Sigma_{p+q=n,p>0}\,(-1)^p\,a\tau_p^-\delta_p\otimes a\tau_q^+ \\
&= \Sigma_{p+q=n,p>0}\,(-1)^p\,a\tau_p^-\delta_p\otimes a\tau_{q+1}^+\delta_0 \\
&= \Sigma_{1\leqslant p\leqslant n}\,(-1)^p\,a\tau_p^-\delta_p\otimes a\tau_{n+1-p}^+\delta_0,
\end{aligned}$$

$$\begin{aligned}
B &= \Sigma_{p+q=n,q>0}\,(-1)^p\,a\tau_p^-\otimes a\tau_q^+\delta_0 \\
&= \Sigma_{p+q=n,p<n}\,(-1)^p\,a\tau_{p+1}^-\delta_{p+1}\otimes a\tau_q^+\delta_0 \\
&= \Sigma_{1\leqslant p\leqslant n}\,(-1)^{p+1}\,a\tau_p^-\delta_p\otimes a\tau_{n+1-p}^+\delta_0 = -A.
\end{aligned}$$

Finally, (8.77) amounts to $\tau_{n-1}\partial_n(a)$, applying (8.72) and (8.73):

$$\begin{aligned}
\tau_{n-1}\partial_n(a) &= \tau_{n-1}(\Sigma_{0\leqslant i\leqslant n}(-1)^i a\delta_i) \\
&= \Sigma_{0\leqslant i\leqslant n}\Sigma_{p+q=n-1}\,(-1)^i\,a\delta_i\tau_p^-\otimes a\delta_i\tau_q^+ \\
&= \Sigma_{p+q=n-1}\Sigma_{0\leqslant i\leqslant p}\,(-1)^i\,(a\delta_i\tau_p^-\otimes a\delta_i\tau_q^+) \\
&\quad + \Sigma_{p+q=n-1}\Sigma_{1\leqslant i\leqslant q+1}\,(-1)^{p+i}\,(a\delta_{p+1}\tau_p^-\otimes a\delta_{p+i}\tau_q^+) \\
&= \Sigma_{p+q=n-1}\Sigma_{0\leqslant i\leqslant p}\,(-1)^i\,(a\tau_{p+1}^-\delta_i\otimes a\tau_q^+) \\
&\quad + \Sigma_{p+q=n-1}\Sigma_{1\leqslant i\leqslant q+1}\,(-1)^{p+i}\,(a\tau_p^-\otimes a\tau_{q+1}^+\delta_i) \\
&= \Sigma_{p+q=n-1}\Sigma_{0\leqslant i<p}\,(-1)^i\,(a\tau_p^-\delta_i\otimes a\tau_q^+) \\
&\quad + \Sigma_{p+q=n}\Sigma_{1\leqslant i\leqslant q}\,(-1)^{p+i}\,(a\tau_p^-\otimes a\tau_q^+\delta_i).
\end{aligned} \tag{8.78}$$

(c) The cup product can be obtained composing three arrows:

- first, the bilinear mapping induced by tensoring

$$\mathrm{Hom}(S_p(X),R)\times\mathrm{Hom}(S_q(X),R)\;\longrightarrow\;\mathrm{Hom}(S_p(X)\otimes S_q(X),R\otimes R),$$
$$(\lambda,\mu)\mapsto\lambda\otimes\mu\qquad(\text{for }\lambda\colon S_p(X)\to R,\;\mu\colon S_q(X)\to R),$$

- second, the homomorphism induced covariantly by the product in R

$$\mathrm{Hom}(S_p(X)\otimes S_q(X),R\otimes R)\;\longrightarrow\;\mathrm{Hom}(S_p(X)\otimes S_q(X),R),$$

- third, the homomorphism induced contravariantly by $\tau_{pq}X$

$$(\tau_{pq}X)^\sharp\colon\mathrm{Hom}(S_p(X)\otimes S_q(X),R)\;\longrightarrow\;\mathrm{Hom}(S_{p+q}(X),R).$$

8.5.11 Solutions of 5.5.7

(a) It is a trivial consequence of the analogous formula for the cup product, in 5.5.2(d), taking into account that the differential commutes with the homomorphisms $p^{\sharp i}$.

(b) The diagonal $\Delta\colon X \to X \times X$ gives a composite

$$S^p(X;R) \times S^q(X;R) \;\longrightarrow\; S^n(X \times X; R) \to S^n(X;R),$$
$$(\lambda,\mu) \mapsto \Delta^\sharp(\lambda \times \mu) \qquad (p+q=n),$$

(8.79)

which coincides with the cup product, because

$$\Delta^\sharp(\lambda \times \mu) = \Delta^\sharp p^{\sharp 1}(\lambda) \smile \Delta^\sharp p^{\sharp 2}(\mu) = (p_1\Delta)^\sharp(\lambda) \smile (p_2\Delta)^\sharp(\mu) = \lambda \smile \mu.$$

8.5.12 Solutions of 5.6.7

(a)–(c) Obvious consequences of Theorem 5.6.6, taking into account that, for free abelian groups A, B of finite dimension

$$\dim(A \oplus B) = \dim(A) + \dim(B), \qquad \dim(A \otimes B) = \dim(A)\dim(B).$$

(d) In fact $\chi(X) = \beta X(-1)$, and it is sufficient to apply formula (5.115).

(e) The Poincaré polynomial of the circle is $\beta\mathbb{S}^1(t) = 1 + t$, whence

$$\beta\mathbb{T}^n(t) = (1+t)^n, \qquad H_k(\mathbb{T}^n) \cong \mathbb{Z}^{\binom{n}{k}} \qquad (0 \leqslant k \leqslant n),$$

(8.80)

and $H_k(\mathbb{T}^n) = 0$ for $k > n$ (as for any manifold of dimension n).

The identity $\binom{n}{k} = \binom{n}{n-k}$ corresponds to the formula $\beta_k(M) = \beta_{n-k}(M)$ in (5.56), for a compact orientable manifold of dimension n.

8.6 Exercises of Chapter 6

8.6.1 Solutions of 6.1.2

(a) We have two homotopies h, k with fixed endpoints, between consecutive paths

$$h\colon a_0 \simeq_2 a_1\colon x \simeq x', \qquad k\colon b_0 \simeq_2 b_1\colon x' \simeq x''.$$

Concatenating them, on the first variable, we obtain a 2-homotopy $H\colon$ $a_0 * b_0 \simeq_2 a_1 * b_1\colon x \simeq x''$

$$H(s,t) = \begin{cases} h(2s,t), & \text{for } 0 \leqslant s \leqslant 1/2, \\ k(2s-1,t), & \text{for } 1/2 \leqslant s \leqslant 1, \end{cases}$$

(8.81)

$$H(-,i) = a_i * b_i, \qquad H(0,t) = x, \qquad H(1,t) = x'' \qquad (t \in \mathbb{I},\ i = 0,1),$$

whose continuity follows from an obvious closed cover of $\mathbb{I} \times \mathbb{I}$.

(b) We take the iterated concatenations of three consecutive paths $a\colon x \simeq x'$, $b\colon x' \simeq x''$, and $c\colon x'' \simeq x'''$

$$d = a * (b * c), \qquad d' = (a * b) * c.$$

(8.82)

The first path is parametrised on $\mathbb{I} = [0, 1/2] \cup [1/2, 3/4] \cup [3/4, 1]$, moving with double velocity on the first subinterval, and four-tuple velocity on the remaining two. The second is parametrised on $\mathbb{I} = [0, 1/4] \cup [1/4, 1/2] \cup [1/2, 1]$, moving with four-tuple velocity on the first two subintervals, then with double velocity on the last.

Each of these paths is a reparametrisation of the other; for instance $d' = df \colon \mathbb{I} \to X$, where the (invertible) function $f \colon \mathbb{I} \to \mathbb{I}$ turns the previous second covering into the first

$$f(t) = \begin{cases} 2t, & \text{for } 0 \leqslant t \leqslant 1/4, \\ t + 1/4, & \text{for } 1/4 \leqslant t \leqslant 1/2, \\ (t+1)/2, & \text{for } 1/2 \leqslant t \leqslant 1. \end{cases}$$

Now any 2-homotopy $h \colon \mathrm{id} \simeq_2 f \colon \mathbb{I} \to \mathbb{I}$ gives a 2-homotopy $dh \colon d \simeq_2 d' \colon \mathbb{I} \to X$ (by whisker composition). For instance we can take at each $s \in \mathbb{I}$ the affine linear path from s to $f(s)$

$$h \colon \mathbb{I} \times \mathbb{I} \to \mathbb{I}, \qquad h(s,t) = (1-t).s + t.f(s) \qquad (s, t \in \mathbb{I}), \tag{8.83}$$

which is constant at $s = 0, 1$.

(c) For a path $a \colon x \simeq x'$, the paths $e_x * a$ and $a * e_{x'}$ are reparametrisations of a by (non-invertible) functions $f, g \colon \mathbb{I} \to \mathbb{I}$

$$c = e_x * a = af, \qquad\qquad d = a * e_{x'} = ag, \tag{8.84}$$

$$f(t) = \max(0, 2t - 1) \qquad\qquad g(t) = \min(2t, 1).$$

Therefore a 2-homotopy $h \colon \mathrm{id} \simeq_2 f$ (resp. $k \colon \mathrm{id} \simeq g$) gives a 2-homotopy $ah \colon a \simeq_2 e_x * a$ (resp. $ak \colon a \simeq_2 a * e_{x'}$). As in (b), these 2-homotopies $\mathbb{I} \times \mathbb{I} \to \mathbb{I}$ can be defined by affine linear combinations in the second variable.

(d) For a path $a \colon x \simeq x'$, a 2-homotopy $h \colon e_x \simeq_2 a * a^{\sharp}$ can be obtained as

$$h(s,t) = \begin{cases} a(2st), & \text{for } s \leqslant 1/2, \\ a(2(1-s)t), & \text{for } s \geqslant 1/2. \end{cases} \tag{8.85}$$

Similarly we have a 2-homotopy $k \colon e_{x'} \simeq_2 a^{\sharp} * a$.

8.6.2 Solutions of 6.1.3

(a) A 2-homotopy $h\colon a \simeq_2 a'\colon x \simeq x'$ in the space X gives a 2-homotopy $fh\colon fa \simeq_2 fa'\colon fx \simeq fx'$ in Y.

(b) Composing with $f\colon X \to Y$ preserves path-concatenation and trivial loops (of X).

(c) Obvious:

$$(gf)_*([a]) = ([gfa]) = g_*(f_*([a])), \quad (\operatorname{id} X)_*([a]) = [(\operatorname{id} X)a] = [a].$$

(d) There is an obvious functor $F\colon \mathsf{Gpd} \to \mathsf{Set}$ that takes a small groupoid G to the quotient of the set $\operatorname{Ob} G$ of its objects, modulo the equivalence relation that relates two objects x, x' if there is in G an arrow $x \to x'$. The composite $F\Pi_1$ coincides with Π_0.

Viewing a groupoid as a representation of a 1-dimensional space, this functor F can be written as $\Pi_0\colon \mathsf{Gpd} \to \mathsf{Set}$. Similarly, a small category can be viewed as an elementary form of a *directed* space [G1].

Letting $\operatorname{Ob}\colon \mathsf{Gpd} \to \mathsf{Set}$ be the forgetful functor of objects, there is a chain of adjunctions $\Pi_0 \dashv D \dashv \operatorname{Ob} \dashv C$, where $D(X)$ is the discrete groupoid on the set X, and $C(X)$ is the indiscrete one.

8.6.3 Solutions of 6.1.5

(b) Obviously (i) $\Leftrightarrow$ (iv) $\Leftrightarrow$ (v) $\Leftrightarrow$ (ii).

(c) A consequence of (b).

The additional property (iv′) is also equivalent to the previous ones, once we know that all groups $\pi_1(X, x)$ are isomorphic when X is path connected (see 6.3.1).

(d) A consequence of the previous points: the corresponding properties of the fundamental groupoid (namely: being equivalent to $\mathbf{0}$, or having at most one arrow between two given objects, or being equivalent to $\mathbf{1}$) are invariant up to categorical equivalence.

8.6.4 Solutions of 6.1.9

(b) After the suggested choice of the arrows $\lambda_x\colon x \to P(x)$, the functor $P\colon \Pi_1 X \to \Pi_1 X|A$ is defined on vertices, and we complete it on arrows, letting

$$P(\alpha\colon x \to y) = (\lambda_x^{-1}\alpha\lambda_y\colon P(x) \to P(y)). \tag{8.86}$$

(Recall that, in these groupoids, we are writing the composition in the natural order of concatenation.)

We constrain the previous choice requiring that, if $x \in A$, then $\lambda_x = [e_x]$ (the identity of x). In this way PJ is the identity. Moreover, the family $\lambda = (\lambda_x)$ forms a natural transformation $\mathrm{id}\,\Pi_1 X \to JP$, which is invertible (because all its components are).

8.6.5 Solutions of 6.2.3

(a) For a loop $a\colon x_0 \simeq x_0$, the pointed map

$$h = \varphi(a \times \mathbb{I})\colon \mathbb{I} \to X \times \mathbb{I} \to Y$$

is a 2-homotopy $h\colon fa \simeq ga$ of loops at $y_0 \in Y$, which gives $[fa] = [ga]$.

We can also deduce this result from Theorem 6.1.4 on fundamental groupoids: the commutative square (6.7) gives the following relation in the group $\pi_1(Y, y_0)$

$$[fa]\varphi_*(x_0) = \varphi_*(x_0)[ga],$$

and $[fa] = [ga]$, because $\varphi_*(x_0) = [e_{y_0}]$ is the class of the trivial loop at y_0.

8.6.6 Solutions of 6.3.1

(a) Suppose that $\pi_1(X, x_0)$ is commutative and $c\colon x_0 \simeq x_1$ in X. Any other path $d\colon x_0 \simeq x_1$ in X can be expressed, up to 2-homotopy, as $d = u * c$, where u is a loop at x_0. Then, for every $a \in \Omega(X, x_0)$: $\hat{d}[a] = [c]^{-1}[u]^{-1}[a][u][c] = \hat{c}[a]$.

Conversely, the transition isomorphism $\hat{u}\colon \pi_1(X, x_0) \to \pi_1(X, x_0)$ determined by any loop u at x_0 must be the identity, which means that $[u]$ commutes with any element of $\pi_1(X, x_0)$.

8.6.7 Solutions of 6.3.3

(a) The loop $a \in \Omega(X, x_0)$ is a cycle of X, and its homology class is invariant up to loop homotopy (Exercise 2.2.8(c)), which is wider than homotopy with fixed basepoint. The property $h([a][b]) = h[a * b] = [a]_H + [b]_H$ is proved in Exercise 2.2.8(e).

(b) For a loop $a \in \Omega(X, x_0)$, the loop $c^{\sharp} * a * c$ is loop homotopic to a (by Lemma 6.3.6); therefore $h'\hat{c}[a] = h'[a] = [a]_H = h[a]$.

(c) For a pointed map $f\colon (X, x_0) \to (Y, y_0)$ we have:

$$h(Y, y_0)\,\pi_1(f)[a] = [fa]_H = H_1(f)\,h(X, x_0)[a].$$

(d) It is sufficient to verify that $[G, G]$ is closed under the inner automorphism $\hat{g}(x) = g^{-1}xg$, for every $g \in G$. In fact $\hat{g}$ takes every commutator $[x, y]$ to the commutator $[\hat{g}(x), \hat{g}(y)]$.

(e) The commutativity of $G/[G, G]$ is obvious. As to the universal property, the homomorphism $f\colon G \to A$ takes the subgroup $[G, G]$ to 0, and induces a homomorphism g, determined as $g[x] = f(x)$.

8.6.8 Solutions of 6.4.1

(a) To construct $*_i G_i$, we first form an *alphabet* $X = \Sigma_i |G_i|$, the disjoint union of the underlying sets, and then the set W of finite words $(x_1, x_2, ..., x_n)$ on this alphabet X, including the empty word e. Multiplying words by concatenation, the set W is a monoid. (It is actually the free monoid on the set X, as we have seen in 1.6.7(c).)

Each word $w \in W$ has a 'reduced form' $R(w)$, obtained by replacing every occurrence x_h, x_{h+1} of two elements in the same group G_i with their product in this group, and taking out any unit-element of some G_i. The result does not depend on the sequence of reductions (this seemingly obvious point requires in fact a complex proof). We say that $w \sim w'$ when $R(w) = R(w')$.

The quotient set $*_i G_i = W/\!\sim$ has a well defined product: $[w][w'] = [ww']$; indeed the concatenation ww' is equivalent to $R(w)R(w')$, because $R(ww')$ can be obtained by iterated reductions, as $R(R(w)R(w'))$.

The product is associative (it already is in W). The unit is the class of the empty word. The inverse of the class $[x_1, x_2, ..., x_n]$ is obviously the class $[x_n^{-1}, ..., x_2^{-1}, x_1^{-1}]$.

(b) First we have a unique mapping of sets $h\colon X \to |H|$ which coincides with f_i on each subset $|G_i|$ (by the universal property of the sum of sets, in 1.6.2(a)). Then we extend it to the monoid W of finite words on X, letting $k(x_1, x_2, ..., x_n) = h(x_1)h(x_2)...h(x_n)$. Finally, this homomorphism of monoids $k\colon W \to H$ is invariant under the equivalence relation $\sim$ leading to G, and the induced homomorphism $f\colon G \to H$ satisfies the required conditions $fu_i = f_i$ (for all i).

Its uniqueness is attested by the fact that G is spanned by the images of all injections.

(d), (e) For a mapping $f\colon X \to |H|$ with values in a group H, we consider the homomorphism

$$g\colon *_{x \in X} \mathbb{Z} \to H, \qquad g(x_1^{k_1} x_2^{k_2} ... x_p^{k_p}) = f(x_1)^{k_1} f(x_2)^{k_2} ... f(x_p)^{k_p}.$$

It does satisfy $g\eta = f$, and is determined by this property since the elements $\eta(x)$, for $x \in X$, generate the domain of g.

8.6.9 Solutions of 6.4.2

(a) Obvious: in this case the commutativity conditions $u_1 f = u_2 g$ and $v_1 f = v_2 g$ are automatically satisfied, and the new universal property coincides with 6.4.1(i*).

(b) The quotient modulo K is precisely what is required to produce a commutative square.

(c) As in 2.3.6(b).

8.6.10 Solutions of 6.4.4

(a) We use the open cover (U, V) of $\mathbb{S}^n$ already used in 2.4.2 for the Mayer–Vietoris sequence.

The subspaces $U = \mathbb{S}^n \setminus \{\underline{s}\}$ and $\mathbb{S}^n \setminus \{\underline{n}\}$ are contractible, and their fundamental group is trivial (by 6.3.2(c)). The intersection $U \cap V \simeq \mathbb{S}^{n-1}$ is also path connected (for $n \geqslant 2$), and the van Kampen theorem applies. This is sufficient to say that $\pi_1(\mathbb{S}^n) = 0$.

(b) As in 2.5.1, we realise the figure-eight space $X = \mathbb{S}^1 \vee \mathbb{S}^1$ as two circles in $\mathbb{R}^2$ tangent at the origin, and we use the same open cover (U, V) used there to compute $H_1(X)$.

Now $\pi_1(U, 0)$ and $\pi_1(V, 0)$ are infinite cyclic groups, generated by the homotopy classes $[a]$ and $[b]$ of the following loops at the origin, around the left and the right circle (represented in figure 6.24)

$$a(t) = (-1 + \cos 2\pi t,\ \sin 2\pi t), \quad b(t) = (1 - \cos 2\pi t,\ -\sin 2\pi t). \quad (8.87)$$

Since $\pi_1(A)$ is the trivial group, the van Kampen Theorem gives a canonical isomorphism $\pi_1(U, 0) * \pi_1(V, 0) \to \pi_1(X, 0)$.

Therefore $\pi_1(X, 0)$ is the free group generated by two elements, the homotopy classes $[a]$ and $[b]$. By 6.4.1(e), each element can be uniquely written as a product

$$[a]^{h_1} [b]^{k_1} [a]^{h_2} [b]^{k_2} \dots [a]^{h_n} [b]^{k_n}, \quad (8.88)$$

with *non-zero* integral exponents, except $h_1, k_n \in \mathbb{Z}$.

Similarly, the fundamental group of a bouquet of n circles $\mathbb{S}^1 \vee \dots \vee \mathbb{S}^1$ is the free group generated by n elements.

(c) Suppose that $\pi_1(X, x_0)$ has two elements $[a]$, $[b]$ which do not commute; for instance, this is the case of the eight-figure space. Then the loops a and

$a' = b^\sharp * a * b$ are not 2-homotopic; but we know that they are homotopic as free loops, by Lemma 6.3.6.

(d) We use the same open cover (U, V) of P_n, as in the solution of Exercise 2.5.8(a)

$$U = P_n \setminus \{0\} \simeq \mathbb{S}^1, \qquad V = \operatorname{int} \mathbb{D}^2 \simeq \{*\},$$
$$A = U \cap V = \operatorname{int} \mathbb{D}^2 \setminus \{0\} \simeq \mathbb{S}^1. \tag{8.89}$$

The van Kampen Theorem gives a pushout of groups

$$\begin{array}{ccc}
\mathbb{Z} & \xrightarrow{\ u\ } & \mathbb{Z} \\
\downarrow & & \downarrow{\scriptstyle v} \\
0 & \longrightarrow & \pi_1(P_n, x_0)
\end{array}$$

where the homomorphism u is the multiplication by n (as in the solution of Exercise 2.5.8(a), for the homomorphism $h_1 \colon H_1(A) \to H_1(U)$). The group $\pi_1(P_n, x_0)$ is thus the cokernel of u in Gp (and in Ab), namely the abelian group $\mathbb{Z}_n$.

8.6.11 Solutions of 6.5.2

(a) For $n = 0$, $\mathbb{I}^0$ is the singleton $\{0\} = \mathbb{R}^0$, and $\underline{\partial}\mathbb{I}^0 = \emptyset$; thus $\Omega_0(X, x_0)$ can be identified with the underlying set $|X|$, pointed at x_0. Relative homotopies are paths $x \simeq x'$, and $\pi_0(X, x_0) = (|X|/\!\simeq, [x_0])$ is the set of path components of X, pointed at the class of the basepoint x_0, as defined in (1.70).

(b) For $i = 1, ..., n$, two relative cubes $a, b \in \Omega_n(X, x_0)$ are always consecutive in direction i (all their faces being constant at x_0); all the faces of their concatenation $a *_i b$ are constant at x_0.

Each operation induces a group structure $[a] +_i [b] = [a *_i b]$ on $\pi_n(X, x_0)$, with the same proof as in the case $n = 1$ (although we shall see that these structures coincide). The identity is (always) the class $[u_n]$ of the constant map $\mathbb{I}^n \to X$ at x_0. The inverse $-_i[a]$ is the class of the i-reversed cube

$$(t_1, ..., t_i, ..., t_n) \mapsto (t_1, ..., 1 - t_i, ..., t_n).$$

(c) This property holds more generally for four cubes $a, b, c, d \colon \mathbb{I}^n \to X$ whose i-faces and j-faces agree, as in the following picture (we only represent the two directions $i < j$ of the operations)

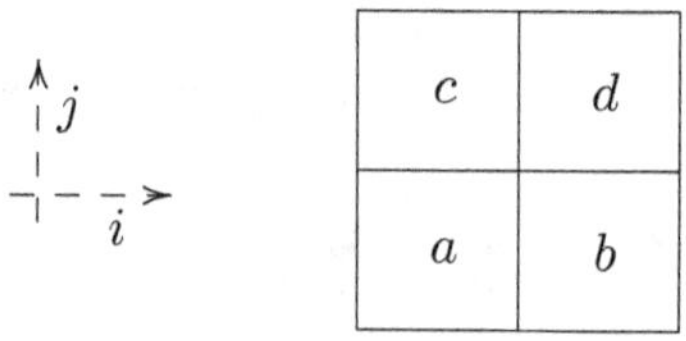

$$\partial_i^+ c = \partial_i^- d,$$
$$\partial_j^- c = \partial_j^+ a,$$
$$\partial_j^- d = \partial_j^+ b,\qquad (8.90)$$
$$\partial_i^+ a = \partial_i^- b.$$

Then both pastings $(a *_i b) *_j (c *_i d)$ and $(a *_j c) *_i (b *_j d)$ give the same n-cube $f : \mathbb{I}^n \to X$, defined as follows (omitting the other variables t_k, for $k \neq i, j$)

$$f(t_i, t_j) = \begin{cases} a(2t_i, 2t_j), & \text{for } t_i, t_j \leqslant 1/2, \\ b(2t_i - 1, 2t_j), & \text{for } t_i \geqslant 1/2,\ t_j \leqslant 1/2, \\ c(2t_i, 2t_j - 1), & \text{for } t_i \leqslant 1/2,\ t_j \geqslant 1/2, \\ d(2t_i - 1, 2t_j - 1), & \text{for } t_i, t_j \geqslant 1/2. \end{cases}$$

(e) First, the operations coincide, because

$$x \bullet y = (x + e) \bullet (e + y) = (x \bullet e) + (e \bullet y) = x + y. \qquad (8.91)$$

Second, this operation is commutative, because

$$x \bullet y = (e + x) \bullet (y + e) = (e \bullet y) + (x \bullet e) = y + x = y \bullet x. \qquad (8.92)$$

Concretely, if $a, b \in \Omega_n(X, x_0)$ and $[a] = x$, $[b] = y$ in $\pi_n(X, x_0)$, the identities (8.91) show how we should deform the cube $a *_i b$ into $a *_j b$

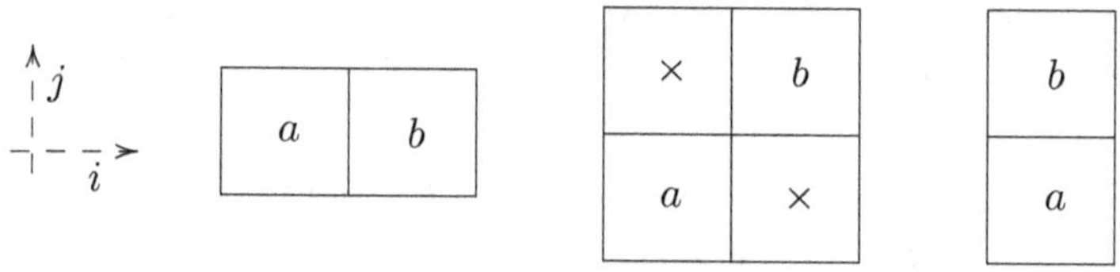

$$(8.93)$$

The cross-marked squares represent the constant cube $u_n : \mathbb{I}^n \to X$ at x_0, whose class is the identity of $\pi_n(X, x_0)$. A similar deformation transforms $a *_j b$ into $b *_i a$.

References

[AHS] J. Adámek, H. Herrlich, G.E. Strecker, Abstract and concrete categories. The joy of cats, John Wiley & Sons, 1990.

[Bo1] F. Borceux, Handbook of categorical algebra 1, Cambridge Univ. Press, 1994.

[Bo2] F. Borceux, Handbook of categorical algebra 2, Cambridge Univ. Press, 1994.

[Bo3] F. Borceux, Handbook of categorical algebra 3, Cambridge Univ. Press, 1994.

[Bou1] N. Bourbaki, Algebra I, Chapters 1–3, Hermann and Addison–Wesley, 1974.

[Bou2] N. Bourbaki, General Topology I, Chapters 1–4, Springer, 1989.

[Bou3] N. Bourbaki, Algebra II, Chapters 4–7, Springer, 1990.

[Br1] R. Brown, Groupoids and van Kampen's theorem, Proc. London Math. Soc. 17 (1967), 385–401.

[Br2] R. Brown, Elements of modern topology, McGraw-Hill, 1968.

[Br3] R. Brown, Topology and groupoids, Third edition of Elements of modern topology, BookSurge, LLC, Charleston, SC, 2006.

[CE] H. Cartan, S. Eilenberg, Homological algebra, Princeton Univ. Press, 1956.

[Co] A. Connes, Noncommutative Geometry, Academic Press, 1994.

[Di] J. Dieudonné, A history of algebraic and differential topology, 1900–1960, Birkhäuser, 1989.

[Do] A. Dold, Lectures on Algebraic Topology, Springer, 1972.

[EM1] S. Eilenberg, S. Mac Lane, General theory of natural equivalences, Trans. Amer. Math. Soc. 58 (1945), 231–294.

[EM2] S. Eilenberg, S. Mac Lane, Relations between homology and homotopy groups of spaces, Ann. of Math. 46 (1945), 480–509.

[EM3] S. Eilenberg, S. Mac Lane, Relations between homology and homotopy groups of spaces II, Ann. of Math. 51 (1950), 514–533.

[EM4] S. Eilenberg, S. Mac Lane, Acyclic models, Amer. J. Math. 75 (1953), 189–199.

[EK] S. Eilenberg, G.M. Kelly, Closed categories, in Proc. Conf. Categorical Algebra, La Jolla 1965, Springer, 1966, pp. 421–562.

[ES] S. Eilenberg, N. Steenrod, Foundations of algebraic topology, Princeton Univ. Press, 1952.

[EZ1] S. Eilenberg, J.A. Zilber, Semi-simplicial complexes and singular homology, Ann. of Math. 51 (1950), 499–513.

[EZ2] S. Eilenberg, J.A. Zilber, On products of complexes, Amer. J. Math. 75 (1953), 200–204.

[Fu] L. Fuchs, Abelian groups, Springer, 2015.

[Fr] P. Freyd, Abelian categories, An introduction to the theory of functors, Harper & Row, New York 1964. Republished in: Reprints Theory Appl. Categ. 3 (2003).

[GaZ] P. Gabriel, M. Zisman, Calculus of fractions and homotopy theory, Springer 1967.

[G1] M. Grandis, Directed Algebraic Topology, Models of non-reversible worlds, Cambridge Univ. Press, 2009.
Available at: http://www.dima.unige.it/~grandis/BkDAT_page.html

[G2] M. Grandis, Homological Algebra, The interplay of homology with distributive lattices and orthodox semigroups, World Scientific Publishing Co., 2012.

[G3] M. Grandis, Homological Algebra in strongly non-abelian settings, World Scientific Publishing Co., 2013.

[G4] M. Grandis, Manifolds and local structures, A general theory, World Scientific Publishing Co., 2021.

[G5] M. Grandis, Category Theory and Applications: A textbook for beginners, Second Edition, World Scientific Publishing Co., 2021.

[GM] M. Grandis, L. Mauri, Cubical sets and their site, Theory Appl. Categ. 11 (2003), No. 8, 185–211.

[Gt] A. Grothendieck, Sur quelques points d'algèbre homologique, Tôhoku Math. J. 9 (1957), 119–221.

[Ha] A. Hatcher, Algebraic Topology, Cambridge University Press, 2002.
Available at: https://www.math.cornell.edu/~hatcher/AT/ATpage.html

[He] H. Herrlich, Topologische Reflexionen und Coreflexionen, Lecture Notes in Math. 78, Springer, 1968.

[Hi] P. Hilton, Correspondences and exact squares, in: Proc. of the Conf. on Categorical Algebra, La Jolla 1965, Springer, 1966, 255–271.

[HiW] P.J. Hilton, S. Wylie, Homology theory: An introduction to algebraic topology, Cambridge University Press, 1960.

[Hu] S.T. Hu, Homotopy Theory, Academic Press, 1959.

[Hur] W. Hurewicz, On duality theorems, Bull. Amer. Math. Soc. 47 (1941), 562–563.

[Jo1] P.T. Johnstone, The point of pointless topology, Bull. Amer. Math. Soc. 8 (1983), 41–53.

[Jo2] P.T. Johnstone, The art of pointless thinking: a student's guide to the category of locales, in: Category theory at work (Bremen, 1990) Heldermann, Berlin 1991, pp. 85–107.

[KP] K.H. Kamps, T. Porter, Abstract homotopy and simple homotopy theory, World Scientific Publishing Co., 1997.

[Ka] D.M. Kan, Adjoint functors, Trans. Amer. Math. Soc. 87 (1958), 294–329.

[Ke] J.L. Kelley, General topology, Van Nostrand, 1955.

[Kl] G.M. Kelly, Basic concepts of enriched category theory, Cambridge Univ. Press, 1982.

[LoR] E. Lowen-Colebunders, G. Richter, An elementary approach to exponential spaces, Appl. Categ. Structures 9 (2001), 303–310.

[M1] S. Mac Lane, Homology, Springer, Berlin 1963.

[M2] S. Mac Lane, Categories for the working mathematician, Springer, 1971.

[M3] S. Mac Lane, Categories for the working mathematician, Second Edition, Springer, 1998.

[MaM] S. Mac Lane, I. Moerdijk, Sheaves in geometry and logic. A first introduction to topos theory, Springer, 1994.

[Mas1] W.S. Massey, Algebraic topology: An introduction, Harcourt, Brace & World, 1967.

[Mas2] W.S. Massey, Homology and cohomology theory: an approach based on Alexander-Spanier cochains, Marcel Dekker, Inc., New York, 1978.

[Mas3] W.S. Massey, Singular homology theory, Springer, 1980.

[Mat] M. Mather, Pull-backs in homotopy theory, Can. J. Math. 28 (1976), 225–263.

[May] J.P. May, Simplicial objects in algebraic topology, Van Nostrand Co., 1967.

[Mi] B. Mitchell, Theory of categories, Academic Press, 1965.

[Mu] J.R. Munkres, Topology: a first course, Prentice-Hall, 1975.

[Po] H. Poincaré, Analysis situs, J. de l'École Polytechnique 1 (1895), 1–123.

[Sp1] E.H. Spanier, Cohomology theory for general spaces, Ann. Math. 49 (1948), 407–427.

[Sp2] E.H. Spanier, Algebraic topology, Mc Graw–Hill,1966.

[Sv] M. Spivak, A comprehensive introduction to Differential Geometry, Vol. I, Second Edition, Publish or Perish, Wilmington, 1979.

[Vi] J.W. Vick, Homology theory. An introduction to algebraic topology, Second Edition, Springer, 1994.

Index